The Six Pillars of Calculus

Biology Edition

Pure and Applied
UNDERGRADUATE TEXTS · 60

The Six Pillars of Calculus

Biology Edition

Lorenzo Sadun

AMERICAN
MATHEMATICAL
SOCIETY
Providence, Rhode Island USA

2020 *Mathematics Subject Classification.* Primary 00-01, 26A06, 92-10, 92D30, 92D25, 00A71, 26-01.

For additional information and updates on this book, visit
www.ams.org/bookpages/amstext-60

Library of Congress Cataloging-in-Publication Data

Names: Sadun, Lorenzo Adlai, author.
Title: The six pillars of calculus : biology edition / Lorenzo Sadun.
Description: Providence, Rhode Island : American Mathematical Society, [2023] | Series: Pure and applied undergraduate texts, 1943-9334 ; volume 60 | Includes bibliographical references and index.
Identifiers: LCCN 2022038762 | ISBN 9781470469962 (paperback) | ISBN 9781470472436 (ebook)
Subjects: LCSH: Biomathematics–Textbooks. | AMS: General – Instructional exposition (textbooks, tutorial papers, etc.). | Real functions – Functions of one variable – One-variable calculus. | Biology and other natural sciences – Genetics and population dynamics – Epidemiology. | Biology and other natural sciences – Genetics and population dynamics – Population dynamics (general). | General – General and miscellaneous specific topics – Theory of mathematical modeling. | Real functions – Instructional exposition (textbooks, tutorial papers, etc.).
Classification: LCC QH323.5 .S228 2023 | DDC 570.1/51–dc23/eng/20220921
LC record available at https://lccn.loc.gov/2022038762

To my students, and to everybody who has helped me understand calculus better. (But I repeat myself.)

Contents

Instructors' Guide and Background

The Purpose of This Book. In 1952, George B. Thomas was asked to revise a publisher's existing calculus textbook. Instead, he wrote his own. The many editions of Thomas's *Calculus and Analytic Geometry*, and the numerous subsequent adaptations of his work, have dominated university classrooms ever since. The names on the textbooks have changed over time, with each new author bringing his or her own distinctive style. (I am especially a fan of James Stewart's precision and attention to detail.) However, the overall blueprint is the same. As a result, today's calculus classrooms largely resemble those of the 1990s, of the 1970s, and even of the 1950s.

For some students (including the author in his youth), and especially for those who wish to pursue careers in math, physics, and engineering, this approach has worked well. But for increasingly many other students, it has failed. To those students, calculus is a weed-out course where success depends mostly on memorizing a large number of unmotivated definitions, formulas, and problem-solving techniques, and applying them without any regard for what the variables and functions actually represent. Lacking any unifying principle, these elements are quickly forgotten when the semester is over. Even successful alumni often look back at their calculus classes as painful and useless ordeals.

It doesn't have to be that way! Calculus is built on a handful of very simple ideas that everybody can and should understand. Calculus is connected to every aspect of day-to-day life, from making business decisions to addressing environmental problems to building bridges that don't fall down, in ways that should interest and inspire everybody. Calculus can also be enjoyable, if students are brought into the process of discovery instead of just being handed results. This text, and the course at the University of Texas that it was written for, are an attempt to reclaim calculus and make it comprehensible, memorable, practical, and even fun.

So what's so different about this text?

- Instead of presenting theorems first and applications later, concepts are developed by tackling realistic problems. For instance, we begin with the SIR system of coupled differential equations that model the spread of a disease, and in particular the Covid pandemic, long before we ever define what a derivative is! Instead of starting with formal definitions of limits and derivatives, we generate "rate equations" from an informal understanding of what "rate of change" means. This leads naturally to linear approximations and to solving difficult problems one step at a time. Students discover that getting an answer through many small steps requires more work than taking a few big steps, but gives more accurate results. Exact answers emerge as the limit of this process, motivating the concept of a derivative.

- Everything in the book is connected explicitly to six central ideas that I call the *Six Pillars of Calculus*:
  (1) Close is good enough (approximation and limits)
  (2) Track the changes (derivatives)
  (3) What goes up has to stop before it comes down (optimization and turning points)
  (4) The whole is the sum of the parts (integration)
  (5) One step at a time (the Fundamental Theorem of Calculus)
  (6) One variable at a time (the key to multi-variable calculus)
  Making sense of these six ideas is both a goal in itself and an organizing principle for every other concept. Having such an overall conceptual framework to build on makes learning easier and forgetting harder.

- We stress multiple ways of looking at each concept. Is a derivative a limit of a difference quotient, a rate of change, the slope of a tangent line, the best linear approximation to a function, or the gain on a tuning knob? To pass a quiz, a student might memorize one of these things, but the real power of a derivative comes from it being all of these at the same time.

- We emphasise "when" and "why" over "how". Recognizing that a real-world phenomenon involves a derivative or an integral is much more important than being able to compute that derivative or integral from a formula.

- We also put a heavy emphasis on numerical approximations and programming. Difference quotients approximate derivatives, but we can just as well use derivatives to approximate difference quotients. Riemann sums approximate integrals, and integrals approximate Riemann sums. By studying approximations on their own terms, and not just as stepping stones toward the definitions of derivatives and integrals, students begin to grasp what derivatives, integrals, and infinite series are trying to accomplish.

    Programming also serves a pedagogical role. The best way to learn anything is to explain it to somebody else. That somebody else can be a computer, and the required explanation can be a MATLAB program. Students are not necessarily expected to write their own MATLAB programs from scratch, but they are expected to modify examples in the book to fit different circumstances.

- We use differential equations and modeling as recurring themes. Many, if not most, quantities in the world, from the number Covid-19 patients to the concentration of CO_2 in the atmosphere to the rabbit population of Australia, are governed by differential equations. The goal is not just to teach students calculus, but to help them recognize calculus in the world around us.

- We embrace heuristic reasoning. While no lies are told and everything in this book *can* be justified rigorously, most of the time we simply present an intuitive explanation and stop there. In particular, we skip over the formal machinery of limits almost completely, relying instead on a basic "what we are getting closer and closer to" notion that Newton, Leibniz, and Euler would have immediately recognized.

- We place integrals before anti-derivatives. In most calculus classes, anti-derivatives and the Fundamental Theorem of Calculus are presented to students before they fully understand what integrals actually are; they come away with the impression that integrals are just anti-derivatives. Later on, double integrals make no sense to them, because you can't take a double anti-derivative with respect to x and y. Here, integrals are developed as generalized sums, period. This allows us to apply integration to a host of problems that bear little superficial resemblance to the area under a curve or a distance traveled. Only after we learn the Fundamental Theorem of Calculus do we turn to the anti-derivative, both as the simplest way to compute most integrals, and as an important and useful concept in its own right.

This book was designed for a 1-semester course at the University of Texas. Naturally, when fitting what has been traditionally been a 2-semester sequence into one semester, something has to give. As noted above, we skip much of the formal structure of limits. We give short shrift to continuity and differentiability, merely presenting a few examples of discontinuous or nondifferentiable functions to build intuition. We also omit many of the calculational techniques that appear in traditional calculus texts. You will find no mention of l'Hôpital's rule, of implicit differentiation, of logarithmic differentiation, of integration by trigonometric (trig) substitution, or of integration by partial fractions. In an age when you can pick up your phone to look up derivatives and integrals online, knowing how to compute a complicated derivative or integral by hand is much less important than it used to be.

Origins. This book did not emerge from a vacuum. Much of its structure and terminology (especially Chapters 2 and 3) is based on the 1994 text *Calculus in Context*, by James Callahan, David Cox, Kenneth Hoffman, Donal O'Shea, Harriet Pollatsek, and Lester Senechal, who compose the Five Colleges Calculus Project. That in turn was based on a course developed by Michael Sutherland and Kenneth Hoffman at Hampshire College.

Calculus in Context stresses the role of calculus as "a language and a tool for exploring the whole fabric of science", with differential equations as a unifying theme. Fortunately for us, the Five Colleges Calculus Project had the support of the National

Science Foundation, with the condition that the results be made available for others to adopt and adapt. I am grateful for the opportunity (and the permission) to do so.

At the University of Colorado, Eric Stade developed his own course on Calculus, Stochastics and Modeling (Math 1310). This course was aimed at biology students, using *Calculus in Context* as the primary text, but also used a large number of home-grown worksheets and mini-projects that walk students through the difficulties. When Bill Wolesensky and I learned about Math 1310, we immediately wanted to teach a similar course at the University of Texas.

As a result, in 2014 Bill and I developed a one-semester calculus course for biologists (M408R) at the University of Texas at Austin that was explicitly modeled on Math 1310. Eric Stade generously shared all of his course materials, which we adapted for use by Texas students. Over time, we (mostly Bill) developed our own materials, and our course developed its own identity, with less emphasis on stochastics and differential equations and more emphasis on the Six Pillars, while maintaining *Calculus in Context*'s commitment to learning through tackling hard, realistic problems.

M408R was so successful that the McCombs School of Business asked us to develop a similar course for business students. At that point, it was clear that we would need our own textbook, as none of the existing business calculus books followed our approach. I took on the task of writing this text, first an edition for business students and then an edition for biology students. Eventually we intend to develop versions of the course, and editions of this text, for students in physics, engineering, and pure math.

Chapter-by-Chapter Summary. Chapter 1 addresses the question "What is calculus?" The short answer is "the study of things that change". A longer answer involves the Six Pillars. This is the shortest chapter in the book, but is also the most important, as it frames everything that comes after.

In **Chapter 2** we start predicting the future with mathematical models. We develop the SIR equations to model the spread of disease, and particularly the spread of Covid-19. Knowing where things stand today allows us to compute how quickly things are changing, which allows us to make projections for the near future. Repeating the process allows us to predict the more distant future. Studying when the derivative of the number of infected individuals is zero allows us to understand the conditions for the peak of an epidemic.

In **Chapter 3** we take up the 1st Pillar of Calculus (*Close is good enough*) in a more systematic way. After discussing approximate methods for solving algebraic equations, we turn to functions and ask how quickly they change. We study linear functions, where the rate of change is constant. We then approximate nonlinear functions by linear functions over short stretches where the rate of change is nearly constant. Euler's method for solving differential equations then emerges as a natural consequence of linear extrapolation. This approximation, and others introduced in this chapter, get better and better as the time step is reduced, which sets up the concept of a derivative. We close with the bisection method for finding approximate roots of a function and the amazingly accurate Babylonian method for approximating square roots.

In **Chapter 4** we turn to the 2nd Pillar (*Track the changes*) and define derivatives. First we approximate rates of changes with forward, backward and centered difference quotients. We also consider derivatives with respect to variable other than time, such as the amount of extra blood that the heart can pump per change in pulse, and we develop the approximation

$$(0.1) \qquad f(x) \approx f(a) + f'(a)(x - a),$$

which we call the *microscope equation,* because any smooth graph looks straight if you zoom in enough. We define the derivative $f'(a)$ as (1) the rate of change of the function at a, (2) the slope of the tangent line, (3) the multiplier in the best linear approximation near a, and (4) a limit of difference quotients, and we see why these are all the same. We look at the derivative *function* $f'(x)$ and we see how the value of $f'(x)$ relates to how fast $f(x)$ is changing. Using algebra, we compute the derivatives of a few simple functions, like x^2 and $1/x$.

Chapter 5, on computing and using derivatives, is similar to a chapter in a standard calculus text. We obtain the derivatives of polynomials, exponentials, sines, and cosines from the definitions. We then develop the product, quotient, and chain rules. By looking at where a derivative is zero, we see how to maximize or minimize a function, invoking the 3rd Pillar (*What goes up has to stop before it comes down*). We use the signs of $f'(x)$ and $f''(x)$ to understand the shape of a graph and develop the first and second derivative tests for maxima and minima. We finish with Newton's Method as an example of the power of linear approximation.

In **Chapter 6** we use differential equations to understand many common functions. Exponentials arise as solutions to $dy/dt = ry$, which describes everything from money earning interest to population growth to the cooling of a cup of coffee. Logarithms appear when we solve problems involving exponentials. Sines and cosines arise as ratios of sides of triangles, as locations of points on the unit circle, as solutions to the differential equations that describe the rotation of a wheel, and in simple harmonic motion. More complicated models of growth and oscillation include the logistic model of limited growth (also of the zombie apocalypse) the Lotka-Volterra model of interactions between predators and prey, related models of mutualism and competition, and models of chemical reactions.

Chapter 7 is about integration as a process of computing a bulk quantity by breaking it into pieces, estimating the pieces, and adding up the pieces. This is the 4th Pillar: *The whole is the sum of the parts.* Mostly we work with Riemann sums and definite integrals, but we also consider the accumulation function $A(x) = \int_0^x f(s)\,ds$, and note that $\int_a^b f(x)\,dx = A(b) - A(a)$.

Chapter 8 is about the Fundamental Theorem of Calculus. First we define the three key quantities: definite integrals, accumulation functions, and anti-derivatives. We have already seen how accumulation functions and definite integrals are related. The second Fundamental Theorem of Calculus (FTC2) relates anti-derivatives and definite integrals. If $F'(x) = f(x)$, then $\int_a^b f(x)dx = F(b) - F(a)$. This is proved directly with the 5th Pillar of Calculus: *One step at a time.* We divide the change in $F(x)$ from $x = a$ to $x = b$ into pieces, estimate each piece as $f(x)\Delta x$, and add up the pieces to get $\int_a^b f(x)\,dx$. The first Fundamental Theorem of Calculus (FTC1) states that the accumulation function is an anti-derivative, and is proved both directly and as a corollary of FTC2. Only after the meaning of integration and the importance of FTC2 are established do we take up anti-derivatives, turning many of our known facts about derivatives into facts about anti-derivatives.

Chapter 9 is about methods of integration. The first half of the chapter is about finding anti-derivatives through integration by substitution and integration by parts. (We do *not* treat trig substitutions, partial fractions, or any but the simplest integrals of trig functions.) The second half of the chapter is about numerical integration, in particular the trapezoidal rule and Simpson's rule. The two halves are grouped together in a single chapter to emphasize that integrals are generalized sums, and that anti-derivatives are (mostly) just a tool for computing integrals.

In **Chapter 10** we consider functions of two or more variables and study them with the 6th and final pillar: *One variable at a time.* The main theme is that multi-variable calculus is not fundamentally different from single-variable calculus. The main ideas are the same, but sometimes we need to do things several times, once for each variable. For instance, standard practice is to experiment with varying one control parameter at a time. This leads naturally to partial derivatives, linear approximations for functions of multiple variables, and the multi-dimensional chain rule. Double integrals are the same as single integrals, in that we break up the region of integration into small pieces, estimate the contribution of each piece, and add up the pieces. By organizing our pieces, summing first over one variable and then over the other, we convert a double integral into an iterated integral on which FTC2 can be applied.

Finally, in **Chapter 11** we return to the 1st Pillar with sequences and series. We define an infinite sum as the limit of finite sums. Saying that a certain quantity *is* an infinite sum is just a shorthand way of saying that it is *approximated* better and better by finite sums. We see that the geometric series $\sum_{n=0}^{\infty} r^n$ converges to $1/(1-r)$ when $|r| < 1$ and diverges otherwise. We then consider power series. By manipulating the power series $\sum_{n=0}^{\infty} x^n = 1/(1-x)$ we obtain series expansions for $\ln(1-x)$ and $\tan^{-1}(x)$. We then consider Taylor polynomials and Taylor series and Taylor's theorem with remainder, and we use Taylor polynomials to approximate functions and their integrals. We compute the power series of several common functions and derive Euler's formula $e^{ix} = \cos(x) + i\sin(x)$. Finally, we develop the main tests for convergence, in particular the root and ratio tests, and apply them to power series to obtain radii of convergence for Taylor series.

Dependencies and Possible Course Structures. The course M408R that we teach at the University of Texas does *not* cover the entire book. We usually cover Chapters 1–9. The remaining chapters are provided in case we (or you!) encounter an exceptionally strong class or can devote more than one semester to the material. For instance, at a school that uses quarters rather than semesters, Chapters 1–6 would make a coherent one-quarter course on differential calculus, to be followed by a second quarter covering Chapters 7–11.

The first eight chapters are laid out in a very definite sequence. It is hard to imagine skipping any of these chapters without completely losing the flow. The first two chapters are short and set the stage for the ideas that follow. Chapter 3 develops the concept of successive approximation that drives almost all of calculus. Chapter 4 develops the concept of the derivative, and Chapter 5 shows how to use derivatives. Exponential functions and trig functions are introduced early, but they only really come into their own in Chapter 6. Chapter 7 explains what integrals are, and Chapter 8 explains the Fundamental Theorem of Calculus. There is no getting around those core subjects.

As a result, almost nothing in Chapters 1, 4, 7, or 8 can really be skipped. However, it is certainly possible to skip some topics in Chapters 2, 3, 5, and 6 without losing continuity. In Chapter 2, it isn't necessary to cover extensions of the SIR model (Section 2.5) or business applications of the SIR model (Section 2.6). This material has considerable real-world significance, but it is included mostly to whet students' interest in modeling. Section 3.7, on the bisection and Babylonian methods of solving algebraic equations, can also be skipped. If you are downplaying the use of computers (an option that is *not* recommended!), then you may find Section 3.6, on implementing Euler's method by computer, to be superfluous.

Chapter 5 contains three applications of derivatives, for optimization (Section 5.4), graph sketching (Section 5.5) and Newton's method (Section 5.6). Most instructors include the first two and skip the third, but the choice is yours.

In Chapter 6, many students will not need the supplementary material (A Crash Course in Trigonometry), and especially well-prepared students may already know the material in Section 6.2 (Exponential Functions and Logarithms). If you are not interested in population dynamics or chemistry, then most of Section 6.5 and all of Section 6.6 can be skipped.

Chapters 9, 10, and 11 are stand-alone introductions to methods of integration, to multi-variable calculus, and to series. Depending on the interests and background of your students, it is reasonable to include any subset of these topics (including the empty subset).

In Chapter 9, the analytic and numerical sections are independent of one another. On the analytic side, I strongly recommend covering at least integration by substitution; integration by parts is optional. For the numerical methods, it is reasonable to include the trapezoidal rule and skip Simpson's rule. If the rest of Chapter 9 is skipped, then Section 9.1 (Integration by Substitution) can be combined with Section 8.5 (Computing Anti-Derivatives).

Within Chapter 10, it is possible to do partial derivatives without doing double integrals, or to do double integrals without partial derivatives. Within the sections on partial derivatives, it is possible to skip the discussion of higher partial derivatives, the "mixed partials are equal" theorem, and quadratic approximations.

Chapter 11 is aimed at a more sophisticated audience than the rest of the book, covering a lot of ground in relatively few pages. The first four sections are essential to understanding what series and power series are and how to use them. Sections 11.5 and 11.6 are about tests for convergence, determining when and where the material of the first four sections can be meaningfully applied. The bulk of a traditional treatment of sequences and series is devoted to convergence; testing for convergence is a wonderful exercise in open-ended mathematical thinking. However, many successful scientists and engineers have gotten away with cavalier attitudes towards convergence, especially since the use of Taylor *polynomials* (as opposed to series) only requires differentiability, not analyticity.

Course development is a collaborative process. If you have questions on how to use this book in your course, and especially if you have feedback on what did and didn't work with your students, please contact me directly.

Acknowledgments. This book would not have been possible without the efforts of Bill Wolesensky, who has been my partner in developing the courses M408R (Calculus for Biologists) and now M408Q (Calculus for Business) at the University of Texas. The structure of this book reflects the structure of those courses, which in turn reflects Bill's ideas as much as my own.

This book would also not have been possible without the pioneering work of the Five Colleges Calculus Project (James Callahan, David Cox, Kenneth Hoffman, Donal O'Shea, Harriet Pollatsek, and Lester Senechal). Not only are the early chapters of this book consciously modeled on their wonderful text *Calculus in Context*, but the courses at the University of Colorado and the University of Texas that directly led to the book you are holding would never have been designed without *Calculus in Context*. The Five Colleges Calculus Project in turn was supported by the National Science Foundation. I am grateful both for the inspiration that *Calculus in Context* provided, and for permission to adapt material from that book.[1] In particular, many of the exercises in this book are modeled on exercises in *Calculus in Context*. Even more are inspired by *Calculus in Context*'s general approach to the subject. Several computer programs in this book, including SIREulers, Riemann, and the unnamed implementation of the Babylonian method, are MATLAB implementations of similar programs in *Calculus in Context*.

Special thanks go to Eric Stade. As already noted, his course on *Calculus, Stochastics and Modeling* (Math 1310) at the University of Colorado was an essential link between *Calculus in Context* and M408R/Q. The initial version of M408R ran primarily on adaptations of Math 1310 course materials that Eric generously shared, and the current course still bears Eric's fingerprints. Bill Wolesensky and I are very happy to enjoy the view from the shoulders of Eric Stade and the Five Colleges Calculus Project.

I am also grateful to Keenan Kidwell and Alex Macedo, who have been teaching M408Q and M408R, semester after semester, and have provided invaluable feedback on what works and what doesn't.

This book would have been *nearly* impossible without the efforts of Anita Sadun. Not only has she provided me with love, inspiration, and moral support through 30+ years of marriage, but she's a darn good illustrator and editor! Nearly half of the figures in this book are Anita's handiwork. I also thank my brother, Alfredo Sadun, who helped me put a realistic biological spin on many of the key ideas.

[1]Source: Wikimedia Commons, authors: Callahan, James; Cox, David; Hoffman, Kenneth; O'Shea, Donal; Pollatsek, Harriet; and Senechal, Lester, "Calculus in Context" (2008). Open Educational Resources: Textbooks, Smith College, Northampton, MA., licensed under the Creative Commons Attribution 4.0 International (CC BY 4.0) (https://creativecommons.org/licenses/by/4.0/) license.

Development of this book was supported by the Texas Institute for Discovery Education in Science (TIDES), in consultation with faculty in the Department of Mathematics, notably Mike Starbird and Mark Daniels.

Finally, I thank everybody who helped me to realize (well into my career!) that teaching is about engaging with *students*, and not just engaging with the material. This includes faculty peers and mentors such as Jane Arledge, Jennifer Austin, Sarah Eichhorn, Sacha Kopp, and Elizabeth Stepp, but especially includes the students of the University of Texas.

Lorenzo Sadun
Austin, Texas
sadun@math.utexas.edu
July 2022

Chapter 1

What is Calculus?
The Six Pillars

Calculus was invented by Isaac Newton and Gottfried Leibniz in the mid-17th century to solve real problems: To understand the orbits of planets. To understand what happens when an apple falls from a tree. To design lenses and clocks and all sorts of machines. Even to make money betting at dice. Over time, people realized how much calculus had to say about almost *everything* in the world. If it's real and it's changing, we can understand it better with calculus.

But as a 21st-century student learning calculus, it's easy to get lost in a maze of formulas, computational techniques, obscure theorems, and contrived problems. You can lose sight of what it's all for.

So what is calculus, anyway? A good working definition is

> *Calculus is the study of things that change.*

- If a plant absorbs solar energy at a rate of 1000 watts, and if the sun is up for 12 hours each day, then the plant absorbs

$$1000 \text{ watts} \times 12 \text{ hours} = 12,000 \text{ watt-hours} = 12 \text{ kilowatt-hours}$$

  of energy each day. That's simple arithmetic. But if the amount of solar radiation changes over the course of the day, depending on the angle of the sun, then figuring out the plant's total energy intake involves adding up the *variable* intakes for the different hours of the day. That's calculus.

- With every beat, your heart delivers oxygen-rich blood to your organs and muscles. You need more oxygen when you exercise, so your heart beats faster. However, your heart is also less efficient when it beats faster, so if it beats too fast it will deliver less oxygen. At what heart rate do your muscles get the most oxygen? That's calculus.

- If a lake has 1000 fish, and if the fish population is growing at 100 fish/year, then it is easy to predict how many fish there will be next year (1100), the year after (1200) or the year after that (1300). But what if population growth slows down as the lake becomes more crowded? Understanding how that changes the answer requires calculus.

Fortunately, calculus isn't black magic. Almost everything we do boils down to six simple ideas that we call the **Six Pillars of Calculus**:

(1) **Close is good enough.** In algebra, we look for exact answers to problems. If somebody asks you to solve $x^2 - 3x + 2 = 0$, you might factor that as $(x-2)(x-1) = 0$, or you might apply the quadratic formula, or you might just try different numbers and see that $x = 1$ and $x = 2$ work. What you probably don't say is "There must be a solution somewhere between 0.9 and 1.1 ..., make that 0.99 and 1.01 ..., no, make that 0.999 and 1.001." But that's exactly the sort of reasoning we use in calculus! When faced with problems that can't be solved right away, we simplify things and look for an approximate answer. Then we try to improve that answer, and improve it some more, until we get something that is fairly accurate. If the calculations are too grungy to do by hand (and they often are), we program a computer to do our dirty work for us. Eventually, we get an answer that's good enough for whatever real-world task we have in mind. Then we stop and congratulate ourselves on a job well done.

Once in a while, we need the exact answer. When that happens, we approach that exact answer as the **limit** of better and better approximations. Almost all of the important formulas of calculus come from this limiting process.

(2) **Track the changes.** We can often tell more about something by looking at the way that it's changing than by asking where it is right now.

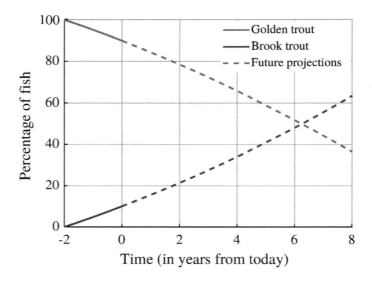

Figure 1.1. Which subspecies of trout is winning?

The Californian golden trout is the state freshwater fish of California. However, it is endangered due to competition from invasive species like brook trout that were introduced (either accidentally or on purpose) into its habitat. Suppose that a stream has 90% golden trout and 10% brook trout, but that 2 years ago it was almost all golden trout and that last year it had 95% golden trout and 5% brook trout, as shown in Figure 1.1. Which subspecies of trout is doing better?

If we only looked at current numbers, we might think that the golden trout were doing fine, as they outnumber the brook trout almost 10:1. However, if we look at the rate of change, we see that the brook trout population has doubled in the past year, while the golden trout population has declined. Within 10 or 20 years, the brook trout are likely to drive the local golden trout population to extinction.

Graphically, the rate of change of a quantity is closely related to the **slope** of the graph of that quantity. We're going to spend a lot of time studying slopes—both how to find them and what to do with them. In mathematical terms, this is called computing a **derivative**.

Figure 1.2. A schematic graph of blood flow vs. heart rate. The overall shape and scale of the graph are realistic, but the details are not medically accurate.

(3) **What goes up has to stop before it comes down.** The amount of blood that a typical 40-year-old person's heart pumps as a function of how fast it beats follows a curve similar to that shown in Figure 1.2. If it is currently beating at 60 beats per minute (bpm), then it can pump more blood by pumping faster, which is what happens when the person starts exercising. This part of the curve slopes upwards, so more beats mean more blood. However, if the heart is currently beating at 220 bpm, then it needs to slow down.[1] This part of the curve slopes downwards, so more beats means less blood. Peak blood flow is at 180 bpm, where the curve is flat and its slope is zero. By studying the rate of change of the curve and figuring

---

[1] Heart rates that high are very dangerous and can trigger fibrillation and cardiac arrest. The top of the curve is considered "maximum heart rate" (MHR) and athletes are generally advised to keep their heart rates below 85% of MHR.

out where that rate of change is zero, we can determine how much our hearts can
(and can't) do.

(4) **The whole is the sum of the parts.** How much carbon dioxide (CO_2) will be
in the atmosphere next year? That's a complicated question, because CO_2 enters
and leaves the atmosphere in many ways, including plants converting CO_2 into
sugar, that sugar decomposing back into CO_2 when the plant is eaten or rots, vol-
canic activity, and of course the burning of fossil fuels. By adding up all of these
processes, we can find the net emission for each year: the amount of CO_2 put into
the atmosphere minus the amount taken out.

Let's break things down year by year. The amount of CO_2 in the atmosphere
next year is the amount there is now, plus this year's net emission. Similarly, this
year's CO_2 is last year's CO_2 plus last year's net emission. Working backwards,
year by year, we see that the CO_2 content of the atmosphere this year is the content
a million years ago, plus the sum of all the net emissions in the last million years
(roughly half of which have been added since 1950).

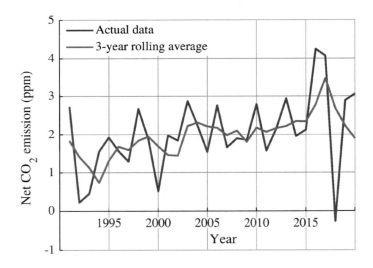

Figure 1.3. Global net CO_2 emissions by year, 1991–2020

If we plot the net emissions as a function of time, as in Figure 1.3, the sum of
all those values is the same as the area under the curve. We're going to spend a
lot of time talking about area, but it isn't because we're obsessed with geometry.
It's because most bulk quantities, like the total power used by the city of Austin
in 2007 or the total amount of water carried by the Colorado River in 2011, can be
treated exactly like CO_2 concentration or area. Break the quantity you're studying
into little pieces, estimate each piece, and add up the pieces. This process is called
integration.

Note that this is very different from how integration is often taught. In most
calculus classes, students are taught that integration is in some sense the opposite
of differentiation. That isn't exactly wrong, as we'll see with the 5th Pillar, but it

misses the point. Integration is about adding up the pieces, which is why it applies to a host of problems. Anti-derivatives are a great tool for actually computing integrals, but they don't explain why such-and-such quantity is represented by such-and-such integral. For that, we need the 4th Pillar.

(5) **One step at a time.** A famous Chinese proverb says, "A journey of a thousand miles begins with a single step". The journey then continues with about two million additional steps. By understanding what happens at each step, we can understand the entire journey.

Instead of asking what the concentration of CO_2 *is*, we can ask how much it *changed* in a short period of time, like a year. Geometrically, the change per year is the slope of the concentration curve, shown in Figure 1.4.[2] The steeper the curve is, the faster the concentration is changing.

In the 1990s, net emissions were typically between 1 and 2 parts per million (ppm) per year, while in the 2010s they were typically between 2 and 3 ppm per year. You can see that the concentration curve is steeper (has a higher slope) in the 2010s than in the 1990s.

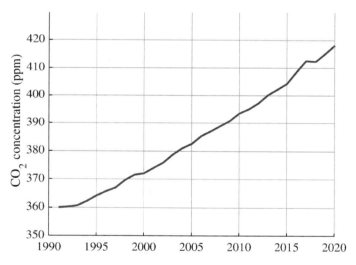

Figure 1.4. Peak concentration of CO_2 by year, 1990–2020

How are Figures 1.3 and 1.4 related? The new emissions in one year is the same as the change in the concentration from that year to the next. This means that the concentration is a running total of the net emissions and corresponds to the **area** under the net emission curve. The net emissions is the rate at which the concentration is changing and corresponds to the **slope** of the concentration curve. Mathematically, the concentration is the **integral** of the net emission and the net emission is the **derivative** of the concentration. The **Fundamental Theorem of Calculus** (FTC) relates derivatives and integrals in general. With it, we

---

[2]The concentration of CO_2 varies by season. The graph shows the peak value in May of each year. On May 8, 2018, the detectors measured a concentration of 418.74, more than 6 ppm higher than on any other day that year. That outlier is not included in the graph.

can use what we already know about derivatives to understand integrals. In fact, it is the single most powerful tool we have for evaluating integrals.

The FTC comes from thinking about changes one step at a time, but there is more to "one step at a time" than just the fundamental theorem. For instance, suppose that a herd of wild animals, currently containing 1000 individuals, is growing at 6% per year. How many animals will there be in 30 years? Using our initial population and the growth law, we can estimate the net births minus deaths in the first year, and from that we can compute the population a year from now. From that, we can estimate the net births minus deaths in the second year and get our population two years from now. Continuing the process, step by step, we can accurately project the population of the herd far into the future.

Finally, "one step at a time" is great slogan for how to approach calculus in general. Many problems are way too complicated to be done all at once. By breaking big problems into sequences of smaller problems, and by solving these smaller problems one at a time, we can accomplish wonders.

(6) **One variable at a time.** Many functions involve two or more input variables. The boiling point of water depends on our elevation and on how much salt we put in the water. In the summer, the heat index depends on temperature and humidity. In the winter, the wind chill factor depends on temperature and wind speed. The price of widgets depends on supply and demand. The value of an oil field depends on how much oil it produces and on the price of oil.

To understand functions of two (or more) variables, we always hold everything but one variable fixed and study just that variable. Asking how a change in temperature changes the heat index is a question about a function of just one variable that we already know how to answer. Similarly, we can figure out how changing the humidity affects the heat index. Putting the two answers together, we can understand how changing both temperature and humidity affects the heat index.

It's tempting to say "that's all there is!", but that isn't really true. Over the centuries, lots of really smart people have cooked up lots of really smart ways to solve lots of really hard problems. Now, with the help of computers, we can solve even more problems. In the next ten chapters, we're going to follow in the footsteps of these masters and learn some of their results. Yes, there will be formulas to memorize and techniques to practice and algorithms to implement on computers. And yes, it will take work to really absorb everything.

But hopefully this voyage of discovery won't be a mystery. Every new formula or technique or algorithm will lead straight back to the Six Pillars, so it will be connected to every other formula or technique or algorithm. If you remember the pillars, you'll have a framework for organizing all the little details. Over time, you'll forget many of those details, and that's OK. As long as you remember the pillars and are willing to look up the details as needed, you'll be able to use calculus for your whole life, not just in your classes and in your job, but in understanding the wild and complicated world we live in.

Chapter 2

Predicting the Future: The SIR Model

In December 2019, a strange form of pneumonia started affecting residents of Wuhan, China. The virus causing this disease was quickly identified and the disease was named COVID-19, for "**CO**rona**VI**rus **D**isease 20**19**".[1] By April 2020, Covid-19 had spread around the world and had mutated into a more transmissible form, leading to drastic lockdowns to control the spread. Schools closed or went online, most people were told to stay home except for essential errands like buying food, and everybody was told to practice "social distancing". Despite these measures, Covid-19 continued to spread, and by mid-2022 had killed at least 6 million people worldwide (with some estimates being twice that number) including over a million Americans. By the time you read this, the death toll will almost certainly be even higher. Throughout the pandemic, everybody wanted to know what would happen next, which kept mathematical modelers (including the author) very busy.

In this chapter we will use four of the Pillars of Calculus to get a handle on the spread of disease.

- It's impossible to keep track of the behavior of every person in the world, so we devise a simplified model to describe average behavior. *Close is good enough!*

- This involves understanding the rate at which people become sick and the rate at which they recover. *Track the changes!*

- Once we understand where things stand and how fast they are changing, we can make realistic predictions about what is likely to happen in the future, as well as what we can do to improve that trajectory. Our projections are only accurate for a short time, so we combine a lot of short-term projections to get a long-term projection. *One step at a time!*

---

[1] After a while, people dropped the "19", stopped capitalizing the whole word, and just wrote "Covid" or "covid".

- Even without these long-term projections, we can understand how the epidemic will peak by comparing the factors that increase infection with those that decrease infection. *What goes up has to stop before it comes down!*

We will begin with a fairly simple model, called the SIR model, that works very well with diseases ranging from ebola to measles to the common cold. We'll see how to set up the model, well see how to use the model to give quantitative predictions of how an epidemic will run, and we'll see how to analyze the model theoretically. Covid has features that make it slightly harder to model, but we'll return to Covid at the end of the chapter and see how modelers tweaked the SIR model to understand the pandemic.

2.1. Worried Sick

Suppose that you are the principal of a school where a few students have just come down with measles. You've heard that measles is very contagious and you're worried about the disease spreading. Do you need to shut the school down? Or is it enough just to tell parents to keep their sick kids at home? In order to make sense of the situation, you will need to run different scenarios and forecast the number of sick children for the next month or two. You hire a mathematical consultant (your Friendly Author), and together we attack the problem.

The first thing we do is define our quantities. Let t denote time, measured in days, and let $I(t)$ be the number of infected children at time t. We call t and I **variables** because they change. The time t is our **input variable**, while I is an **output variable**. The graph of I as a function of t will look something like the plot in Figure 2.1. There is an initial phase where more and more people get sick, a moment of peak infection, and a long tail as everybody recovers. We want the rise to be as slow as possible, the peak to be as low as possible, and the decline to be as fast as possible.

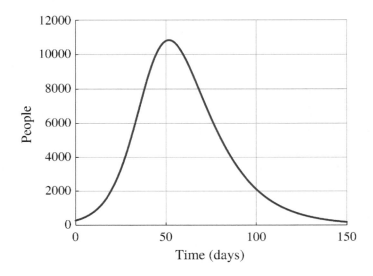

Figure 2.1. Infected children as a function of time

Before we try to control I, we need to understand what makes I change over time. That is, we need to make a mathematical **model** for what is going on. The model needs to take into account all the *important* features of what's happening in the world, while being simple enough to be solvable. We are going to ignore a *lot* of details, because *close is good enough!*

Once we have our model, we have to **analyze** it. In this step, we don't care where our equations came from. We just have to solve them. Maybe we can find a formula for the answer. More likely, we can't find a formula, but we can run the model on a computer to generate accurate predictions. Once again, *close is good enough!*

Finally, we need to **interpret** our results. Math can tell us that such-and-such variable will have such-and-such value at such-and-such time, but we need to understand biology to say what that means for the health of our school.

In other words, predicting the future is a three-step process:

(1) **Model** our system mathematically. Define appropriate variables and write down some equations that describe how these variables change with time. This step requires real-world understanding as well as math.

(2) **Solve** the model to determine what each variable will be at some future time. This step is 100% math, using techniques that we are about to develop.

(3) **Interpret** the results. Take our mathematical results and make real-world sense of them.

We'll tackle these one at a time, which is a lesson in itself. *One step at a time.* We need to break hard problems into bite-sized tasks, and then do each task in turn.

2.2. Building the SIR Model

To understand how germs spread, we divide our population of students (and maybe teachers and staff) into three groups, as in Figure 2.2:

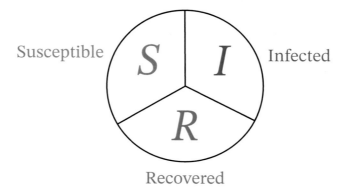

Figure 2.2. Three categories of people

- **Susceptible** people, or *Susceptibles*. These are people who might get sick if they're exposed to the disease.
- **Infected** people, or *Infecteds*: These are people who have already gotten sick. Not only are they sick, but they are spreading germs to everybody they meet.
- **Recovered** people, or *Recovereds*: These are people who can't get sick. Maybe they used to be sick and then recovered. Maybe they were vaccinated. Maybe they have some natural immunity that protects them. For whatever reason, they're no longer in the game.

Next we define our variables. Let $S(t)$ be the number of Susceptibles at time t, let $I(t)$ be the number of Infecteds at time t, and let $R(t)$ be the number of Recovereds at time t. Obviously, there are differences among the people in each category, but we're going to ignore those differences. This is a model, not a complete description of reality! Instead, we're going to talk about the behavior of the *average* Susceptible, the *average* Infected, and the *average* Recovered. We then ask two questions:

(1) At what rate do Susceptibles get exposed and become Infecteds?

(2) At what rate do Infecteds clear their bodies of germs (but not antibodies!) and become Recovereds?

From these rates, we can figure out the rates S', I', and R' at which the quantities S, I, and R are changing, and from those we can predict the future.

Losing Patients Through Recovery. Diseases typically don't last forever. Measles lasts about two weeks.[2] Of course the duration can vary from one person to the next, but because it takes an average of 14 days to recover from measles, it's a good approximation to say that about 1/14 of all the measles patients (Infecteds) will recover each day. That is,

(2.1)   today's change in the Recovered population $\approx \dfrac{I(\text{today})}{14}$.

Likewise,

(2.2)   tomorrow's change in the Recovered population $\approx \dfrac{I(\text{tomorrow})}{14}$,

and in general

(2.3)   the change in the Recovered population on day $t \approx \dfrac{I(t)}{14}$.

Note the units in this equation. $I(t)$ and $R(t)$ are numbers of *people*, and the number of new R's on any given day is also measured in people. But the *rate R'* at which R is changing is measured in people/day. That is,

$$R'(t) = \frac{I(t)}{14 \text{ days}}.$$

The numerator has units of people, the denominator has units of days, and the ratio has units of people/day.

---

[2]There are several stages to measles, including incubation, a period of high fever, and a period including the telltale rash, all spread over a total of two to three weeks. Because the period when you can infect others lasts about two weeks, we're going to simplify things and just say that the disease lasts two weeks.

This is an example of a **rate equation**. A rate equation describes the rate at which something is changing in terms of other data. In this case, it describes R' in terms of I.

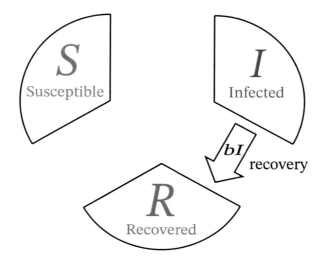

Figure 2.3. With the passage of time, Infecteds become Recovereds.

More generally, if T is the average length of time that people stay sick, then we expect

(2.4) $$R' = \frac{I}{T}.$$

If we define $b = 1/T$, then we can write our rate equation without fractions;

(2.5) $$R' = bI;$$

see Figure 2.3. R' has units of people/day, I has units of people, and b has units of (1/day)s. The parameter b is called the **recovery coefficient**, because it says how quickly people recover.

The number b is called a **parameter**. It isn't a variable, because it doesn't change during the epidemic. However, it may take on different values for different diseases or different populations.[3] By adjusting the value of b, we can use the same rate equation to model ebola, the common cold, seasonal flu, and many other diseases as well.

Gaining patients: Transmission. Recovery is good. If a city has 21,000 people sick with measles, then people recover at a rate of

(2.6) $$\frac{21{,}000 \text{ people}}{14 \text{ days}} = \frac{1500 \text{ people}}{\text{day}}.$$

As long as fewer than 1500 people get sick each day, the epidemic will subside.

To understand transmission, let's look at things from the perspective of a single Susceptible, who we'll call Joe. Joe will come in contact with a certain fraction p of

---

[3]For instance, measles is more dangerous in adults over 20 and in children under 5 than in teenagers or children over 5. The average duration depends on the age of the patient.

the Infecteds each day, for a total of pI contacts. The fraction p is typically very small, but the Infected population I can be very large. The more Infecteds there are, the more chances there are of Joe being exposed to the disease. Every time that Joe comes in contact with an Infected, there is a probability q that he will breathe in enough germs to become sick himself. That is, there is a probability pqI per day of Joe becoming infected.

We don't actually care about p and q separately, as all that matters is their product. We define $a = pq$, and call a the **transmission coefficient**. Like b, this is a parameter, not a variable. Different diseases in different communities will have different values of a, but we can use the same reasoning for all of them.

Finally, we look at the entire population of Susceptibles, and not just at Joe. If there are S Susceptibles, each of whom has a probability aI (per day) of becoming an Infected, then each day there will be approximately aSI Susceptibles who become Infecteds. That is,

$$(2.7) \qquad\qquad S' = -aSI;$$

see Figure 2.4. S and I have units of people and S' has units of people/day, so a must have units of $1/(\text{people} \times \text{days})$. Also note the minus sign in our equation! The more Susceptibles become Infecteds, the fewer Susceptibles are left.

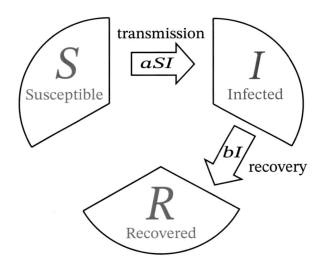

Figure 2.4. Through contact with Infecteds, Susceptibles become Infected.

What about I? Completing the model. So far we have figured out how the numbers of Recovereds and Susceptibles change, but what about the number of Infecteds? We have

$$
\begin{aligned}
I' &= +(\text{Rate at which people get sick}) \\
&\quad - (\text{Rate at which they recover}) \\
&= aSI - bI.
\end{aligned}
$$

(2.8)

Another way to see this is that $S + I + R$ is the total number of people out there, and that doesn't change. Because $S' + I' + R' = 0$, we must have $I' = -S' - R'$.

Putting everything together, we have a system of three rate equations. Together, they're called the **SIR model**:

$$(2.9) \qquad \begin{aligned} S'(t) &= -aS(t)I(t), \\ I'(t) &= aS(t)I(t) - bI(t), \\ R'(t) &= bI(t), \end{aligned}$$

where

- t is the time. Depending on the setting, you may want to measure t in days, weeks, months, or years.

- $S(t)$ is the number of Susceptibles at time t, measured in people.

- $I(t)$ is the number of Infecteds at time t, measured in people.

- $R(t)$ is the number of Recovereds at time t, measured in people.

- b is the recovery coefficient, which is the reciprocal of the average time T that the disease lasts. The units of b are the reciprocal of whatever units we are using for time. This parameter can vary from disease to disease.

- a is the transmission coefficient. This can depend both on the disease and on the community. If there are outbreaks in several different places, b will be more or less the same in each place, but a will typically be bigger in smaller communities (where each Susceptible knows a greater fraction of the Infecteds) and smaller in larger communities. The units of a are $1/(\text{time} \times \text{people})$.

2.3. Analyzing the Model Numerically

Now that we have our model, let's use it to predict the future. Suppose that a disease that lasts an average of 15 days is spreading across a college campus of 50,000 people. We measure t in days, we suppose that $a = 0.000004/\text{person-day}$ and $b = \frac{1}{15}/\text{day}$, and we start with $S(0) = 40,000$, $I(0) = 2100$, and $R(0) = 7900$. What will S, I, and R be in two days? In five days? In ten days? For that matter, what were the values yesterday? A week ago? A month ago?

A naive approach is to use the SIR equations to compute S', I', and R' once and for all:

$$(2.10) \qquad \begin{aligned} S' &= -0.000004(40,000)(2100) &= -336 \text{ people/day}, \\ I' &= 0.000004(40,000)(2100) - 2100/15 &= 196 \text{ people/day}, \\ R' &= 2100/15 &= 140 \text{ people/day}. \end{aligned}$$

If we have 2100 Infecteds on day 0, and if I is growing at a rate of $I' = 196$ people/day, then we should expect $2100 + 196 = 2296$ Infecteds tomorrow. We should expect $2296 + 196 = 2100 + 2(196) = 2492$ Infecteds the day after tomorrow. After a week, we should expect $2100 + 7(196) = 3464$ Infecteds, after a month we should have

$2100 + 30(196) = 7980$ Infecteds, and after a year we should have $2100 + 365(196) = 73,640$ Infecteds. In general, after t days we should have

$$(2.11) \qquad I(t) \approx I(0) + I'(0)t = 2100 + 196t.$$

This is called a **linear approximation**. Instead of finding the exact equation for $I(t)$, we found the equation of the line that has the right value and the right slope at $t = 0$. We then approximate the value of the true $I(t)$ function by the value of the linear function $2100 + 196t$.

Before we move on, let's recall some basic facts about things that change at a constant rate. If we are driving at a constant speed of 57 miles per hour and pass milepost 253 at 3:00, where will we be at 4:00? At 5:00? At time t?

From 3:00 to time t is $t - 3$ hours. As rate × time = distance, and as we are going at 57 mph, we will travel $57(t - 3)$ miles in that time. Adding that to our starting point at milepost 253, we will find ourselves at milepost

$$(2.12) \qquad x(t) = 253 + 57(t - 3)$$

at time t. (At least if we're traveling in the direction where the mile markers are increasing. Otherwise we might find ourselves at milepost $253 - 57(t - 3)$.)

The same idea works for any quantity that is changing at a constant rate, not just for position. If a quantity Q is changing at rate Q', and if Q starts at a value Q_0 at time t_0, then what will Q be at time t? Because it grows at rate Q' for time $(t - t_0)$, it will increase by $Q' \times (t - t_0)$. Adding that to our starting value of Q gives

$$(2.13) \qquad Q(t) = Q_0 + Q' \times (t - t_0).$$

This is the equation of a straight line in **point-slope form**. The linear approximation (2.11) is a special case of this, with I instead of Q, and with starting time $t_0 = 0$. We'll have a lot more to say about equations of lines and linear approximations in Chapter 3.

Returning to the SIR model, we can use the linear approximation (2.11) to study the past as well as the future. According to this approximation, yesterday we had around $2100 - 196 = 1904$ Infecteds, a week ago we had $2100 - 7(196) = 728$ Infecteds, and a month ago we had $2100 - 30(196) = -3780$ Infecteds.

How much do you trust those numbers? You should take them with a grain (or more) of salt. In particular, the estimate $I(-30) \approx -3780$ is absurd. You can't have a negative number of Infected people! You also can't have $I(365) = 73,640$, since that's more than the entire population.

In making those estimates, something went seriously wrong. You should get into the habit of always asking yourself whether answers make sense. Reality check! Before reading on, take a minute to think about what went wrong with the methods we used to get our negative answer and our overly positive answer.

The problem with our linear approximation is that it assumes that the rate of change of I is 196 people/day, always was 196 people/day, and always will be 196 people/day. In reality, we know that $I' = 196$ people/day *today*, and it's realistic to assume that I' won't change much in the next few days, but in a few weeks, or a few months,

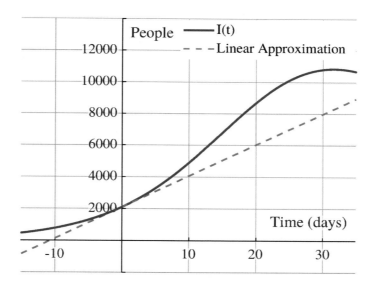

Figure 2.5. The linear approximation is negative when $t < -10.75$, but the actual $I(t)$ isn't.

it could change by a lot. This means that we can trust our linear approximation when $t = 1$ or $t = 2$ or $t = -1$ or $t = -2$, but we shouldn't trust it when $t = 30$ or 365 or -30. The actual situation is shown in Figure 2.5. The linear approximation tracks $I(t)$ very closely for $-2 < t < 2$, but it does not account for the curvature in the graph of $I(t)$. As t gets more and more negative, the graph of $I(t)$ flattens out and stays positive, while the linear approximation goes negative. Meanwhile, as t gets more and more positive, $I(t)$ goes up and then down, while the linear approximation just keeps on growing.

To do better than our naive approximation, we need a way of estimating S' and I' and R' not just right now, but also in the future and the past. Here's a way to estimate $S(10)$, $I(10)$, and $R(10)$.

First, we need to decide how far we can trust our linear approximation. Since I seems to be changing at about 10% each day, trusting it for a couple of days is reasonable. We're going to pick a time interval of two days, which we'll call Δt days, and predict the future $\Delta t = 2$ days at a time.

(1) Use the initial values $S(0)$, $I(0)$, and $R(0)$, together with the SIR equations, to estimate $S'(0)$, $I'(0)$, and $R'(0)$.

(2) Use a linear approximation to estimate $S(2) \approx S(0) + 2S'(0)$, as well as $I(2) \approx I(0) + 2I'(0)$ and $R(2) \approx R(0) + 2R'(0)$.

(3) Use the values of $S(2)$, $I(2)$, and $R(2)$, together with the SIR equations, to estimate $S'(2)$, $I'(2)$, and $R'(2)$.

(4) Use a linear approximation to estimate $S(4) \approx S(2) + 2S'(2)$, etc.

(5) Use those values and the SIR equations to approximate $S'(4)$, etc.

(6) Lather, rinse, repeat. At each time t, plug the estimated values of $S(t)$, $I(t)$, and $R(t)$ into the SIR equations to get estimated values of $S'(t)$, $I'(t)$, and $R'(t)$. Then

Table 2.1. Projecting forward with $\Delta t = 2$

t	$S(t)$	$I(t)$	$R(t)$	$S'(t)$	$I'(t)$	$R'(t)$
0	40,000	2,100	7,900	-336	196	140
2	39,328	2,492	8,180	-392	226	166
4	38,544	2,944	8,512	-454	257.5	196.5
6	37,636	3,459	8,905	-520.5	290	230.5
8	36,595	4,039	9,366	-591.5	322	269.5
10	35,412	4,683	9,905			

use a linear approximation to compute $S(t + \Delta t) \approx S(t) + \Delta t\, S'(t)$, etc. Table 2.1 shows the results for the first ten days.

Likewise, we can go backward in time, using a small negative time step. We can take $\Delta t = -2$ days and go from $t = 0$ to $t = -2$ to $t = -4$, etc., all the way back to to $t = -30$. The results are shown in Table 2.2, with some times skipped and the values of S', I', and R' omitted to save space. As you can see, $I(-30)$ is *not* negative.

Table 2.2. Projecting backward with $\Delta t = -2$

t	$S(t)$	$I(t)$	$R(t)$
0	40,000	2,100	7,900
-6	41,683	1,109	7,208
-12	42,591	562	6,848
-18	43,055	278	6,667
-24	43,286	136	6,577
-30	43,400	66	6,534

This method still doesn't give exact answers, but *close is good enough!* If we want to compute $I(10)$ with greater accuracy, we can use the same algorithm with $\Delta t = 1$ instead of $\Delta t = 2$, for an answer of $I(10) = 4783$ instead of 4683. That requires ten iterations instead of five. If we want even more accuracy, we can take $\Delta t = 0.1$ and do 100 iterations, getting $I(10) = 4876$, or take $\Delta t = 0.01$ and do 1000 iterations, getting $I(10) = 4886$. If we're willing to do the extra work, or if we program a computer to do the extra work for us, we can have as much accuracy as we want. (However, our model is only an approximation of the real world, so even if we can solve our model to great accuracy, that doesn't necessarily mean that we can predict the future with that much accuracy.)

This is the 5th Pillar of Calculus: *One step at a time.* Every big change is made up of many little changes. If we can understand each little change, we can put the pieces together to understand the big change.

2.4. Theoretical Analysis: What Goes Up Has to Stop Before It Comes Down

So far we have used the 1st, 2nd, and 5th Pillars of Calculus. By making approximations and tracking the changes in our variables S, I, and R and by putting a lot of short-term linear approximations together, we figured out how to obtain good projections of the future, and we were able to use those same projections to understand the past. Now we're going to tackle the question:

What is happening when I(t) reaches its peak?

The key fact is that *the sign of I′ tells you whether I is increasing or decreasing.* Whenever $I'(t) > 0$, $I(t)$ must be increasing. Whenever $I'(t) < 0$, I must be decreasing. At the very top of the curve, when $I(t)$ has stopped increasing and hasn't yet started decreasing, $I'(t)$ transitions from positive to negative. At that instant of time, we must have $I'(t) = 0$.

Let's figure out what is happening at that time. The SIR equations tell us that

$$(2.14) \qquad I'(t) = aS(t)I(t) - bI(t) = I(t)(aS(t) - b).$$

Since $I(t)$ is always positive, the sign of $I'(t)$ is the same as the sign of $aS(t) - b$. As long as $S(t) > b/a$, $I'(t)$ will be positive and $I(t)$ will increase. But eventually we will run out of Susceptibles. When $S(t)$ drops below b/a, $I'(t)$ will become negative and $I(t)$ will start to decrease.

The number b/a is called a **threshold**. In our example, the threshold was 16,667. This is *not* the value of $I(t)$ at the peak. Rather, it is the value of $S(t)$ when $I(t)$ hits its peak. In Figure 2.6, $I(t)$ hits its peak value of 10,870 when $t = 31.6$, which is when $S(t)$ passes through $16,667$.

If the threshold b/a is large and there aren't many Susceptibles to begin with, the outbreak will fizzle from the start. If b/a is small and $S(0)$ is large, there will be a

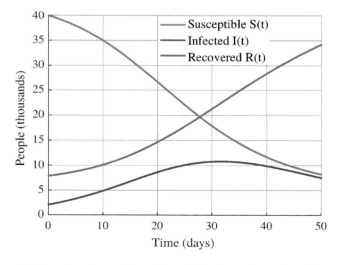

Figure 2.6. Over time, the number of Susceptibles drops, the number of Recovereds rises, and the number of Infecteds rises, reaches a peak, and then falls.

long period of growth before the epidemic peaks. As a public health consultant, our considerations for controlling the outbreak are the following:

(1) Make b as big as possible. Equivalently, make the duration $T = 1/b$ of the disease as small as possible. The less time that you're sick, the less time you have to infect others. Unfortunately, this is usually very difficult, especially with viral diseases. For instance, anti-viral drugs can sometimes reduce the length of the flu, but only by a small amount. A better approach is through testing and quarantining. Even if somebody hasn't *recovered* yet, we can try to *remove* them from the general population until they stop being infectious. That doesn't help the patient any, but when it comes to stopping an epidemic, removal is as good as recovery. For this reason, many people like to say that R stands for Removed rather than Recovered.

(2) Make a as small as possible. Do everything possible to minimize contact between Infecteds and Susceptibles. Stay home if you're sick. Wash your hands. Wear a mask. Use hand sanitizer. If the outbreak is severe, public health officials may need to take more drastic measures, like closing schools, canceling large public events, or issuing shelter-in-place orders.

(3) Reduce the number of Susceptibles through **vaccination,** which directly converts Susceptibles into Recovereds. That is, into people who aren't sick, won't get sick, and can't spread the disease. If we can push the starting value of S below the threshold, then we can prevent the epidemic from even starting. This is called **herd immunity**. People often think that it's important to get a shot to protect themselves, and it is. But it's even more important for them to get that shot to protect the rest of us!

2.5. Covid-19 and Modified SIR Models

The SIR model does a very good job of describing epidemics in well-mixed populations where the underlying conditions are changing slowly if at all. However, with just two parameters, it can't capture all of the details of a global crisis like the Covid-19 pandemic. In this section we'll go over several ways that modelers modified the SIR model to deal with Covid.

Scaled models and the replication number R_0. One problem with the SIR model is that the parameters depend on the size of the city where the outbreak is happening. Typically, the larger the population, the smaller a fraction of the population that each person knows or meets and the smaller the transmission coefficient a will be. However, the product of a and the total population tends to be the same in different cities. For this reason, it's useful to write a scaled version of the model.

Let $T = S + I + R$ be the total population. This number is of course constant, or at least approximately constant. We let $s = S/T$, $i = I/T$ and $r = R/T$ be the fractions of the population that are susceptible, infected, or recovered, so that $s + i + r = 1$. Since

$s' = S'/T$, $i' = I'/T$ and $r' = R'/T$, we can write rate equations for these quantities.

(2.15)
$$\begin{aligned} s' &= & -aSI/T & = & -aTsi, \\ i' &= & (aSI - bI)/T & = & aTsi - bi, \\ r' &= & bI/T & = & bi. \end{aligned}$$

Modelers usually use the Greek letters β and γ in this scaled model, with

(2.16)
$$\beta = aT; \qquad \gamma = b.$$

This makes the scaled SIR equations:

(2.17)
$$\begin{aligned} s' &= & -\beta si, \\ i' &= & \beta si - \gamma i, \\ r' &= & \gamma i. \end{aligned}$$

Conditions do vary from place to place, with β being larger in cities that are more crowded and where people are less careful, but β varies much less than a and T vary separately. As a result, it's sensible to talk about the values of β and γ for each disease.

In the early stages of an epidemic, when s is close to 1, people are getting sick at rate $\beta si \approx \beta i$ and recovering at rate γi. That is, there are β/γ people getting sick for every person who recovers. This ratio is called the **basic replication number**

(2.18)
$$R_0 = \frac{\beta}{\gamma},$$

and represents the average number of new people that each sick person infects. If $R_0 > 1$, then the epidemic grows. If $R_0 < 1$, then the epidemic fizzles out.[4] For Covid, R_0 was originally estimated to be between 2 or 3. However, the speed at which it spread through the USA and Europe suggests that R_0 was actually higher. Later variants evolved to be even more infections, with values of R_0 around or even above 10.

Controlling an epidemic then amounts to getting s and R_0 as low as possible. In the short term, public health measures like wearing masks, closing schools, and staying home as much as possible can reduce β, and so can reduce R_0. Testing helps, too. Sick people don't have to infect others. If they can be identified and quarantined (or removed) from the general population, they can be infect**ed** without being infec**tious**.

However, such efforts can't last forever. Sooner or later people want to return to normal, they start behaving like they did pre-pandemic, and R_0 goes back up. The only long-term solution is reducing the fraction s of Susceptibles to below $1/R_0$, either through vaccination (best case) or natural infection (worst case).

The SEIR model. If Alice is sick and coughs on Bob, then Bob might get sick. If Bob then coughs on Carol, then Carol might get sick. However, if Bob coughs on Carol immediately after meeting Alice, then Carol is safe. It takes time for Alice's viruses to grow in Bob's body to the point that he can infect Carol.

---

[4]It's unfortunate that the basic replication number uses the same letter as the number of recovered individuals. However, since we're looking at r rather than R, and mostly care about s and i, this isn't really a problem.

To take this into account, modelers divide the population into four groups instead of three:

- **Susceptible** people (S, making a fraction $s = S/T$ of the population) are healthy but can become sick if they meet an infectious person.
- **Exposed** people (E, making a fraction $e = E/T$) are incubating the disease, but aren't sick enough to infect anybody else yet.
- **Infectious** people (I, making a fraction $i = I/T$) are a danger to everybody they meet.
- **Removed** people (R, making a fraction $r = R/T$) have recovered, have been quarantined, or have died. In any case, they aren't going to get sick again and they aren't going to infect anybody else.

Susceptible individuals become Exposed at a rate proportional to SI, just as in the SIR model. Exposed individuals become Infectious, and Infectious individuals become Removed, through the passage of time. After rescaling to express everything in terms of fractions of the population, our rate equations become

$$
\begin{aligned}
s' &= -\beta si, \\
e' &= \beta si - \alpha e, \\
i' &= \alpha e - \gamma i, \\
r' &= \gamma i.
\end{aligned}
$$

(2.19)

The new parameter α describes how long the incubation period is. On average, a person who gets infected is Exposed for time $1/\alpha$ and then is Infectious for time $1/\gamma$ before finally becoming Removed.

SEIR models behave a lot like SIR models with the same values of β and γ, except that the time delay causes the onset of the epidemic to grow a bit more slowly. In particular, we still talk about $R_0 = \beta/\gamma$ and the "herd immunity" threshold for starting to recover from the epidemic is still when $s = 1/R_0$.

The SIRS model. Immunity doesn't last forever. Over time, you lose the antibodies you had to a virus. Worse, germs can mutate into new forms that your initial antibodies don't recognize.[5] Either way, being Removed isn't really permanent.

We take this into account by adding a new parameter δ to our model to indicate the rate at which Removed individuals rejoin the ranks of Susceptibles.

$$
\begin{aligned}
s' &= \delta r - \beta si, \\
i' &= \beta si - \gamma i, \\
r' &= \gamma i - \delta r.
\end{aligned}
$$

(2.20)

The average duration of a person's immunity is $1/\delta$. This new model is called the SIRS model, for Susceptible-Infected-Removed-Susceptible. We can similarly turn the SEIR model into an SEIRS model.

---

[5] That's why you need a new flu shot each year. You may still be immune to last year's flu strain, but not to this year's.

Asymptomatic transmission and other compartments. One of the most dangerous things about Covid is its ability to be transmitted by people who don't even know that they are sick, either because they don't yet have symptoms (**presymptomatic transmission**) or because they never got symptoms (**asymptomatic transmission**). At the beginning of the pandemic, roughly a third of Covid cases were asymptomatic. This fraction has increased over time as Covid has evolved. While asymptomatic carriers tended to shed fewer viruses than symptomatic or presymptomatic carriers, they tended to go out a lot more. This made disease control very difficult.

To take asymptomatic transmission into account, modelers divided the Infected category (or the Infectious category for the SEIR model) into two groups, I_s and I_a for "symptomatic" and "asymptomatic". They had different rates β_s and β_a of infecting others and different recovery rates γ_s and γ_a. (They might take the same time to actually recover from the disease, but symptomatic individuals get removed from circulation a lot faster than asymptomatic carriers.) The SIR version of the model would then look like

(2.21)
$$s' = -\beta_s s i_s - \beta_a s i_a,$$
$$i_s' = \mu(\beta_s s i_s + \beta_a s i_a) - \gamma_s i_s,$$
$$i_a' = (1 - \mu)(\beta_s s i_s + \beta_a s i_a) - \gamma_a i_a,$$
$$r' = \gamma_s i_s + \gamma_a i_a,$$

where μ is the fraction of cases that develop symptoms.

You can also write down versions of the SEIR and SIRS models that take asymptomatic transmission into account. These models are more complicated than the basic SIR model, but they can be solved numerically in exactly the same way that we solved the SIR model. It's too complicated to do by hand, but programming a computer to solve equations (2.21) really isn't any harder than programming one to solve the basic model.

In practice, most of the models that were used to understand the spread of Covid had many compartments. Besides sorting people according to whether they were susceptible, exposed, infectious, or removed, there were compartments for being hospitalized, for being in intensive care, and for dying. People were sorted by age and sex, since Covid hit older people and men much harder than it hit younger people and women. To understand patterns of disease spread within a single city, people were also sorted by occupation, race, and zip code, as it turned out that all three factors were important. Once vaccines were developed, there were separate compartments for people who were vaccinated or who had recovered from Covid. Some of the models were very complicated. However, they were still just souped-up versions of SIR, with more details but the same underlying reasoning.

2.6. Same Song, Different Singer: SIR and Product Marketing

Throughout history, scientists have developed mathematical explanations for one purpose, only to find the same mathematical structure over and over again in completely different settings. The physicist Eugene Wigner called this the "unreasonable effectiveness of mathematics". In this section we'll explore how the SIR model can be used in business.

Imagine that you work for a company that has just introduced a hot new phone app. People are learning about it by word of mouth, and more and more people are using your product. Of course, that growth can't go on forever. Eventually people will get tired of your app and will move on to the next great thing. In order to make the most of your product's popularity, you need to forecast usage for the next year or two. Having learned about the SIR model of how biological viruses spread, you decide to use the model to study how other things "go viral".

Instead of having Susceptibles, Infecteds, and Recovereds, you divide the population of potential customers into three groups:

(1) **Potential** users, or *Potentials*: These are people who might use the product in the future if they hear enough good things about it from their friends.

(2) **Active** users, or *Actives*: These are people who have already adopted the product. Not only are they using it, but they are spreading the word about it.

(3) **Rejected** users, or *Rejecteds*: These are people who won't ever use the product. Maybe they just aren't interested. Maybe they used to use the product but got tired of it. Maybe our app doesn't even work on their brand of phone. For whatever reason, they're no longer reachable.

As with diseases, people go from the first category (Susceptible or Potential) into the second (Infected or Active) through contact with people in the second category. The only difference is that transmission happens by word of mouth, either literally talking face to face or via texting or social media. Meanwhile, people go from the second category to the third through the passage of time. Every day, there's a chance that an Active user will get bored with the product or will find out about something better.

We define a parameter a that says how well Active users talk their friends into using the app, and a parameter b that says how quickly people stop using the app. We call a the **transmission coefficient**, exactly as before, and call b the **attrition coefficient**. If we let $S(t)$, $I(t)$ and $R(t)$ denote the numbers of Potential, Active and Rejected users,[6] then the equations are exactly the same as before:

$$(2.22) \qquad \begin{aligned} S'(t) &= -aS(t)I(t), \\ I'(t) &= aS(t)I(t) - bI(t), \\ R'(t) &= bI(t), \end{aligned}$$

Since the equations are the same, the way we solve them is the same, the thresholds are the same, and the trajectories are the same.

If we want, we can make a scaled version of this model, with parameters β and γ and a replication number $R_0 = \beta/\gamma$ that describes how many new customers each Active user attracts before abandoning the product. We can introduce time delays with an SEIR model. We call allow for Rejected users to become interested in the product

---

[6]We could use different letters and call this the PAR model, but we won't.

again with an SIRS model. To a mathematician, the epidemic and the product launch are exactly the same.

There is a big difference, of course, in what we want to see happen. If an epidemic has an initial value of S that is below the threshold, causing the outbreak to fizzle, then we congratulate the health authorities who kept us safe. If a product launch has an initial value of S that is below the threshold, causing the product to fizzle (something we call a **product bust**), then the marketing team is likely to get fired. The math is the same, but the interpretation is completely different.

In marketing, we generally want the opposite of what the health authorities want. We want $I(t)$ to grow as quickly as possible, to reach a peak that is as high as possible, and to die out as slowly as possible. This means that we want to

(1) Make b as *small* as possible. Equivalently, make the usage time $T = 1/b$ as big as possible. This has to do with the quality of our product. The more exciting our product is, and the more we provide updates and other continuing benefits to the users, the lower the attrition coefficient will be.

(2) Make a as *big* as possible. Do everything possible to encourage word-of-mouth communication between Active and Potential users.

(3) Pick a market where $S(0)$ is as *big* as possible. Even the best products can fizzle if they're directed at the wrong market. A lemonade stand will do a lot better in Texas in August than in Minnesota in February.

2.7. Chapter Summary

The Main Ideas.

- *Close is good enough.* Most of the time, we don't need an exact answer. A good simple approximation is often more useful than a complicated formula.

- Mathematical models help us to understand real-world problems.
    - Take what we know about our problem.
    - Throw out the unimportant details and keep the main features.
    - Express those features mathematically.
    - Use math to **analyze** the model.
    - **Interpret** the results. Turn numbers and equations into real-world conclusions. Without this last step, the rest is useless.

- *Track the changes.* Many useful models involve **rate equations** that describe how fast something is changing in terms of its current state. Many of the same equations show up in different settings. The SIR equations model the spread of disease. The same equations also model market penetration.

- Once you know how fast something is changing, you can use a **linear approximation** to predict its future or explore its past. This is accurate for a short time, but can't be trusted over long time intervals.

- *One step at a time.* If you don't trust a linear approximation to predict what will happen next year, just use it to predict what will happen tomorrow. Then use those results to predict what will happen the day after tomorrow, then the day after that, and so on.

- *What goes up has to stop before it comes down.* You can learn a lot by studying the point when something stops moving: that is, when the rate of change equals zero.

Expectations. You should be able to:

- Model epidemics with a particular set of rate equations (the SIR model).
- Explain how the parameters in the SIR model relate to properties of the disease and the community where it's spreading.
- Use the SIR equations, together with a linear approximation, to predict future infection rates.
- Iterate this process to generate a table of values for several times.
- Relate the sign of I' to whether an epidemic is getting better or worse, and relate the size of I' to how fast this is happening.
- Use the same tools to analyze a product launch.

2.8. Exercises

Rate of Change

2.1.   A farmer is monitoring the number of fire ant mounds on his property. Suppose that there are 1200 mounds in March and 1250 mounds in May of the same year. Let $F(t)$ be the number of fire ant mounds at time t.
   (a) What is the change in the number of fire ant mounds between March and May? Denote your answer by ΔF. What are the units for ΔF?
   (b) What is the change in time between these two observations? Denote your answer by Δt. What are the units for Δt?
   (c) At what rate did the number of mounds change from March to May? Call your answer m_1. What are the units for m_1?
   (d) Assuming that the number of mounds is a linear function of time, how many mounds should we expect in August? What value of Δt are you using to find your answer? Use the notation and answers from parts (a), (b), and (c) to express your answer.
   (e) Again assuming a linear relationship, how many mounds were there in January of the same year? What value of Δt are you using to find your answer?
   (f) Find a linear formula for the number of mounds t months after March (that is, $t = 0$ means March).

2.2.   In Exercise 2.1 you developed a formula for the number of fire ant mounds t months after March. Suppose that you wish to update your model using additional data. You observe that model you developed in Exercise 2.1 worked well through August, but that the *change* in the number of mounds between the months of August and September was 100. (That is, there were 100 more mounds in September than in August.)

(a) What is the per month rate of change between August and September? Denote your answer by m_2.

(b) How does m_2 compare to the value of m_1 you found in Exercise 2.1(c)? What does this say about the seriousness of the fire ant infestation?

(c) Use m_2 to find an updated linear formula for predicting the number of mounds for October through December. (This is similar to the formula you found in Exercise 2.1(f).) State the correct domain for this expression.

(d) Use the information given in Exercise 2.1 and the result of Exercise 2.2(c) to write a piecewise function that expresses the number of ant mounds for the entire year (January through December), letting $t = 0$ represent the month of March.

(e) Use MATLAB to graph the piecewise function you found in part (d). Label your axes and give your graph an appropriate title. Use different colors for the two pieces of your function.

(f) Carefully explain why using the piecewise function to express monthly ant mounds for the entire year is better than simply using the linear function you found in Exercise 2.1(f) for the entire year.

2.3. Explain the term "rate equation". How is a rate *equation* different from the *rate of change m* associated with a linear function?

2.4. Suppose we have a rate equation $F' = \cdots$ for a quantity $F(t)$. At a starting time t_0, we have $F(t_0) = F_0$, and the rate equation tells us that $F'(t_0) = R_0$.

(a) Explain, using proper notation, how we can approximate the change in $F(t)$ between times $t = t_0$ and $t = t_0 + a$ using a linear approximation.

(b) Explain, using proper notation, how we can approximate $F(t_0 + a)$.

(c) If $F(t)$ is a linear function, would your answers in parts (a) and (b) still be approximations? Explain.

Interpreting SIR Models

2.5. Give approximate answers to these questions about the following graph of SIR model behavior:

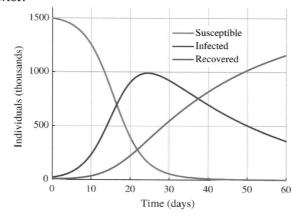

(a) When does the number of Infecteds reach its peak? How many people are infected at that time?

(b) Initially, how many people are Susceptible? How many days does it take for the Susceptible population to be cut in half?

(c) How many days does it take the Recovered population to reach 500,000? How many people recover by day 60?

(d) On what day is the size of the Infected population increasing most rapidly? When is it decreasing most rapidly? How can you tell?

(e) How many people became Infected at some time during the first 30 days? (Note that this is not the same as the number of people who are infected on day 30.) Explain how you found this information.

2.6.   Below are two graphs depicting the spread of disease according to the usual SIR model. The initial values $S(0)$, $I(0)$, and $R(0)$ and the transmission coefficient a are the same for both graphs. However, the two graphs correspond to different recovery coefficients b.

(a) For each graph, indicate which curve is S, which is I, and which is R.

(b) Which graph corresponds to the larger value of b? Explain.

2.7.   Consider an epidemic with the simplified rate equation $S' = -35{,}000$ persons per day. (This is not the usual SIR model.) Suppose the initial Susceptible population is 2,000,000 on December 14.

(a) How many Susceptibles will be left on Christmas Day (December 25)?

(b) When will the Susceptible population vanish entirely?

(c) How many Susceptibles were there on December 10?

(d) When were there 2,200,000 Susceptibles?

2.8. Suppose that a new disease is circulating in Smalltown, population 10,100, and that it evolves according to the usual SIR equations

$$\begin{aligned} S' &= -aSI, \\ I' &= aSI - bI, \\ R' &= bI. \end{aligned}$$

It appears that this disease lasts for an average of 20 days. It is also thought that there were initially 100 Infected people. That is, we had

$$\begin{aligned} S(0) &= 10,000, \\ I(0) &= 100, \\ R(0) &= 0. \end{aligned}$$

One day later, it was observed that there were eight *newly infected* people.

(a) Use the given information to complete the model. That is, find the transmission and recovery coefficients a and b.

(b) Using $\Delta t = 2$, find estimates for $S(4)$, $I(4)$, and $R(4)$.

(c) Find the threshold and explain its significance.

(d) Suppose that upon becoming Infected, patients are quarantined. It still takes time to identify Infected people, but the average length of time that they can infect others drops from 20 days to 10. What effect will this have on the threshold?

2.9. (a) Construct the appropriate SIR model for a disease with the following characteristics:

- Infected people remain sick for an average of 20 days.
- A typical Susceptible meets about 0.01% of the Infecteds each day (because he meets 0.01% of the entire population each day).
- It takes an average of eight contacts with an Infected before a Susceptible becomes Infected.

(b) How large does the Susceptible population need to be for an epidemic to grow?

2.10. A town's public health department has created an SIR model to predict the effects of this year's new strain of flu. The SIR equations are given by

$$\begin{aligned} S' &= -.0000002SI, \\ I' &= .0000002SI - I/15, \\ R' &= I/15. \end{aligned}$$

The initial values at $t = 0$ are thought to be

$$S(0) = 2,000,000, \quad I(0) = 21,000, \quad R(0) = 15,000.$$

(a) Using the initial rates, use the SIR model to find $S(1)$, $I(1)$, and $R(1)$.

(b) Using the SIR equations, calculate the rates of change $S'(1)$, $I'(1)$, and $R'(1)$. Then use these values to determine $S(2)$, $I(2)$, and $R(2)$.

(c) Using the values of $S(2)$, $I(2)$, and $R(2)$ that you computed in part (b), calculate the rates of change $S'(2)$, $I'(2)$, and $R'(2)$. Then estimate $S(3)$, $I(3)$, and $R(3)$.

(d) Go back to the starting time $t = 0$ and to the initial values

$$S(0) = 2{,}000{,}000, \qquad I(0) = 21{,}000, \qquad R(0) = 15{,}000.$$

Recalculate the values of $S(2)$, $I(2)$, and $R(2)$ by using a time step of $\Delta t = 2$. This only requires a single round of calculations, using $S'(0)$, $I'(0)$, and $R'(0)$. How do your answers compare to those computed in part (b)? Which estimates do you think are most accurate, those in part (b) or those in part (d)? Why?

2.11. **Herd Immunity.** Suppose there is a measles outbreak in a small community of 40,000 people, almost all of them Susceptible, with $b = 1/(14 \text{ days})$ and $a = 0.000003$.

(a) What is the threshold in this situation? How many people will get sick before the epidemic reaches its peak?

(b) Now suppose that 10,000 people (out of 40,000 total) are immune to measles, either because they already had the disease or because they were vaccinated. How many will get sick before the epidemic reaches its peak?

(c) If enough people are immunized, the epidemic will fizzle from the start. How many people would have to be immunized for that to happen?

(d) When enough people are immunized to keep outbreaks from spreading, the population is said to have herd immunity. Explain how the number of vaccinations needed for herd immunity is related to the threshold b/a.

(e) If you were a public health official, what would you say to somebody who doesn't want to get vaccinated and who insists that their (not) getting vaccinated is none of your business?

2.12. **Replication Number.** The ratio of how many people get sick to how many recover is $aSI/(bI) = aS/b$. At the beginning of an epidemic, when S is almost the entire population, this is the same as the basic replication number $R_0 = \beta/\gamma$ that we saw in our scaled SIR model. If $R_0 > 1$, then the epidemic will grow, at least at first. If $R_0 < 1$, it will fizzle out.

(a) Suppose that the total population is 40,000. If $R_0 = 1.2$, what is the threshold? How many people will get sick before the epidemic reaches its peak? What if $R_0 = 10$? In general, how are R_0, the total population, and the threshold related?

(b) In a town with total population T and a disease with basic replication number R_0, how many people would need to become immune to achieve herd immunity?

Food for thought: The number of people needed to reach herd immunity is the same as the number of people who have to get sick *before* the epidemic reaches its peak. However, a substantial number of people can still get sick *after* the peak. For instance, suppose that $R_0 = 2$. The peak will be reached when half the population has gotten sick, but nearly 30% of the population will get sick later. Only about 20% will avoid the disease altogether. This is one reason why it is much better to immunize people before an epidemic than to wait for herd immunity during an epidemic.

2.13. **Superspreader Events.** One way for a disease to spread is through big public gatherings such as sporting events, music festivals and political protests. By giving sick individuals the opportunity to interact with a lot of other people, these gatherings effectively increase the transmission coefficient a.[7]

(a) Suppose that we are in the setting of Exercise 2.10, except that a music festival doubles the chance that a Susceptible will become Infected. What is the new transmission coefficient?

(b) Compute the new threshold if the festival of part (a) takes place. Compare your answer to the threshold in the original setting of Exercise 2.10.

(c) Suppose you have a model for a small community where $S(0) = 40,000$. It is estimated that, if the threshold is 30,000, then the hospitals will be able to keep up with the flow of sick patients. However, if the threshold is less than 30,000, then the epidemic will progress so fast that the hospitals won't be able to keep up. If $b = 1/14$ and $a = 0.000002$, will the health care system be able to handle the strain?

(d) Now suppose that the town hosts an athletic competition that increases a to 0.000003. What will the new threshold be? What will happen to the town?

2.14. **Dynamic Energy Budget.** The reasoning we used to build the SIR model, namely tracking the inputs and outputs for each variable to come up with a rate of change, also helps us build models that have nothing to do with disease spread. For instance, plants and animals, in all of their wondrous variety, can be viewed as consumers. Plants must consume water, nutrients, and sunlight to survive. Animals must consume plants or other animals. The energy in the resources they consume is utilized in growth, reproduction, and basal maintenance. (To simplify the model, we have ignored several other energy requirements.) The following is a simple schematic that represents the basic components of a dynamic energy budget.

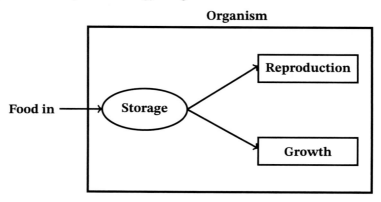

---

[7]An August 2020 motorcycle rally drew almost half a million people to the small town of Sturgis, SD. Within weeks, South Dakota and its neighboring states were leading the country in Covid infections per capita.

(a) Suppose energy (measured in calories) enters storage (S) via a food source at rate p_E (measured in calories per day). Energy leaves storage at rate p_L. A fraction κ of that energy is used for growth G, while a fraction $1 - \kappa$ is used for reproduction R. Write down a system of rate equation for S, G, and R.

(b) Suppose that a fraction m of the energy designated for growth is actually used for organism maintenance. (These are the processes essential to keeping an organism alive, such as respiration, cell repair, etc.) Revise your model to show the energy used for organism maintenance separately from the energy used for growth.[8]

Exponential Growth. In many cases, the rate of change of a quantity is proportional to the quantity itself. We will see many different examples of that in Chapter 6, but the two simplest are money earning interest and population growth. We will examine these in Exercises 2.15–2.17.

If a bank account containing $10,000 earns $150/year in interest, then an account at the same bank containing $20,000 should earn $300/year in interest. Either way, the interest is 1.5%/year of the bank balance.

If a city of 100,000 is growing at a rate of 3000 people per year, then we would expect a similar city of 200,000 (or even the same city a few years later) to grow by 6000 people per year. That is, if $P(t)$ is the population at time t, then the **net growth rate** P' is proportional to P:

$$P' = kP,$$

where k is a constant. In this example, $k = .03$/year, or 3%/year.

2.15.   In the equation $P' = kP$ for the population of a city, the number k is called the **per capita growth rate**, with the Latin phrase "per capita" literally meaning "per head". Explain why the units for k are 1/year.

2.16.   In 2020, the population of Australia was 25 million and was growing at 1.36%/year, while the population of the United States was 330 million and was growing at 0.71%/year. Assume that both countries' populations keep growing at these per capita rates.

(a) Let $A(t)$ and $U(t)$ be the populations of Australia and the United States, respectively. Write down the rate equations that govern the growth of these functions.

(b) What were the net growth rates $A'(2020)$ and $U'(2020)$ in 2020?

(c) In general, does a country with a higher per capita growth rate necessarily have a higher net growth rate?

(d) On average, how many seconds did it take for the population of Australia to increase by 1 in 2020? For the population of the United States? (A year is about 31.6 million seconds.)

Food for Thought: At these per capita growth rates, the populations of Australia and the United States would become equal in 2417, at which time each country would have

---

[8]Plants don't need much energy to keep themselves going, so m is typically small for plants. m is larger for animals, especially warm-blooded animals (like humans) that spend a lot of energy just keeping their body temperatures up.

about 4.3 *billion* people. Exponential growth is great when it comes to money, but it can be scary in other cases.

2.17.   A bank account is earning interest at a constant percentage rate. When the account had $3200, it was growing at a rate of $79/year.
   (a) Write an equation that links the net growth rate (meaning P', not the interest rate) to the bank balance.
   (b) At a later time, the bank balance is growing at $100/year. What is the balance at that time?

Scaling the Time Variable. In a rate equation, the quantity Y' has units, namely whatever units we use for Y divided by whatever units we use for time. If Y is people and t is measured in days, then Y' is measured in people/day. In Section 2.5, we generated a scaled model, where S, I, and R were replaced with dimensionless quantities s, i, and r that are proportional to S, I, and R. In Exercises 2.18–2.20, we will see what happens when we similarly replace the time t with a dimensionless variable.

2.18.   Suppose that $i' = 0.014$/week. That is, i is increasing at a rate of 0.014 per week.
   (a) How fast is i increasing per *day*? How fast is i increasing per *year*? Which is bigger, the growth per year or the growth per day?
   (b) Let t_d, t_w, and t_y be the number of days, weeks, and years since a fixed starting time. How are the variables related? Which is bigger, t_d or t_y?

To make a model dimensionless, we pick a fixed unit of time T. T might be a second, an hour, a day, a year, or 3723.849 years. We then measure time in units of T. That is, we define a new dimensionless variable $\tau = t/T$ that counts how many T's have elapsed since $t = 0$. If T is small, then τ will be big, while if T is big, then τ is small. A very *large* number of nanoseconds equals a very *small* number of centuries. For variables Y and P that change in time, we let $\dot{Y}$ and $\dot{P}$ denote the changes in Y and P per change in τ.

2.19.   How is $\dot{Y}$ related to Y'? (*Hint*: if Y' is constant, how much will Y change by the time that $\tau = 1$?)

2.20.   Suppose that $P' = rP$. Let $T = r^{-1}$.
   (a) Compute $\dot{P}$ in terms of P.
   (b) Compute T when $r = 1\%$/year, when $r = 3\%$/month, and when $r = 0.06$/minute.

Food for Thought: You should discover that the trajectory of $P' = rP$ looks essentially the same for all of these values of r, only we trace out that trajectory at different speeds. If we can understand the solution to $P' = rP$ for one value of the parameter r (say, for $r = 1$), then we can understand the solution to $P' = rP$ for all values of r.

2.21.   Consider the scaled SIR equations (2.17). Let $T = \gamma^{-1}$. Write down a system of equations for $\dot{s}$, $\dot{i}$, and $\dot{r}$.

Food for Thought: The resulting equations still have one dimensionless parameter, namely the basic replication number R_0. The details of an epidemic depend on a lot of different parameters and initial conditions, but the general shape of the curve only depends on R_0.

Chapter 3

Close is Good Enough

In Chapter 2, we used a lot of approximations to understand the spread of disease. First, we made approximations in building the SIR model, pretending that all Susceptibles are the same, all Infecteds are the same, and all Recovereds are the same. Then, when we couldn't find a formula for the solution, we used linear approximations and baby steps to get (approximate) numerical solutions. And that was good enough.

In this chapter we're going to flesh out the idea of linear approximation that we began in Chapter 2. Before we can do that, we need to remember how linear functions behave. Before we can do *that*, we need to understand how functions behave in general. And before we can do *that*, we need to get comfortable with the whole idea of making approximations. So let's get started!

3.1. The Idea of Approximation

Back in algebra class, you learned how to solve cut-and-dried problems like

Find all solutions to $x^2 - 3x + 2 = 0$.

There are several ways to do this. We begin by defining $f(x) = x^2 - 3x + 2$.

(1) We can factor $f(x)$ as $(x - 1)(x - 2)$. If $(x - 1)(x - 2) = 0$, then either $x - 1 = 0$ or $x - 2 = 0$, so our solutions are $x = 1$ and $x = 2$.

(2) We can apply the quadratic formula.

$$(3.1) \qquad x = \frac{-b \pm \sqrt{b^2 - 4ac}}{2a} = \frac{3 \pm \sqrt{(-3)^2 - 4(1)(2)}}{2} = \frac{3 \pm 1}{2} = 1 \text{ or } 2.$$

(3) We can plot a few points to get a feel for how f behaves.

x	$f(x)$
0	2
1	0
2	0
3	2

Because $f(1) = f(2) = 0$, $x = 1$ and $x = 2$ are solutions. A quadratic function can only have two roots, so there are no other solutions, and we're done.

All of these strategies worked because *the problem was very simple.* Unfortunately, the real world rarely gives us such simple problems. It gives us complicated problems with messy numbers that we have to make sense of anyway. Most of the time, there isn't a method like factoring that gives us the answers. There isn't a formula that gives us the answers. The answers are messy, so there's almost no hope of just guessing them.

For instance, let's tackle the problem

Find all solutions to $x^5 + x - 31 = 0$.

We can't factor the polynomial. There is no formula for the roots of a fifth-order polynomial. There aren't any simple roots. So what can we do?

The answer is that we can still plot a few points. As before, we define $f(x) = x^5 + x - 31$ and we get the following table of values.

x	$f(x)$
0	−31
1	−29
2	2
3	215

Because $f(1)$ is negative and $f(2)$ is positive, there must be a root between 1 and 2. What's more, $f(2)$ is a lot closer to 0 than $f(1)$ is, so that root is likely to be close to 2.

So let's try 1.9. We compute $f(1.9) = 1.9^5 + 1.9 - 31 = -4.33901$ and add it to our table.

x	$f(x)$
0	−31
1	−29
1.9	−4.33901
2	2
3	215

Now we know that our root is somewhere between 1.9 and 2, and probably closer to 2 than 1.9 (Why?) So we try 1.96 and 1.97 and get

x	f(x)
0	−31
1	−29
1.9	−4.33901
1.96	−0.11453
1.97	0.64093
2	2
3	215

Now we know that our root is somewhere between 1.96 and 1.97, and probably a lot closer to 1.96 than 1.97.

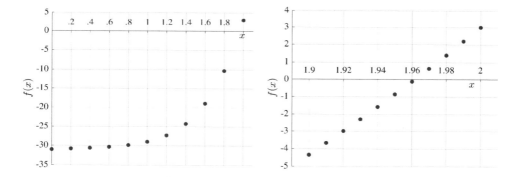

Figure 3.1. The graph of $f(x)$ on two different scales

Graphing our results helps us to visualize what is going on. By plotting points between $x = 0$ and $x = 2$, as in the first graph in Figure 3.1, we see that our root is between 1.8 and 2, and looks to be between 1.9 and 2. By plotting points between 1.9 and 2, as in the second graph in Figure 3.1, we see that our root is between 1.96 and 1.97. Some would say that we still haven't solved the problem, because we still don't know what the answer is *exactly*. But we've actually found the answer to two decimal places. It's 1.96! By plugging in a few more numbers, we can figure out that the answer is between 1.961 and 1.962, in other words that to three decimal places it's 1.961. Then that to four decimal places it's 1.9615. Then that to five decimal places it's 1.96153. No matter how much accuracy we need, this simple method can get it for us.

Lessons from this example

- We didn't use any fancy mathematical tools. Having good tools is useful, of course, and we're going to develop some great tools in the rest of the book, but most math can be done with the tools you learned in middle school.

- We didn't do anything especially clever or elegant. Cleverness and elegance are overrated! In time we'll learn some clever methods that give us more accuracy with less effort, but you don't need to be clever to do good math.

- Having the right mindset *is* important. *Close is good enough.* If you can't find an exact answer, find a way to get an approximate answer. Then find a way to improve that answer, and improve it again, until you have as much accuracy as you need.

- The most important thing is *perseverance*. In this problem, there was no way to find the answer in one step, or even in several steps, but we didn't let that stop us. Instead, we solved the problem to several decimal places, one step at a time.

- Computers aren't clever, and the mathematical tools they have aren't fancy. However, they have incredible perseverance, and will do millions of calculations for us every second. We just have to understand the procedure well enough to ask them to do it. That is, we have to write a program. We'll have a lot to say about using computers to solve rate equations later in this chapter.

The same lessons apply to almost everything in calculus. We almost never can get an answer directly, so we almost always look for an approximate solution. Then we look for ways to improve that approximation and home in on the true solution. More accuracy usually means more work, but with enough perseverance we can obtain as much accuracy as we need. If we're too lazy to do all that work ourselves (and who isn't?), we can ask computers to do it for us, but only after we've understood the procedure well enough to explain it to them.

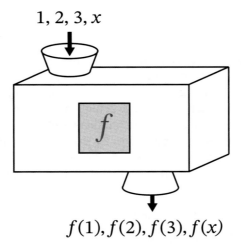

Figure 3.2. A function viewed as a machine for turning inputs into outputs.

3.2. Functions

A **function** is a rule that assigns one and only one output to every possible input, as shown in Figure 3.2.

The inputs can be, well, anything, and so can the outputs. For instance, the function *SC* has states as inputs and cities as outputs, and

(3.2)
$$SC(\text{Alabama}) = \text{Montgomery}$$
$$SC(\text{Alaska}) = \text{Juneau}$$
$$SC(\text{Arizona}) = \text{Phoenix}$$
$$\vdots \quad \vdots$$
$$SC(\text{Wyoming}) = \text{Cheyenne}$$

You may have already guessed that $SC(\text{New York}) = \text{Albany}$, that $SC(\text{Texas}) = \text{Austin}$, and that SC stands for "state capital".

Most of the time, however, our inputs and outputs will be ordinary numbers. (Occasionally the inputs and outputs will be pairs or triples of numbers.) Such a numerical function can be described in four ways:

(1) **As a procedure for computing the output in terms of the input.** For instance, there is a function *SQ* where the output is what you get when you multiply the input by itself.

There is also a function *DVTemp* whose input is a date and whose output is the high temperature in Death Valley National Park on that date. Computing $DVTemp(t)$ for a date in the past is easy: just look up the weather records. Computing $DVTemp(t)$ for a future date is a lot harder and mostly involves waiting for that date.

(2) **As a formula.** We give the input a name (usually x or t) and we try to find a formula for the output. In the case of *SQ*,

(3.3)
$$SQ(x) = \text{What you get when you multiply } x \text{ by itself}$$
$$= x^2.$$

Of course, we could just as well have called the input variable something else and written $SQ(y) = y^2$ or $SQ(t) = t^2$ or $SQ(\text{George}) = \text{George}^2$. (We don't usually use human names for variables, but there's no reason why we couldn't.) It's all the same function that takes the input and squares it.

Warning! Not every function has a formula. There is no formula for $DVTemp(t)$, neither for the past nor the future. Even when looking at the past, the fluctuations in the weather are too complicated to be described by mathematical formulas.

(3) **As a table.** We just make a list of inputs and outputs, as with SC, and write something like

x	$SQ(x)$
-3	9
-2	4
-1	1
0	0
1	1
2	4
3	9

In this case, the table doesn't describe the function completely, as it doesn't tell us what $SQ(7)$ or $SQ(\pi)$ or $SQ(-24)$ are. There are infinitely many possible inputs, and we can only fit a finite number of rows in the table. We have to use our imagination to figure out what $SQ(x)$ is when x isn't on the list.

In many real-world applications, however, we don't know the formula for the function, but we do have measured data. For instance, the function might give a person's weight as a function of time, the number of legal n-letter Scrabble words as a function of n, or (as with $DVTemp(t)$) the high temperature in a certain place as a function of the date. In those cases, the table *is* the function.

(4) **As a graph.** The graph of a function is the set of all ordered pairs (x, y), where x is an input and y is the corresponding output. This gives a curve in the plane that passes the **vertical line test**, which says that a vertical line can only hit the curve once. It contains a lot more information than a table, as it shows what happens for every possible input x, at least within the interval that we are graphing. For

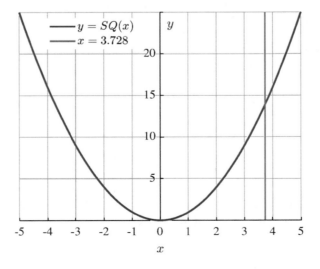

Figure 3.3. The graph of SQ

instance, Figure 3.3 shows what *SQ* does when the input x is between -5 and 5. If we want to know what $SQ(3.728)$ is, we just have to look closely to see where the vertical line $x = 3.728$ hits the graph.

If we have two ordinary functions—let's call them f and g—that take numbers as inputs and outputs, then we can combine them by feeding the output of one function to the other function. We use the symbol $f \circ g$ to mean the function "apply g to the input, and then apply f to that". In other words,

(3.4) $$(f \circ g)(x) = f(g(x)).$$

Be careful with the order. *We read from right to left, applying first g and then f.* Likewise, if we had a function $f \circ g \circ h$, then we would first apply h, then g, then f.

The order matters, as you can see in Figure 3.4. If f is the "multiply by 2" function, described by the formula $f(x) = 2x$, and if g is the "add 7" function, described by the formula $g(x) = x + 7$, then

(3.5)
$$f \circ g(x) = f(x + 7) = 2(x + 7) = 2x + 14,$$
$$g \circ f(x) = g(2x) = 2x + 7.$$

Those are different functions!

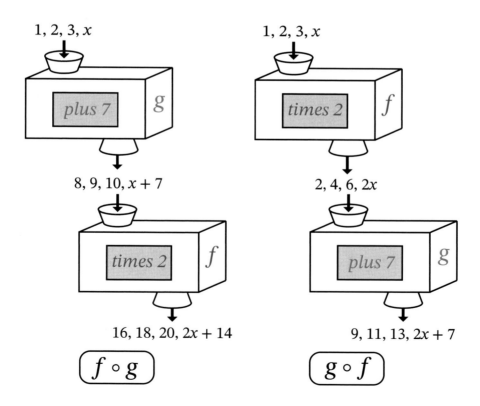

Figure 3.4. The functions $f \circ g$ and $g \circ f$ for $f(x) = 2x$ and $g(x) = x + 7$

3.3. Linear Functions and Their Graphs

If f is a function and x_1 and x_2 are two possible inputs, we are often interested in how fast the output changes as the input goes from x_1 to x_2. Let $y_1 = f(x_1)$ and let $y_2 = f(x_2)$, so (x_1, y_1) and (x_2, y_2) are two points on the curve $y = f(x)$. We use the Greek letter Δ to indicate the change in a quantity, so

(3.6)
$$\Delta y = y_2 - y_1 = \text{ Change in } y$$
$$\Delta x = x_2 - x_1 = \text{ Change in } x.$$

Δ is the Greek equivalent of "d", which stands for "difference". We often call Δy the **rise** and Δx the **run**, even though these quantities may in fact be negative.

The ratio $\Delta y / \Delta x$ measures how fast our function is changing. We call it the **average rate of change (AROC)**, of the function between x_1 and x_2. For instance, in Figure 3.5 the AROC between 1 and 3 is

(3.7)
$$\frac{\Delta y}{\Delta x} = \frac{9 - 1}{3 - 1} = 4,$$

while the AROC between 3 and 5 is

(3.8)
$$\frac{\Delta y}{\Delta x} = \frac{25 - 9}{3 - 1} = 8.$$

The function is increasing faster between 3 and 5 than between 1 and 3.

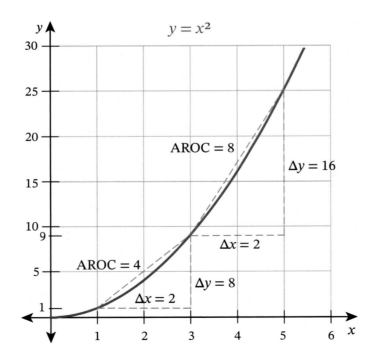

Figure 3.5. The rate of change of a typical function is not constant.

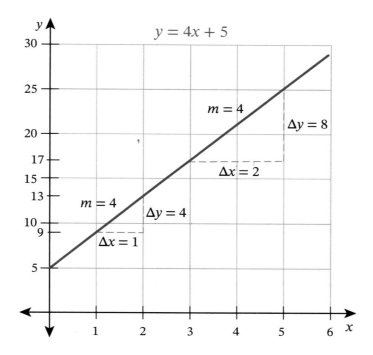

Figure 3.6. The rate of change of a linear function is constant.

A **linear function** is a function whose rate of change is constant, as shown in Figure 3.6. That is, there is a number m such that, for any two points, $\Delta y/\Delta x = m$. For instance, if $y = f(x) = 4x$, then y always changes 4 times as much as x changes. The same is true if $y = f(x) = 4x + 5$. The number m goes by many names: it is the **slope** of the line $y = f(x)$ or the **rate of change** or the **multiplier** of the function f.

Most questions about linear functions have two parts. First, we want to figure out the multiplier from data. Second, we want to use that multiplier to find the equation of the line.

For instance, suppose that we are driving along a highway at constant speed. At 1:00 p.m., we reach mile marker 132. At 3:00 p.m. we reach mile marker 262. How fast are we going? Where will we be at 4:00 p.m.? Where will we be at time t? If our destination is at mile marker 457, when will we get there?

Our input variable is the time t, measured in hours, and $f(t)$ is our position at time t, measured in miles. We are told that $f(1) = 132$ and $f(3) = 262$, so our rate of change is

$$(3.9) \qquad m = \frac{\Delta y}{\Delta t} = \frac{262 \text{ mi} - 132 \text{ mi}}{3 \text{ hr} - 1 \text{ hr}} = \frac{130 \text{ mi}}{2 \text{ hr}} = 65 \text{ miles/hour.}$$

To figure out where we'll be at 4:00 p.m., we compare 4:00 p.m. to 1:00 p.m. We have $\Delta t = 4 - 1 = 3$ (hours), so $\Delta y = 65(3) = 195$ (miles). That is, at 65 miles/hour we will drive 195 miles between 1:00 p.m. and 4:00 p.m., putting us at mile marker 132+195=327. At time t we will have gone $65(t - 1)$ miles since 1:00 p.m., so we will be at mile marker $132 + 65(t - 1)$. That's the equation of our position function:

$$(3.10) \qquad\qquad f(t) = 132 + 65(t - 1),$$

where $f(t)$ is measured in miles and t is measured in hours. If we prefer, we can do a little algebra to rewrite $f(t)$ as $65t + 67$, or as $262 + 65(t - 3)$. The formulas look different, but the function is the same.

So far we have been using m to convert known values of Δt into values of Δy. We can also do the reverse:

$$(3.11) \qquad\qquad \Delta y = m\Delta t, \text{ so } \Delta t = \frac{\Delta y}{m}.$$

When we reach our destination we will have gone

$$(3.12) \qquad\qquad \Delta y = (457 - 132) \text{ miles } = 325 \text{ miles.}$$

That takes us

$$(3.13) \qquad\qquad \Delta t = \frac{\Delta y}{m} = \frac{325 \text{ miles}}{65 \text{ miles/hour}} = 5 \text{ hours,}$$

so we'll reach our destination at 1:00 p.m. + 5 hours = 6:00 p.m.

Let's summarize what we've learned, and apply it to an arbitrary linear function $f(x)$.

- Two points determine the slope of a line. If $f(x_1) = y_1$ and $f(x_2) = y_2$, then

$$(3.14) \qquad\qquad m = \frac{y_2 - y_1}{x_2 - x_1} = \frac{\text{rise}}{\text{run}}.$$

- A point and a slope give us the equation of a line. If our point is (x_1, y_1) and the slope is m, and if (x, y) is any other point, then

$$(3.15) \qquad\qquad y - y_1 = \Delta y = m\Delta x = m(x - x_1),$$

  so

$$(3.16) \qquad\qquad y = y_1 + m(x - x_1).$$

  This is called the **point-slope** form of the equation of the line. See Figure 3.7.

- Some people prefer to use a little algebra to rewrite the equation in **slope-intercept** form, as

$$(3.17) \qquad\qquad y = mx + b,$$

  where $b = y_1 - mx_1$ is the value of y when $x = 0$, also known as the y-**intercept**. In other words, slope-intercept form is what you get when you write point-slope form with $x_1 = 0$. Most algebra classes stress slope-intercept form, but you should get used to using point-slope form, *which is much more useful for calculus.*

- The multiplier m is a conversion factor for getting Δy from Δx, but *it is also a conversion factor for getting Δx from Δy.*

$$(3.18) \qquad\qquad \Delta y = m\Delta x, \qquad \Delta x = \frac{\Delta y}{m}.$$

  To figure out when a linear function reaches a certain value y, first compute $\Delta y = y - y_1$, then $\Delta x = \Delta y/m$, and finally $x = x_1 + \Delta x$.

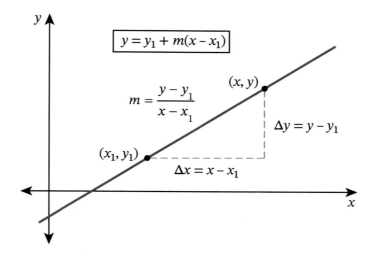

Figure 3.7. The point-slope form of the equation of a line

3.4. Linear Approximations and Microscopes

Linear functions are perfect for describing quantities that change at a constant rate. They are also great for *approximating* functions that change at a *nearly* constant rate. If a curve doesn't curve much, then a line through any two points will do a good job of approximating the entire curve. This allows us to **extrapolate** and **interpolate** data.

For instance, consider the graph shown in Figure 3.8. As you can see, it is pretty close to a straight line. If I tell you that $f(1) = 1$ and $f(1.01) = 1.0201$, you can figure

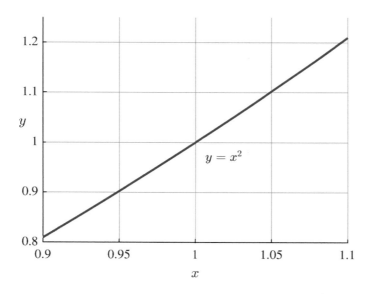

Figure 3.8. When zoomed in on, most graphs look linear.

out the slope of that line.

$$(3.19) \qquad m = \frac{f(1.01) - f(1)}{1.01 - 1} = \frac{.0201}{.01} = 2.01.$$

Once you have the slope, you can write down the equation of the line in point-slope form:

$$(3.20) \qquad y = 1 + 2.01(x - 1).$$

Once you have the equation of the line, you can use it to approximate any point along the curve:

$$(3.21) \qquad f(x) \approx 1 + 2.01(x - 1),$$

as shown in Figure 3.9. For example, $f(1.005) \approx 1 + 2.01(0.005) = 1.01005$. As 1.005 is halfway from 1 to 1.01, we approximate $f(1.005)$ as the value halfway from $f(1)$ to $f(1.01)$. Using known values of a function at two points x_1 and x_2 to estimate $f(x)$ for x between x_1 and x_2 is called **interpolation**.

We can use the same reasoning to estimate $f(x)$ when $x < x_1$ or when $x > x_2$. For instance, $f(1.03) \approx 1 + 2.01(.03) = 1.0603$ while $f(0.98) \approx 1 + 2.01(-.02) = 0.9598$. Extending our line past x_1 and past x_2 is called **extrapolation**.

Notice the advantage of using point-slope form instead of slope-intercept form. In point-slope form, the numbers $x - 1$ are small and easy to multiply by 2.01, and you can see what is going on. The equation $y = 2.01x - 1.01$ describes the same line in slope-intercept form, but it requires you to multiply 2.01 by 1.005 or 1.03 or 0.98, which is more complicated and much more likely to lead to arithmetic errors.

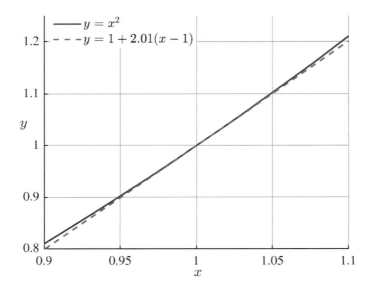

Figure 3.9. Linear approximations work well over short distances.

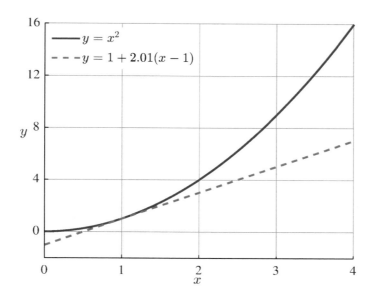

Figure 3.10. Over longer distances, linear approximations aren't nearly as good.

Linear approximations are not exact, of course. In this example, the function was our old friend $f(x) = x^2$. The exact value of $f(1.005)$ was 1.010025, not 1.01005, the exact value of $f(1.03)$ was 1.0609 rather than 1.0603, and the exact value of $f(0.98)$ was 0.9604 instead of 9.9598. Not exact, but very, very close.

So did we just get lucky by studying a function whose graph didn't curve much? Not at all! As you can see in Figure 3.10, the graph $y = x^2$ curves quite a bit, and is much steeper at $x = 3$ than at $x = 1$. The curve hugs the line near $x = 1$, but eventually veers away. As a result, the linear approximation $f(3) \approx 1 + 2(2.01) = 5.02$ doesn't come anywhere close to the true value of $f(3) = 9$.

However, $y = x^2$ doesn't curve much *between* $x = 0.9$ and $x = 1.1$. Most curves have the same property: if we zoom in on them enough, they begin to look straight. Linear approximations can be used for all sorts of functions, but we should only use them over short distances.

3.5. Euler's Method and Compound Interest

In the 1700s, the Swiss mathematician Leonhard Euler (pronounced "Oiler") invented a method for solving rate equations. We already used his method on the SIR equations in Chapter 2. Now we're going to apply Euler's method to a financial example, and then we'll develop the method in general.

Suppose that we invest $1000 in a fund that pays 6% interest per year. How much money will we have after t years?

Our contract with the bank says that our investment is supposed to grow at a rate that is 6% of whatever is there. In other words, our rate equation is

(3.22) $y' = 0.06y.$

Because we started with $1000, our initial condition is $y(0) = 1000$. Putting these together, we have $y'(0) = 0.06(1000) = 60$ dollars/year. The equation of the line through $(0, 1000)$ with slope 60 is

(3.23) $y(t) = 1000 + 60t.$

This equation says that the bank gives us $60/year forever, which is what would happen if the account earned **simple interest**. According to this **linear approximation** (shown by the dashed line in Figure 3.11), we should have $1060 after one year, $1120 after two years, $1600 after ten years, and $2800 after thirty years.

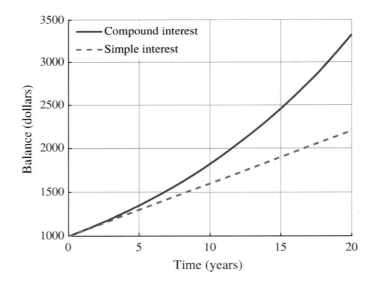

Figure 3.11. The actual solution to $y' = 0.06y$ and the linear approximation.

But that's not how investments actually work. In the second year, we will have more than $1000 in our fund, because we will have the $1000 that we started with, plus the interest we earned in the first year. That means that we should earn more than $60 in interest in the second year. We should earn even more the third year, and so on. This is called **compound interest**, as shown by the solid line in Figure 3.11.

To take that into account, we can update our estimate of y' every year. That is, we can

(1) Use $y(0) = 1000$ and the rate equation $y' = 0.06y$ to compute $y'(0) = 60$.

(2) Use the linear approximation $y(t) \approx y(0) + y'(0)t$ to approximate $y(1)$, getting $y(1) \approx \$1060$.

(3) Use $y(1) \approx 1060$ and the rate equation to estimate $y'(1) \approx 63.6$.

(4) Use the linear approximation $y(t) \approx y(1) + y'(1)(t - 1)$ to approximate $y(2) \approx 1123.6$.

(5) Repeat the process, one year at a time, for as many years as we hold the investment.

Applying this for 5 years generates the data shown in Table 3.1. The time step is $\Delta t = 1$ year. The first column is the time, and the second column is our bank balance. The third is the rate at which we are earning interest, which is 0.06 times the second column. The fourth is how much interest we earn in one time step (which looks like the third column because $\Delta t = 1$, but if we were working 2 years at a time, this column would be twice the third column.) The fifth is our balance at the end of the year, which also gives us our starting balance the next year.

Table 3.1. Computing our balance after 5 years of annual compounding

t	y	y'	$y'\Delta t$	$y + y'\Delta t$
0	1000	60	60	1060
1	1060	63.6	63.6	1123.6
2	1123.6	67.4	67.4	1191.0
3	1191.0	71.5	71.5	1262.5
4	1262.5	75.7	75.7	1338.2
5	1338.2			

In fact, this is how banks used to compute interest, with updates done yearly. This interest is said to be **compounded annually** or **compounded yearly**. If instead we calculated interest every six months, our interest would be **compounded semi-annually**. As you can see from Table 3.2, semi-annual compounding is better than annual compounding, but the difference is fairly small. Monthly compounding is a little bit better than semi-annual compounding, and daily compounding is a tiny bit better than monthly compounding. These are all approximations to the solution of the rate equation, which is called **continuous compounding**. But they're very good approximations! If we used our procedure with a time step of one month to estimate the result of 20 years of continuously compounded interest, we would only be off by \$9.90.

More generally, suppose that we have a rate equation

$$(3.24) \qquad\qquad y' = f(y, t),$$

Table 3.2. 20-year earnings from $1000 invested at 6% interest with simple, annual, semi-annual, monthly, and continuous interest

Year	Simple	Annual	Semi-annual	Monthly	Continuous
0	1000.0	1000.0	1000.0	1000.0	1000.0
5	1300.0	1338.2	1343.9	1348.9	1349.9
10	1600.0	1790.8	1806.1	1819.4	1822.1
15	1900.0	2356.6	2427.3	2454.1	2459.6
20	2200.0	3207.1	3262.0	3310.2	3320.1

where $f(y, t)$ is some fixed function of y and t. (In the banking example, we had $f(y, t) = 0.06y$, but it could just as well have been $f(y, t) = 0.06y + t$.) We start at a particular time t_0, with initial condition $y = y_0$. There is a later time t_f, and we want to estimate $y(t_f)$.

(1) Pick a number of steps N, which makes our **time step** $\Delta t = (t_f - t_0)/N$.

(2) Make a table similar to Table 3.1 with $N + 1$ rows and with at least three columns, labeled t, y and y'. It's convenient (but not necessary) to have two more columns, labeled $y'\Delta t$ and $y + y'\Delta t$. Enter t_0 and y_0 in the first row.

(3) Fill out the table one row at a time. Once we have t and y, the rate equation tells us y', and we can then figure out $y'\Delta t$ and $y + y'\Delta t$.

(4) Start on the next row. The new t is the previous t plus Δt. The new y is the previous y plus the previous $y'\Delta t$. This is our linear approximation.

(5) Repeat until the entire table is filled out. Then read off our final answer $y(t_f)$. If we want to see the whole progression, we can plot the y column against the t column.

(6) If we want more accuracy, we can do it all over again with N larger and Δt smaller.

3.6. The SIR Model by Computer

We now revisit the SIR model that we discussed in Chapter 2.

If this book were written in 1940, there would be nothing more to say. We have a nice model describing the spread of disease. While we know how to solve rate equations by Euler's method in principle, in practice it takes a lot of work. Computing 10 years of compound interest with $\Delta t = 1$ requires 10 turns of the crank. Doing it with $\Delta t = 1/2$ requires 20 turns, and doing it with $\Delta t = 0.1$ requires 100 turns. Solving the SIR equations, with three different variables to keep track of, is even messier. Who has the patience for all that?

Fortunately, we live in the 21st century, and we have computers to do our chores for us. With the right instructions, a computer can do the calculation with $\Delta t = 0.1$

or even $\Delta t = 0.01$ in the blink of an eye. We just need to understand the SIR rate equations, and Euler's method for solving them, well enough to give the computer its instructions—in other words, we need to write a program.

We begin with a clear step-by-step description of the algorithm that we can turn into computer code.

(1) Specify the parameters of our model, namely the transmission coefficient a and the recovery coefficient b.

(2) Specify our initial conditions, that is, the values of S, I, and R at time $t = 0$.

(3) Decide how far into the future we are going to run. We'll call the final time t_f. It's also possible to set a negative value of t_f, in which case we're using the SIR model to understand the past.

(4) Decide how many steps we are going to take. Call this number N. Our **step size** will then be $\Delta t = t_f/N$.

(5) Generate a list of times $t_0 = 0$, $t_1 = \Delta t$, $t_2 = 2\Delta t$, up through $t_N = N\Delta t = t_f$. Note that there are $N + 1$ different times that we care about, not N.

(6) Make a table with 7 columns and $N + 1$ rows. Label the columns t, S, I, R, S', I', and R'. (Some people like to add three more columns labeled $S'\Delta t$, $I'\Delta t$, and $R'\Delta t$, and maybe even three after that for $S + S'\Delta t$, $I + I'\Delta t$, and $R + R'\Delta t$.) Put the list of times in the first column. Fill in the second, third, and fourth entries of the first row with $S(0)$, $I(0)$, and $R(0)$.

(7) Use the SIR equations to fill out the rest of the first row. That is, compute $S'(0) = -aS(0)I(0)$, $I'(0) = aS(0)I(0) - bI(0)$, and finally $R'(0) = bI(0)$.

(8) Use a linear approximation to figure out the values of S, I, and R at the next time. We'll call these S_{new}, I_{new} and R_{new}, and call the previous values S_{old}, etc. This involves adding $S'_{old}\Delta t$, $I'_{old}\Delta t$, and $R'_{old}\Delta t$ to S_{old}, I_{old}, and R_{old}. Write down the new values on the next row.

   If we aren't interested in keeping track of S', I', and R', we can combine steps (7) and (8), writing

$$\begin{aligned} S_{new} &= S_{old} - aS_{old}I_{old}\Delta t, \\ I_{new} &= I_{old} + (aS_{old}I_{old} - bI_{old})\Delta t, \\ R_{new} &= R_{old} + bI_{old}\Delta t. \end{aligned}$$

(9) Repeat steps (7) and (8) (or the combined step) N times. At this point, the first N rows of our table will be filled in completely, and the last row will feature t, S, I, and R. Congratulations! We're done!

Now let's see how that looks with an actual program, called SIREulers. The programs in this book are all written in MATLAB, which includes some simple graphing routines for displaying our output. However, if you prefer to work with another programming language, that's fine, too.

```
function [S, I, R, t] = SIREulers(tmax, N, a, b, S0, I0, R0)
        dt = tmax/N;                                    % Compute Δt
        S = zeros(1, N+1);                              % Start with empty
        I = zeros(1, N+1);                              % Vectors for
        R = zeros(1, N+1);                              % S, I, and R.
        S(1) = S0;                                      % Initialize S
        I(1) = I0;                                      % Initialize I
        R(1) = R0;                                      % Initialize R
        for n = 1:N                                     % Start the loop
            S(n+1) = S(n) - a*S(n)*I(n)*dt;             % Rate eq.for S
            I(n+1) = I(n) + (a*S(n)*I(n) - b*I(n))*dt;  % Rate eq. for I
            R(n+1) = R(n) + b*I(n)*dt;                  % Rate eq. for R
        end                                             % End the loop
        t = linspace(0,tmax,N+1);                       % A vector of times
        plot(t, S, 'red', t, I, 'blue', t, R, 'green')  % Graph the results
        title('Plot of SIR model behavior')            % Add a title
        ylabel('Individuals'), xlabel('Days')          % Add labels
        legend('Susceptible', 'Infected', 'Recovered') % Add a legend.
```

There are a few differences between this program and the algorithm above.

(1) The program is written to be flexible. Instead of having lines that specify t_f, N, a, b, $S(0)$, $I(0)$, and $R(0)$, we type in these parameters when we use the program. In fact, most of the data in Tables 2.1 and 2.2, from Section 2.3, where we studied the SIR model with $a = 0.000004$, $b = 1/15$, $S(0) = 40,000$, $I(0) = 2100$, and $R(0) = 7900$, was generated with SIREulers.[1] Most of the data for Table 2.1 was generated by typing

   SIREulers(10,5,0.000004, 1/15, 40000, 2100, 7900),

and the data for Table 2.2 was generated by typing

   SIREulers(-30,15,0.000004, 1/15, 40000, 2100, 7900).

To run it again with $N = 10$ or $N = 100$ or $N = 10,000$ steps, just replace "5" or "15" with "10" or "100" or "10000". By picking a big enough value of N, we can get as much accuracy as we want. The figures in Chapter 2 were generated using SIREulers with N in the thousands; the run times were negligible.

(2) The program doesn't generate tables directly. Rather, it generates arrays. In MATLAB arrays are indexed starting at 1, so we have to make our index n run from 1 to $N + 1$ instead of from 0 to N. We just have to remember that $S(n)$, $I(n)$, and $R(n)$ represent the values of S, I, and R at time $t_{n-1} = (n-1)\Delta t$ and not at time t_n. In particular, our final answers are $S(N + 1)$, $I(N + 1)$, and $R(N + 1)$.

---

[1]To use this program we must work in a directory that contains the program file SIREulers.m. You are welcome to copy and use SIREulers.m, but it is not a built-in part of MATLAB.

(3) The program combines steps (7) and (8) and doesn't actually keep track of S', I' and R'. (To create Table 1 of Section 2.3, those values were calculated after the fact.)

(4) The program doesn't just compute numbers. The last five lines of the program *graph* the output, with S, I, and R shown as differently colored lines on the same chart. As SIREulers saves the vectors S, I, R, and t, all of the graphing can be done after we have run the program. For instance, if we only wanted to graph S we could just type `plot(t,S)`, and similarly for I and R. Graphing after the fact also allows us to vary the colors and format of the graph, for instance generating the graphs of Figure 3.12.

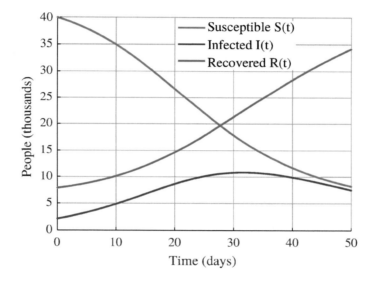

Figure 3.12. Graphs generated by SIREulers

3.7. Solving Algebraic Equations

We close this chapter with two approximation methods for computing roots of algebraic equations. One is an elaboration of what we did at the beginning of the chapter to solve $x^5 + x - 31 = 0$. The other is a very efficient algorithm for computing square roots.

The bisection method. There is a common guessing game where one person thinks of a whole number between 1 and 100 and the second person tries to guess it. After each guess, the first person says whether the number is higher or lower than the guess. Here is a typical run:

Alice: Guess my number.
Bob: 50
Alice: Lower!
Bob: 25

Alice: Lower!
Bob: 12
Alice: Higher!
Bob: 18
Alice: Higher!
Bob: 22
Alice: Lower!
Bob: 20
Alice: Right!

At each step, Bob guessed a number in the middle of the range of possible answers (sometimes rounding up or down). When Alice told him whether he needed to go higher or lower, that range was cut in half and he guessed in the middle again. 50 is in the middle of 1–100, 25 is in the middle of 1–49, 12 is in the middle of 1–24, 18 is in the middle of 13–24, 22 is in the middle of 19–24, and 20 is in the middle of 19–21. By shrinking the interval each time, Bob homed in on Alice's number. No matter what number Alice picked, Bob was guaranteed to get the answer in at most 7 guesses.

Now suppose that we have a polynomial $p(x)$ and points a and b such that $p(a)$ and $p(b)$ have opposite signs. Polynomials are continuous and can't jump from positive to negative, so there has to be a **root** (that is, a point where $p(x) = 0$) somewhere between a and b. We then guess the midpoint, $c = (a + b)/2$, and compute $p(c)$. From the sign of $p(c)$ we know whether to go higher or lower. Like Bob, we narrow our interval and keep guessing in the middle until our interval is so small that any point is close enough.

For instance, suppose that $p(x) = x^5 + x - 31$. We already know that $p(1) = -29 < 0$ and $p(2) = 3 > 0$, which gives us a starting point. The result of our game is shown in Table 3.3.

Table 3.3. Solving $x^5 + x - 31 = 0$ by the bisection method

Lower	Upper	Guess	p(Guess)	Adjustment
1.00000	2.00000	1.50000	−21.906	Higher!
1.50000	2.00000	1.75000	−12.837	Higher!
1.75000	2.00000	1.87500	−5.9507	Higher!
1.87500	2.00000	1.93750	−1.7596	Higher!
1.93750	2.00000	1.96875	0.54566	Lower!
1.93750	1.96875	1.95312	−0.2655	Higher!
1.95312	1.96875	1.96093	−0.0449	Higher!
1.96093	1.96875	1.96484	0.24921	Lower!

In eight steps, we went from "somewhere between 1 and 2" to "somewhere between 1.96093 and 1.96484", an interval that is a factor of 256 smaller. Every ten steps improves our accuracy by a factor of $2^{10} = 1024$, in other words by about three decimal places. This is true not only when we solve $p(x) = x^5 + x - 31 = 0$, but when we find a root of any polynomial.

The Babylonian Method. More than 3000 years ago (scholars differ on the exact date), the ancient Babylonians invented a very efficient method for computing square roots using only addition, subtraction, multiplication, and division. Suppose they were trying to find the square root of a number a. They started with a guess, which we'll call x. Because

$$(3.25) \qquad x \cdot \frac{a}{x} = a = \sqrt{a} \cdot \sqrt{a},$$

one of x and $\frac{a}{x}$ is bigger than $\sqrt{a}$ and the other is smaller. Either way, the average of x and $\frac{a}{x}$ is likely to be closer to $\sqrt{a}$ than x was. They then replaced x with the average of x and $\frac{a}{x}$ and repeated the process as many times as necessary to get the accuracy they wanted.

In fact, this procedure homes in on the true square root amazingly quickly. Table 3.4 shows what happens when we use this method to find $\sqrt{5}$ with the initial guess $x = 2$.

Table 3.4. Computing $\sqrt{5}$ with the Babylonian method gives 12 decimal places of accuracy in just four steps.

x	$5/x$	Average
2	2.5	2.25
2.25	2.222222222222	2.236111111111
2.236111111111	2.236024844720	2.236067977916
2.236067977916	2.236067970840	2.236067977500
2.236067977500	2.236067977500	2.236067977500

The numbers in Table 3.4 were generated with a very simple MATLAB program:

```
a=5              % We're finding the square root of 5
x=2              % That's our initial guess
for j=1:4        % Run the loop four times
    y=a/x        % Compute a/x
    x=(x+y)/2    % Replace x with the average
end              % Complete the loop
```

To compute the square root of a different number, just change the first two lines.

You might think that the Babylonian method is obsolete, as nowadays we have calculators and computers with built-in square root functions. But how do you think the square root function on your calculator actually works? It uses the Babylonian method!

Several thousand years after the Babylonians, Isaac Newton modified the Babylonian method to find roots of all sorts of functions, not just quadratic polynomials. We'll learn about Newton's method in Section 5.6.

3.8. Chapter Summary

The Main Ideas.

- Only very easy problems can be solved quickly and exactly. Most of the time we need to find approximate answers and then improve the approximations.
- A function is a rule for turning an input into an output. It can be understood as
  - An algorithm,
  - A formula,
  - A table, or
  - A graph.
- Linear functions are functions that change at a constant rate m, called the *slope*, *multiplier* or *rate of change*. The formulas

$$m = \frac{\Delta y}{\Delta x}, \qquad \Delta y = m\Delta x, \qquad \Delta x = \frac{\Delta y}{m}$$

are mathematically equivalent, but they carry three different lessons:
  - You can compute the slope from two points as rise/run.
  - You can predict Δy from Δx. This gives us the equation of our line in *point-slope* form: $y = y_1 + m(x - x_1)$.
  - You can predict Δx from Δy.
- You can use lines to approximate curves, *interpolating* and *extrapolating* from given data to estimate other values.
- Euler's method is a "follow your nose" method for finding an approximate solution to a rate equation. More accuracy requires more work.
- With a simple program, we can get a computer to do that work for us.
- For a very, very long time, people have been using approximations to solve algebraic equations. The bisection method and the Babylonian method are simple and accurate.

Expectations. You should be able to:

- Translate back and forth between different representations of a function.
- Find the slope of a line from the coordinates of two points.
- Find the equation of a line in point-slope form from a slope and the coordinates of one point.
- Manipulate that equation to answer questions like "where will you be at such-and-such a time?" and "when will you reach such-and-such a place?"
- Use linear approximations to approximate functions. Extrapolate and interpolate from nearby data points.
- Implement Euler's method by hand, generating a table of (approximate) values of y and y' as a function of time.
- Explain the steps in a computer program that uses Euler's method for a certain rate equation, and modify the program to handle slightly a different rate equation.

3.9. Exercises

Understanding Functions and Composite Functions

Exercises 3.1 and 3.2 refer to the following functions:

$$h(s) = \frac{2s}{s+3}$$
$$f(x) = x^2$$
$$g(u, v) = u + 2v$$
$$q(x, y) = 3$$
$$r(z) = \begin{cases} 2 & \text{if } z < 0 \\ z^2 + 3 & \text{if } 0 \le z < 5 \\ 5 - z & \text{if } 5 \le z \end{cases}$$

3.1.   Evaluate each function.
   (a) $h(3)$
   (b) $(f \circ h)(2)$
   (c) $g(3, 5)$
   (d) $q(\pi, -3)$
   (e) $h(f(3))$
   (f) $r(r(r(-2)))$

3.2.   **True or False?** If true, explain why. If false, provide a counterexample.
   (a) If $a > b$, then $f(a) > f(b)$.
   (b) For all real numbers a and b, $f(a + b) = f(a) + f(b)$.
   (c) For all real numbers c and d, $g(c, d) > 0$.
   (d) For all real numbers c, $(f \circ h)(c) = (h \circ f)(c)$.

3.3.   Graph $y = |x|$. Does this relationship between x and y describe a function? Why or why not?

3.4.   **Sales Tax.** The Texas state sales tax rate is 6.25%. That is, the sales tax T (in dollars) is proportional to the price P of an object, also in dollars, and the constant of proportionality is $k = 6.25\% = 0.0625$. Write a formula that expresses the sales tax as a linear function of the price. Use your formula to compute the tax on a computer that costs $1200.00.

3.5.   Suppose you work at a cell phone store. Your weekly salary is $400 plus a 4% commission on total sales over $8000. Let x represent your total sales (in dollars) in a particular week.
   (a) In words, briefly describe what the function $f(x) = 0.04x$ is computing.
   (b) In words, briefly describe what the function $g(x) = x - 8000$ is computing.
   (c) Suppose that you sold enough cell phones that week to earn a commission. Which of the following would correctly compute your earnings for that week:

   $$S_1(x) = 400 + (f \circ g)(x) \qquad \text{or} \qquad S_2(x) = 400 + (g \circ f)(x)?$$

   Explain.

3.6. Suppose that you want to grow your own vegetables in raised garden beds. You go to the local farm supply store and buy wood, fertilizer, soil, and other supplies for the beds. The store will deliver your purchases to your house for a fee. You pay for your purchases, the sales tax, and the delivery fee. The cost of the supplies, before taxes, is x. The delivery fee is \$30. Sales tax is 7.5%.

(a) Write a function $g(x)$ for the total cost *after* taxes, *not* counting the delivery fee. (Remember that x is the price of your purchases.)

(b) Write a function $f(x)$ for the total cost *before* taxes, *including* the delivery fee.

(c) Calculate and interpret $(f \circ g)(x)$ and $(g \circ f)(x)$.

(d) If sales taxes are *not* charged on delivery fees, which of the compositions in part (c) represents your actual cost? Explain.

Linear Functions and Slope

3.7. Graph the function $f(x) = 0.8x - 2$ on the interval $-5 \le x \le 5$.

(a) Read from your graph the value of $f(x)$ when $x = -2$ and when $x = 2$. What is the difference between these function values? Call this difference Δy. What is the difference between the two x-values? Call this difference Δx. Compute the slope using Δx and Δy. According to the equation of the line, which was given in slope-intercept form, what is the slope? Does your computed slope agree with this value? Why or why not?

(b) Rewrite the equation of the line in point-slope form, using as your reference the point $(1, -1.2)$.

3.8. Graph the function $f(x) = 0.8(x + 2) - 3.6$ on the interval $-5 \le x \le 5$.

(a) Read from your graph the value of $f(x)$ when $x = -2$ and when $x = 2$. What is the difference between these function values? Call this difference Δy. What is the difference between the two x-values? Call this difference Δx. Compute the slope using Δx and Δy. According to the equation for the line, what is the slope? Does your computed slope agree with this value?

(b) Rewrite the equation of the line in slope-intercept form.

3.9. A car is driving at a constant speed of 64 miles/hour. At 2:00 p.m. it passes mile marker 150.

(a) How far will the car travel in the next half hour? Where will it be at 2:30?

(b) How far will the car travel between 2:00 p.m. and t o'clock? Where will it be at t o'clock?

(c) Express the position of the car at t o'clock as a linear function *in point-slope form*. Then plug in $t = 2.25$ to figure out the location of the car at 2:15 p.m.

(d) Rewrite this function in *slope-intercept form*. Then plug in $t = 2.25$ to figure out the location of the car at 2:15 p.m.

Food for Thought: Which was easier, plugging $t = 2.25$ into point-slope or slope-intercept form? If you did your calculations with a calculator, it probably didn't matter much. However, if you did them by hand, you probably found multiplying the slope in part (c) by 1/4 to be a lot easier than multiplying the slope in part (d) by $2\frac{1}{4}$. In general, point-slope is much better for understanding a linear function close to your reference point, which is what we usually care about, while slope-intercept is better for understanding that function near $t = 0$ (noon in this case).

3.10.   A car is driving at constant speed. At 3:00 p.m. it is at mile marker 145. At 4:00 p.m. it is at mile marker 201.
  (a) How fast is the car moving?
  (b) Write a function, *in point-slope form*, for the location $x(t)$ of the car at t o'clock, using 3:00 p.m. as your reference time.
  (c) Use this function to determine where the car is at 3:15 p.m.
  (d) Now rewrite your function in *slope-intercept form*. Use this function to determine where the car was at 2:45 p.m.

3.11.   You should be able to answer all parts of this exercise without ever finding the equations of the functions involved.
  (a) Suppose that $f(x)$ is a linear function with multiplier $m = 1/2$. If $f(-2) = 2$, what is $f(-2.1)$? $f(-2.016)$? $f(10)$?
  (b) Suppose that $h(x)$ is a linear function with multiplier $m = -3$. If $h(1) = 4$, for what value of x is $h(x) = -5$? $h(x) = 19$? $h(x) = 97$?
  (c) Suppose that $g(x)$ is a linear function with $g(3) = 3$ and $g(6) = 4.5$. What is $g(3.1)$? What is $g(3 + a)$?

3.12.   Suppose that $w = 150 - 12s$. How much does w change when s changes from 4 to 10? From 4 to 4.5? From 4 to 4.1? Let Δs denote a change in s and Δw denote the corresponding change in w. Is $\Delta w = m\Delta s$ for some constant m? If so, what is m?

3.13.   Consider two quantities x and y.
  (a) Explain what it means for y to be a linear function of x.
  (b) If y is a linear function of x, is x necessarily a linear function of y? Either provide an argument for why it always is, or an example in which it is not.
  (c) In the following table, x and y are linear functions of each other. Fill in the rest of the table.

x	-5	-2		3	6	
y	8		-0.5	-4		-10

  (d) Based on the table in part (c), express Δy as a function of Δx. Then find a formula for y in terms of x.

3.14.   **Thermometers**. Temperatures are usually measured in degrees Fahrenheit in the United States and in degrees Celsius elsewhere. The Fahrenheit reading F and the Celsius reading C are linear functions of each other, based on two reference points:
  • 0 degrees Celsius is the freezing point of water, which is 32 degrees Fahrenheit.
  • 100 degrees Celsius is the boiling point of water, which is 212 degrees Fahrenheit.
  (a) Use the freezing and boiling points of water to write F as a linear function of C.
  (b) Express C as a linear function of F.

(c) Which represents a larger change in temperature, a degree Celsius or a degree Fahrenheit?

(d) Suppose that an item heats up or cools down. Express ΔF (the change in the Fahrenheit reading) as a linear function of ΔC, and vice-versa.

(e) Find a temperature that has the same reading on both scales.

A True Story: On a spring day in 1975, a young boy from Nebraska was riding with his father to Canada. As they neared the border, they picked up a radio station from Winnipeg. The weather report came on and said that the temperature was 5 degrees. The boy and his father were shocked that it could be so cold in the spring. After a commercial break, the disc jockey came on and said, "That's 5 degrees Celsius, or 41 degrees Fahrenheit". Canada had switched its weather forecasts a few weeks earlier from Fahrenheit to Celsius. It took a while for Canadians (not to mention Americans) to get used to the change.

Falling Bodies. Ignoring air resistance, the velocity of a falling body increases in proportion to the increase in the time that the body has been falling. The constant of proportionality is approximately 32 feet per second squared, or 32 ft/s^2.

3.15.   A ball is falling at a velocity of 48 feet/sec after 1 second.

(a) How fast is it falling after 3 seconds?

(b) Express the change in the ball's velocity Δv as a linear function of the change in time Δt.

(c) Express the ball's velocity $v(t)$ as a linear function of time.

3.16.   If d and v denote the distance fallen (in feet) and the velocity (in feet per second) of a falling body, then the motion can be described by the following equations:

$$d' = v, \qquad v' = 32.$$

The first equation says that velocity is the rate of change of distance. The second equation is what we discussed in Exercise 3.15.

Suppose that $d(0) = 0$ and $v(0) = 16$. Using a step size of 1 second, estimate $d(3)$ and $v(3)$.

3.17.   **Depreciation.** A farmer buys a new combine (used for harvesting crops) for $380,000. He assumes that the value of the combine will drop, or **depreciate**, at a constant rate, and that it will be worth $100,000 in ten years. Let $V(t)$ represent the value of the combine t years after it was purchased.

(a) Find an expression relating ΔV to Δt. Then find an expression relating V to t.

(b) What will the combine be worth after seven years?

(c) When will the value of the combine fall below $80,000?

(d) Graph V for $0 \le t \le 20$ and indicate on your graph the answers for parts (b) and (c).

(e) This method of estimating the value of a used piece of equipment is called **straight-line depreciation**. Explain why this is a sensible name.

3.18. Using MATLAB, graph the function $f(x) = 1/x$ on the interval $2 \le x \le 4$.
   (a) Determine the equation of a line between the points $(2, f(2))$ and $(4, f(4))$. Write this equation in the form $h(x) = m(x - 2) + b$.
   (b) Using the MATLAB "hold on" command, graph $h(x)$ on the same set of axes that you graphed $f(x)$ on.
   (c) Compare the values $f(2)$ and $h(2)$. Do you think $h(x)$ would be a good approximation for $f(x)$ at $x = 3$? Regardless of your answer, how could you create a line that would provide a better approximation for $f(x)$ at $x = 3$?

3.19. **Euler's method**. For the rate equation $y'(t, y) = 2t + y - y^2$ and initial condition $y(0) = 2$, use Euler's method and given step size to answer the following.
   (a) For step size $\Delta t = 2$, approximate $y(2)$.
   (b) For step size $\Delta t = 1$, approximate $y(2)$.
   (c) For step size $\Delta t = 0.5$, approximate $y(1)$.
   (d) How does the $y(1)$ computed in part (c) compare to the $y(1)$ computed in the intermediate step for part (b)? Which of the two estimates for $y(1)$ is more accurate? Explain why.

3.20. **Radioactivity**. During the 1986 Chernobyl nuclear disaster, an estimated 27 kg of radioactive Cesium-137 was released into the environment. Cesium-137 has a half-life of 30 years, meaning that the amount $C(t)$ in the environment satisfies the rate equation

$$C' = -0.0231C,$$

where time is measured in years. Using a step size of 10 years, estimate how much Cesium-137 will remain in 2036, on the 50th anniversary of the disaster.

Food for Thought: Not understanding radioactivity can be deadly. During Russia's 2022 invasion of Ukraine, Russian soldiers who had never heard of the Chernobyl disaster were sent to camp and dig trenches in the most contaminated part of the Chernobyl Exclusion Zone, the so-called Red Forest. Their digging unearthed contaminated soil that had previously been buried. Many of those soldiers died of radiation poisoning.

The Bisection and Babylonian Methods

3.21. Use the bisection method to solve $x^3 + x^2 = 20$ to at least two decimal places, starting with the fact that $2^3 + 2^2 < 20 < 3^3 + 3^2$.

3.22. Use five steps of the bisection method to estimate the positive solution to $x^2 = 10$ (in other words, estimate $\sqrt{10}$), starting with the fact that $3^2 < 10 < 3.2^2$.

3.23. In the setup of Exercise 3.22, how many steps would it take to get eight-digit accuracy?

3.24. Use the Babylonian Method to estimate $\sqrt{10}$ to at least eight-digit accuracy, starting with the initial guess $x_0 = 3$. (This should only take three steps.)

Chapter 4

Track the Changes

So far, we've talked a lot about rates of change, but we never really defined what "rate of change" actually means. We tiptoed around the concept, which was the best we could do with the tools we had. But now we have the tools to take a closer look at the **derivative** of a function. That's the precise mathematical object that corresponds to the vague idea of "rate of change".

4.1. Atmospheric Carbon

Carbon dioxide (CO_2) plays an essential role in life on earth. We produce CO_2 in our cells, converting sugar and oxygen to CO_2, water, and the energy that keeps us alive. Plants do the opposite, converting CO_2, water, and sunlight to sugar and oxygen. CO_2 is also produced whenever we burn anything containing carbon, such as when we use fossil fuels like gasoline, natural gas, and coal. CO_2 keeps our planet warm, trapping heat via the **greenhouse effect**. Visible light from the sun passes right through our atmosphere and heats the earth. The earth radiates its extra heat back into space as infrared light, but the atmosphere is *not* transparent to infrared light. Instead, some of that infrared light is absorbed and reflected back to the earth by CO_2 and other **greenhouse gasses**. The more such gasses are in the atmosphere, the more infrared light is reflected back to the ground and the warmer the earth gets.

Global temperatures have been rising in recent decades, with record heat in many places. Extreme weather events such as hurricanes have become more frequent. At the same time, CO_2 concentrations in the atmosphere have been going up. There are many possible causes of climate change, but most scientists believe that increasing concentrations of CO_2 and other greenhouse gasses are the main cause.

So how fast have CO_2 levels been increasing? CO_2 concentrations vary by season, with a peak typically in May. Table 4.1 shows those peak concentrations, in parts per

Table 4.1. Peak CO_2 concentrations by year in parts per million

Year	Concentration	Year	Concentration	Year	Concentration
1994	362.25	2004	381.04	2014	402.18
1995	364.18	2005	382.58	2015	404.31
1996	365.74	2006	385.34	2016	408.56

million (ppm), in three time periods: one in the 1990s, one in the 2000s, and one in the 2010s.

In 2015, the concentration of CO_2 was 404.31 ppm. If we let $C(t)$ be the concentration (measured in ppm) at time t (measured in years), then $C(2015) = 404.31$. There are three ways to estimate the rate $C'(2015)$ at which that concentration was growing, as shown in Figure 4.1.

(1) We can compare the concentration in 2015 to the concentration in 2016. The concentration went up by $\Delta C = 408.56 - 404.31 = 4.25$ in time $\Delta t = 2016 - 2015 = 1$, so we estimate $C'(2015) \approx 4.25$. ΔC is called a **forward difference**, and the ratio $\Delta C / \Delta t$ is called a (forward) **difference quotient**.

(2) We can compare the concentration in 2015 to the concentration in 2014. As before, we compute $\Delta C = C(2015 + \Delta t) - C(2015)$ and divide by Δt, only now $\Delta t = -1$. Then $\Delta C = 402.18 - 404.31 = -2.13$, and we estimate that $C'(2015) \approx 2.13$. Here ΔC is called a **backward difference**.

(3) We can compare the concentration in 2014 to the concentration in 2016. That is,

$$(4.1) \qquad C'(2015) \approx \frac{C(2015 + \Delta t) - C(2015 - \Delta t)}{2\Delta t} = \frac{408.56 - 402.18}{2} = 3.19.$$

ΔC is called a **centered difference** or a **central difference**. Most of the time, centered difference quotients are much better estimates than forward or backward difference quotients. However, lack of data sometimes forces us to use forward or backward differences.

We can do similar computations to compute $C'(2005)$ and $C'(2015)$. Using centered differences, we get

$$(4.2) \qquad C'(1995) \approx 1.745, \qquad C'(2005) \approx 2.15.$$

The concentration grew much faster in the 2000s than in the 1990s, and even faster in the 2010s.

Let's turn our attention to $C'(1995)$. With that rate of change, we can do a linear approximation. The equation of the line through $(1995, 364.18)$ with slope 1.745 is

$$(4.3) \qquad C(t) \approx 364.18 + 1.745(t - 1995)$$

This linear approximation[1] predicts a concentration of $364.18 + 10(1.745) = 381.63$ in 2005, which is reasonably close to the actual value of 382.58. It also predicts a concentration of $364.18 + 20(1.745) = 399.08$ in 2015, which is substantially below the actual

---

[1] Note that this **tangent line** is not the line through $(1994, 362.25)$ and $(1996, 365.74)$. It has the same slope, as we used a centered difference to estimate $C'(1995)$, but the tangent line touches the actual graph at one point, not two.

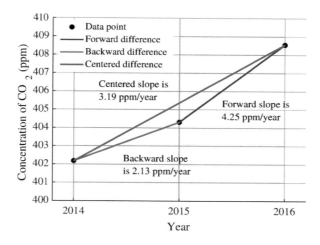

Figure 4.1. Three ways to estimate $C'(2015)$

value of 404.31, and a concentration of 407.81 in 2020, compared to an actual value of 418.32. What happened?

Part of the answer is that our data was limited and noisy. We only used three data points and all scientific measurements are subject to error. If we had used a forward difference quotient or a backward difference quotient, or if we had done our linear approximation starting in 1994 or 1996 instead of 1995, our linear approximation would have had a different slope and would have generated better or worse predictions for 2005, 2015, and 2020.

However, most of the answer is that historical events changed the trajectory. China industrialized, joining the United States and Europe as a major consumer of fossil fuels. Parts of the Amazon were intentionally burned to create farmland. Rising temperatures caused some of the arctic permafrost to melt, releasing CO_2 that had been trapped in the ice. All of these factors, and many others, caused the actual graph of $C(t)$ to bend upwards (sharply!) and leave the linear approximation behind.

4.2. Other Derivatives and Marginal Quantities

In the previous example, our input variable was time, and C' was the rate at which C changed per unit time. There are many other examples where the variable is time. We study the rate of change of position with respect to time and call that **velocity**. We study the rate of change of velocity with respect to time and call that **acceleration**. We study the rate at which population is growing, or the national debt is growing, or the polar ice caps are melting. However, the concept of **rate of change** applies even when the input variable has nothing to do with time.

Pumping blood. The heart of a typical 40-year-old man pumps a little over 4000 ml/min of blood at a pulse rate of 60 beats per minute. How much more blood can it pump at 61 beats per minute, or 65, or 70? How much less would it pump at 59 beats per minute or 55 or 50?

In this case we have a function $H(p)$ that gives the blood flow, in ml/min, as a function of the pulse p. If we measure that $H(59) = 4285$, $H(60) = 4347$ and $H(61) = 4409$, then we can compute $H'(60)$ using either forward, backward, or centered differences. In this case, all three methods give the same result, namely $H'(60) \approx 62$. Every additional beat per minute delivers 62 additional milliliters of blood per minute.

Note that this is *not* the same as asking how much blood flows per heart beat! At 60 beats per minute, the heart delivers 4347 milliliters per minute, or $4347/60 \approx 72.5$ milliliters per beat. But just because the heart is currently delivering 72.5 milliliters per beat doesn't mean it can keep delivering 72.5 milliliters per beat when it beats faster. It can't! The heart gets less efficient as it beats faster, so $H(p)$ is *not* $72.5p$.

Suppose that the man were to begin exercising and his body needed 5000 milliliters of blood per minute. What would his pulse have to be to deliver that much blood? We can understand this by looking at the change in his blood flow, and the change in his pulse, compared to his resting rate. He needs an extra 653 milliliters compared to when his pulse is 60, and each *additional* beat is worth 62 milliliters, so he needs $653/62 = 10.5$ *additional* beats, or a pulse of 70.5 beats per minute.

The **average flow** $\bar{H}(60) = H(60)/60 \approx 72.5$ tells us the overall efficiency of the heart, compared to not beating at all. This is a very important measure of how well the heart is working under normal (resting) conditions. However, it doesn't tell us how much the blood flow *changes* when the pulse changes, or how much the pulse needs to change in order to achieve a certain change in blood flow. To understand changes, we need calculus.

The derivative $H'(60)$ is called the **marginal** flow. It tells us how much *extra* blood flow there is per *extra* pulse. It isn't comparing $H(60)$ to $H(0) = 0$. It's comparing $H(60)$ to $H(61)$ and $H(59)$ and other nearby points. That's what makes it so useful for understanding hearts that are beating at close to 60 beats per minute.

Economics. The word "marginal" is borrowed from economics, where the **marginal cost** of an object is (approximately) the extra cost associated with making one more object. For instance, consider a factory that makes widgets. It costs \$22,000/month just to keep the factory open, plus \$7 for every widget that it produces. If it makes 10,000 widgets per month, how much does it cost to produce a widget?

The *average* cost of a widget is \$9.20/widget, because it costs \$92,000/month to produce 10,000 widgets/month and $\frac{\$92,000/\text{month}}{10,000 \text{ widgets/month}} = \$9.20/\text{widget}$. However, the *marginal* cost is only \$7/widget, because that's how much extra we have to pay for every extra widget we make.

These quantities have different uses. Suppose that the factory sells its widgets for \$9 each. The price is lower than the average cost, so the factory is losing money at a rate of $10,000(9.2 - 9) = 2000$ dollars per month. That's important information! However, it does not mean that the factory should make fewer widgets. The price is higher than the marginal cost; each additional widget costs \$7 but brings in \$9, for an additional profit of \$2/widget. The factory can make more money (or at least lose less money) by making more widgets. If it can make another 1000 widgets per month and sell them for \$9 each, it will break even.

In general, average quantities are important when comparing the current situation to stopping altogether (closing the factory, or replacing a patient's heart with a transplant or artificial pump). However, when considering small changes in the input variable, whether heart rate or production level, marginal quantities are much more important.

In addition to marginal cost, we can talk about **marginal revenue** and **marginal profit**. The revenue from making x widgets isn't really $9x$, as making more widgets causes the supply of widgets to go up and the price of widgets to go down. To manage the factory properly, we need to understand how the price $p(x)$ depends on x, and then compute the revenue function $R(x) = xp(x)$. By taking the derivative of the revenue function, we can calculate the marginal revenue $R'(x)$, which is how much more money can be expected from each additional widget that is made. Profit is the difference between revenue and cost, so the marginal profit $P'(x)$ is the difference between the marginal revenue $R'(x)$ and the marginal cost $C'(x)$. If the marginal profit is positive, or equivalently if the marginal revenue is bigger than the marginal cost, then the factory can make more money by increasing production. If the marginal revenue is less than the marginal cost, then the marginal profit is negative and the factory should decrease production.

4.3. Local Linearity and Microscopes

In the last chapter we saw a linear approximation work very well on the function $SQ(x) = x^2$, approximating x^2 as $1 + 2.01(x - 1)$ for x between 0.9 and 1.1. Earlier in this chapter, we saw a linear approximation do a very poor job of extrapolating CO_2 levels into the mid-2010s from the 1990s. In this section we examine the question of when a linear approximation is justified, and when it isn't. The general rule is:

> **The Microscope Equation:** *If $f'(a)$ exists, then the linear approximation*
>
> (4.4) $$f(x) \approx f(a) + f'(a)(x - a)$$
>
> *is good whenever x is close to a. If x is far from a, or if $f'(a)$ does not exist, then all bets are off.*

To get a feel for how close is close enough, let's look at the function $f(x) = \sin(x)$, shown at four different scales in Figure 4.2, with x measured in radians. On the interval $[-9, 11]$, the graph of $f(x)$ is nowhere near straight. In fact, it undergoes a little more than three full oscillations! Trying to approximate all those wiggles with a straight line is ridiculous.

On the interval $[0, 2]$, however, it only undergoes about a third of a cycle. It mostly goes up, although it heads downhill a little at the end. Approximating this curve with a straight line isn't as absurd as trying it on $[-9, 11]$, but it still isn't a very good fit.

On the interval $[-.9, 1.1]$, however, the sine function only undergoes a 30th of a cycle. In that short stretch, it doesn't have room to change direction very much. For any two numbers x_1 and x_2 between 0.9 and 1.1, $\Delta y/\Delta x$ is somewhere between 0.45 and

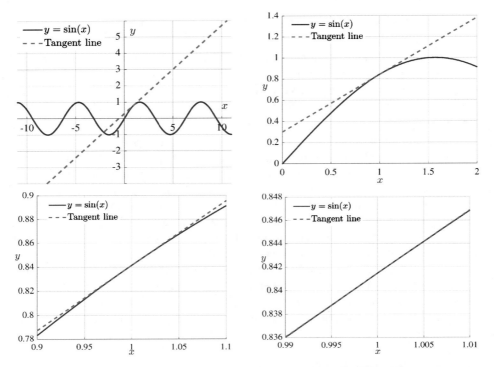

Figure 4.2. The line tangent to $y = \sin(x)$ at $x = 1$ is a terrible approximation on $[-9, 11]$, is poor on $[0, 2]$, is good on $[0.9, 1.1]$ and is indistinguishable from the curve on $[0.99, 1.01]$.

0.62. On the interval $[-0.99, 1.01]$ the fit is even better, and $\Delta y / \Delta x$ only varies between 0.53 and 0.55. Because $f(1) = \sin(1)$ is close to 0.8415, the linear approximation

$$(4.5) \qquad\qquad \sin(x) \approx 0.8415 + 0.54(x - 1),$$

while only reasonably accurate when x is between 0.9 and 1.1, is very accurate when x is between 0.99 and 1.01.

The upshot is that as you zoom in on a smooth curve — *any* smooth curve — the curve looks straighter and straighter, and the linear approximation (4.4) gets better and better. We call equation (4.4) the **microscope equation** because it describes what our graph looks like through a microscope.

Before moving on, let's consider a few functions whose graphs are not smooth, and for which linear approximations don't work.

- Consider the absolute value function $f(x) = |x| = \begin{cases} x & x \geq 0, \\ -x & x < 0. \end{cases}$ The graph of this function between -1 and 1 looks like a V, as seen in Figure 4.3. There is a line $y = x$ that works for $x \geq 0$, and a line $y = -x$ that works for $x \leq 0$, but there isn't any line that works on both sides. No matter how much we zoom in, we still see a V, not a line. The graph does not have a tangent line at $x = 0$, and the function does not have a linear approximation.

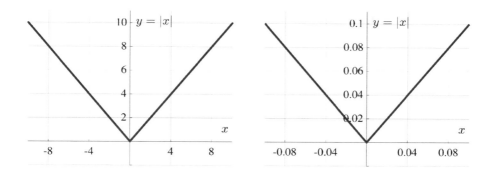

Figure 4.3. The graph of $f(x) = |x|$ is V-shaped at all length scales.

- Next consider the cube-root function $g(x) = x^{1/3}$, shown in Figure 4.4. If we try to compute the slope m at $x = 0$ with a forward difference, we get

(4.6)
$$m \approx \frac{\Delta y}{\Delta x} = \frac{g(0 + \Delta x) - g(0)}{\Delta x} = \frac{\Delta x^{1/3}}{\Delta x} = \Delta x^{-2/3}.$$

The smaller Δx is, the bigger $\Delta y/\Delta x$ is. If $\Delta x = 0.001$, then $\Delta y/\Delta x = 100$. If $\Delta x = 0.000001$, then $\Delta y/\Delta x = 10{,}000$. As Δx gets closer and closer to 0, our approximating line gets steeper and steeper, approaching a vertical line. The same thing happens if we use backward differences or centered differences. The curve simply has a vertical tangent at $x = 0$. A vertical line is not the graph of any function, so there is no linear function that approximates g near $x = 0$.

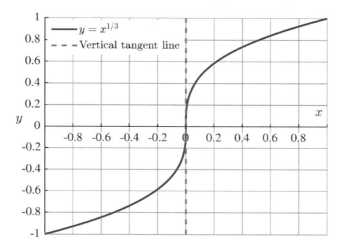

Figure 4.4. The graph of $y = x^{1/3}$ has a vertical tangent at $x = 0$.

- Imagine flipping a fair coin over and over again. For each whole number n, let $h(n)$ be the number of heads in the first n tosses minus the number of tails. This is called a **random walk**. $h(n + 1) - h(n)$ will either be 1 (if the $n + 1$st toss comes up heads) or -1 (if the $n + 1$st toss comes up tails). As there is no way

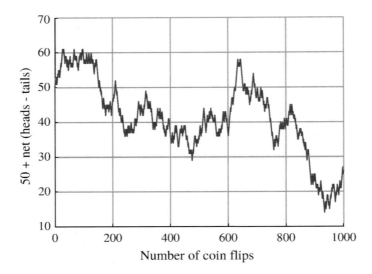

Figure 4.5. A random walk with 1000 coin flips, starting at $h = 50$. The resemblance to a typical stock chart is *not* coincidental.

of predicting what the $n + 1$st toss will be, it is absolutely impossible to use data of $h(1), h(2), \ldots, h(n)$ to predict whether h will go up or down next. The graph of $h(x)$ (with straight lines connecting consecutive integer points), shown in Figure 4.5, is like the graph of $|x|$, only with corners all over the place. It's impossible to find a line that reliably approximates the graph of h near any integer.

That won't stop people from trying, of course. After a run of heads (which happens by chance every once in a while), some people will say "the coin is lucky", and bet on the next toss being heads. Others will say "we've had so many heads that it's bound to be tails next" and bet the other way. Casinos and con artists make a lot of money exploiting that kind of woolly thinking.

The price of a stock from day to day, or a stock average like the Dow Jones, has many of the same mathematical properties as a random walk. There is a slow long-term pattern of growth, a few percent per year, but the short-term fluctuations are completely unpredictable.

Besides pointing out their existence, we don't have much to say about functions with the wild behavior of the last three examples. We are interested in functions that *do* have good linear approximations, and whose graphs *do* have (nonvertical) tangent lines. Such functions are called **differentiable**, and there are plenty of them to keep us busy.

4.4. The Derivative

After several chapters of exploration, we are finally ready to define the **derivative** $f'(a)$ of the function $f(x)$ at $x = a$. There are several different definitions, and it's important to understand why they all describe the exact same thing. If $f'(a)$ satisfies any one of these definitions, then it also satisfies all of the others.

Definition 1. The derivative $f'(a)$ of the function $f(x)$ at $x = a$ is the rate of change of the function at $x = a$.

Definition 2. The derivative $f'(a)$ of the function $f(x)$ at $x = a$ is the slope of the line tangent to the graph $y = f(x)$ at $(a, f(a))$.

Definition 3. The derivative $f'(a)$ of the function $f(x)$ at $x = a$ is the multiplier in the best linear approximation to $f(x)$ near $x = a$. That is, for any constant $c > 0$, no matter how small, the approximation

(4.7) $$f(x) \approx f(a) + f'(a)(x - a)$$

is accurate to within $c|x - a|$ when x is close enough to a.

Definition 4.

(4.8) $$f'(a) = \lim_{x \to a} \frac{f(x) - f(a)}{x - a}.$$

Definition 5.

(4.9) $$f'(a) = \lim_{h \to 0} \frac{f(a + h) - f(a)}{h}.$$

Definitions 4 and 5 use the language and notation of **limits**, and that deserves an explanation. We use the term **limit** to describe what a certain expression is homing in on. The expression "$x \to a$" is read as "x approaches a", while "$\lim_{x \to a}$" (also written "$\lim_{x \to a}$") is "the limit as x approaches a". The statement $f'(a) = \lim_{x \to a} \frac{f(x) - f(a)}{x - a}$ means that $\frac{f(x) - f(a)}{x - a}$ gets closer and closer to $f'(a)$ as x gets closer and closer to a. The statement $f'(a) = \lim_{h \to 0} \frac{f(a + h) - f(a)}{h}$ means that $\frac{f(a + h) - f(a)}{h}$ gets closer and closer to $f'(a)$ as h gets closer and closer to 0.

As we saw in the last section, not every function has a derivative. If $f'(a)$ exists, then we say that f is **differentiable at** a. If $f'(a)$ exists for every number a, we say that $f(x)$ is **differentiable**, period. The study of differentiable functions is called **differential calculus**, and that's what the first half of this book is about. (The second half is about **integral calculus**). For the next few chapters, almost all of the functions that we encounter will be differentiable.

Let's see why all the definitions are really talking about the same thing.

If the function $f(x)$ is changing at rate $m = f'(a)$ near $x = a$, then for any two points x_1 and x_2 near a, $f(x_2) - f(x_1)$ is approximately $m(x_2 - x_1)$. But that means that the ratio

(4.10) $$\frac{\Delta f}{\Delta x} = \frac{f(x_2) - f(x_1)}{x_2 - x_1}$$

is close to m, and so

(4.11) $$f(x_2) - f(x_1) \approx m(x_2 - x_1),$$

with an error that gets smaller and smaller, as a fraction of $x_2 - x_1$, as x_2 and x_1 get closer and closer to a. In particular, if $x_1 = a$ and $x_2 = x$, then

(4.12) $$f(x) \approx f(a) + f'(a)(x - a),$$

which is Definition 3.

If Definition 3 applies, then we have a good linear approximation whose graph has slope $f'(a)$. This graph is the line tangent to the graph of $f(x)$ at $x = a$ (Definition 2). If a line with slope $f'(a)$ is tangent to the graph, then the rate of change of the linear approximation is $f'(a)$, so the rate of change of the original function $f(x)$ must also be $f'(a)$. That is,

(4.13)   Definition 1 $\Rightarrow$ Definition 3 $\Rightarrow$ Definition 2 $\Rightarrow$ Definition 1,

so the first three definitions are all equivalent.

Now suppose that $f'(a)$ satisfies Definition 3. Then, for x close enough to a, $f(x) - f(a)$ must be within $c|x - a|$ of $f'(a)(x - a)$, so $\frac{f(x) - f(a)}{x - a}$ must be within c of $f'(a)$. As x gets closer and closer to a, we can take c smaller and smaller, so the ratio $\frac{f(x) - f(a)}{x - a}$ must be getting closer and closer to $f'(a)$. That is exactly Definition 4.

The reasoning also runs the other way. If $\frac{f(x) - f(a)}{x - a}$ is close to $f'(a)$, then $f(x) - f(a)$ is close (within a small fraction of $x - a$) to $f'(a)(x - a)$, so $f(x) \approx f(a) + f'(a)(x - a)$ is a good linear approximation. The upshot is that Definitions 1–4 are all equivalent.

Finally, Definition 5 is just Definition 4 in a slightly different notation. If we define $h = x - a$ (what we previously called Δx), then $x = a + h$, and looking at what happens as $x \to a$ is the same as looking at what happens when $h \to 0$. In particular,

(4.14) $$\lim_{x \to a} \frac{f(x) - f(a)}{x - a} = \lim_{h \to 0} \frac{f(a + h) - f(a)}{a + h - a} = \lim_{h \to 0} \frac{f(a + h) - f(a)}{h}.$$

Definitions 4 and 5 look like forward difference quotients, but it's a little more complicated than that. When we say $\lim_{x \to a}$ we are considering what happens for all values of x that are close to a. x might be slightly greater than a, in which case we're looking at a forward difference quotient, or x might be slightly less than a, in which case we're looking at a backward difference quotient. Likewise, Definition 5 involves the possibility that h is slightly positive, and also the possibility that h is slightly negative. For the function $f(x)$ to be differentiable at a, both the forward difference quotients and the backward difference quotients have to approach the *same* number $f'(a)$.

What about centered differences? The centered difference quotient

(4.15) $$\frac{f(a + h) - f(a - h)}{2h} = \frac{1}{2}\left(\frac{f(a + h) - f(a)}{h} + \frac{f(a) - f(a - h)}{h} \right)$$

is the average of a forward difference quotient and a backward difference quotient. If $f(x)$ is differentiable at $x = a$, then both the forward and backward difference quotients will be close to $f'(a)$, so their average will also be close to $f'(a)$. Using centered differences is a perfectly good way to compute a derivative.

The trouble is that centered difference quotients sometimes give us an answer even when the function isn't differentiable! Consider the function $f(x) = |x|$ near $x = 0$. The forward difference quotient $f(h)/h$ with $h > 0$ is always 1. The backward difference quotient $f(-h)/(-h)$ (with $h > 0$) is always -1. The centered difference quotient

$$(4.16) \qquad \frac{f(h) - f(-h))}{2h} = \frac{|h| - |-h|}{2h} = \frac{0}{2h}$$

is always 0. It's tempting to just say that $f'(0) = 0$, but that's wrong. The line $y = 0$ is *not* tangent to the curve, and is *not* the graph of a good linear approximation. This glitch is why we don't have a Definition 6 involving centered differences.

Examples of Derivatives. Now that we know what a derivative *is*, let's compute the derivative of the function $f(x) = x^2$ at $x = 1$. Using Definition 5, we have to find the difference quotient $\frac{f(1+h)-f(1)}{h}$ and compute its limit as $h \to 0$. We could plug in numbers like $h = 1, -1, 0.1, -0.1, 0.01, -0.01$ into a calculator, but we can do a lot better using algebra.

$$(4.17) \qquad \frac{f(1 + h) - f(1)}{h} = \frac{(1 + h)^2 - 1^2}{h} = \frac{h^2 + 2h + 1 - 1}{h} = \frac{h^2 + 2h}{h} = h + 2.$$

See Figure 4.6. As h gets closer and closer to 0, the difference quotient gets closer and closer to 2, no matter which way we do the limit. The result is that

$$(4.18) \qquad f'(1) = \lim_{h \to 0} \frac{f(1 + h) - f(1)}{h} = \lim_{h \to 0}(h + 2) = 2,$$

and we're done.

Note that using forward differences ($h > 0$) gives estimated slopes $h + 2$ that are a little too high. That's because the graph of $f(x) = x^2$ is curving up, so it is steeper to the right of $x = 1$ than at $x = 1$ itself. (That's why the slope of our linear approximation in Chapter 3 was 2.01 instead of 2. We estimated $f'(1)$ using a forward difference with $h = 0.01$.) Using backward differences ($h < 0$) gives estimated slopes that are a little too low. If we use centered differences, those errors happen to cancel out:

$$(4.19) \qquad \frac{f(1 + h) - f(1 - h)}{2h} = \frac{(h^2 + 2h + 1) - (h^2 - 2h + 1)}{2h} = \frac{4h}{2h} = 2.$$

In this case we got lucky and the errors canceled completely. Most of the time, using centered differences just makes the error a lot smaller. Forward or backward difference quotients have errors that are typically proportional to h, with one difference quotient being too high and the other too low. Centered difference quotients are typically off by something proportional to h^2. When h is much less than 1, h^2 is a lot smaller than h. If we use $h = 0.001$, we might get an answer accurate to three decimal places using forward or backward difference quotients, while we might get an answer accurate to six decimal places using centered difference quotients. That's why we should use centered differences with numerical data whenever possible.

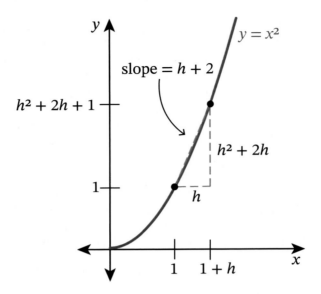

Figure 4.6. Computing the derivative of $f(x) = x^2$ at $x = 1$.

Now let's take the derivative of $f(x) = 1/x$ at $x = 3$. We compute a difference quotient

(4.20)
$$
\begin{aligned}
\frac{f(3 + h) - f(3)}{h} &= \frac{1}{h}\left(\frac{1}{3 + h} - \frac{1}{3}\right) \\
&= \frac{1}{h}\left(\frac{3}{3(3 + h)} - \frac{3 + h}{3(3 + h)}\right) \\
&= \frac{1}{h}\left(\frac{-h}{3(3 + h)}\right) \\
&= \frac{-1}{3(3 + h)}.
\end{aligned}
$$

As $h \to 0$, this gets closer and closer to $-1/9$, so $f'(3) = -1/9$.

If we had used centered differences, we would have computed

(4.21)
$$
\begin{aligned}
\frac{f(3 + h) - f(3 - h)}{2h} &= \frac{1}{2h}\left(\frac{1}{3 + h} - \frac{1}{3 - h}\right) \\
&= \frac{1}{2h}\left(\frac{3 - h}{(3 + h)(3 - h)} - \frac{3 + h}{(3 + h)(3 - h)}\right) \\
&= \frac{1}{2h}\left(\frac{-2h}{9 - h^2}\right) \\
&= \frac{-1}{9 - h^2}.
\end{aligned}
$$

This also approaches $-1/9$ as $h \to 0$. If we're doing things algebraically and taking a limit, it doesn't matter whether we use forward/backward differences or centered differences. Most of the time, the algebra is a little easier using Definition 5 as written, so we'll stick with that.

Table 4.2. Estimating $f'(3)$ with one-sided and centered differences

h	$\frac{-1}{3(3+h)}$	$\frac{-1}{9-h^2}$
-1	-0.1666667	-0.125
-0.1	-0.114943	-0.111234
-0.01	-0.111483	-0.111112
0.01	-0.110742	-0.111112
0.1	-0.107527	-0.111234
1	-0.083333	-0.125

However, if we're working from data and have to stop at a not-so-small value of h, it can make a big difference which method we use, as you can see in Table 4.2.

Both $\frac{-1}{3(3+h)}$ and $\frac{-1}{9-h^2}$ approach $-1/9 \approx -0.111111$ as $h \to 0$. However, $\frac{-1}{9-h^2}$ gets there a *lot* faster! Estimating $f'(1)$ by using $\frac{-1}{9-h^2}$ with $h = \pm 0.01$ is about 300 times more accurate than using $\frac{-1}{3(3+h)}$ with $h = \pm 0.01$. Even using $\frac{-1}{9-h^2}$ with $h = \pm 0.1$ is more accurate than using $\frac{-1}{3(3+h)}$ with $h = \pm 0.01$. *When crunching numerical data, use centered differences whenever possible.*

4.5. A Global View

So far we have been looking at $f'(a)$, the derivative of f at a single point $x = a$. However, f' is also a function in its own right. After all, it's a rule for converting an input a into an output $f'(a)$. In this section we'll look at the derivative function $f'(x)$ and its graph, and see how that relates to the graph of the original function $f(x)$.

Consider the function $f(x) = x^2$, shown in Figure 4.7. At $x = -2$, the graph of $f(x)$ is heading downhill, and the slope of the tangent line is negative, so $f'(-2) < 0$.

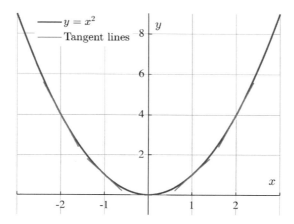

Figure 4.7. The graph of $f(x)$ has a slope that is negative when $x < 0$ and is positive when $x > 0$.

When $x = -1$, the graph is still heading downhill, but not as steeply. That is, $f'(-2) < f'(-1) < 0$. When $x = 0$ the graph is horizontal, so $f'(0) = 0$. When $x = 1$ the graph is heading uphill, so $f'(1) > 0$. When $x = 2$, the graph is heading uphill more steeply, so $f'(2) > f'(1)$.

Putting this all together, we have

(4.22) $$f'(-2) < f'(-1) < f'(0) = 0 < f'(1) < f'(2).$$

Without doing a single computation, we know that the graph of $f'(x)$ looks more-or-less like the line shown in Figure 4.8.

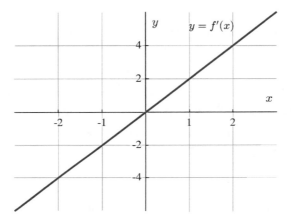

Figure 4.8. The *value* of $f'(x)$ gives the *slope* of the graph of $f(x)$.

Notice that:

- The sign of $f'(x)$ tells you which way the graph of $f(x)$ is heading. When $f'(x) > 0$, the graph of $f(x)$ is heading uphill. When $f'(x) < 0$, the graph of $f(x)$ is heading downhill.

- The size of $f'(x)$ tells you how steeply $f(x)$ is heading uphill or downhill. Steep uphills have $f'(x) \gg 0$, while steep downhills have $f'(x) \ll 0$.

- When $f'(x)$ changes sign from negative to positive, it means that $f(x)$ has stopped going downhill and is about to go uphill. In other words, we're at the bottom of a valley, also called a **local minimum**. At the top of a hill (a **local maximum**, not shown in this example), $f'(x)$ goes from positive to negative.

- If $f(x)$ is differentiable at a local maximum (or minimum), we must have $f'(x) = 0$ there. This is the 3rd Pillar of Calculus:

> *What goes up has to stop before it comes down.*

- The value of $f'(x)$ says absolutely nothing about the value of $f(x)$. It only says how fast the value of $f(x)$ is changing.

Now that we've studied our example qualitatively, let's do some calculations. We can compute $f'(-2)$ from the definition, using algebra:

(4.23)
$$f'(-2) = \lim_{h \to 0} \frac{(-2 + h)^2 - (-2)^2}{h}$$
$$= \lim_{h \to 0} \frac{4 - 4h + h^2 - 4}{h} = \lim_{h \to 0}(-4 + h) = -4.$$

Similar calculations give the values of $f'(-1)$, $f'(0)$, $f'(1)$, and $f'(2)$:

$$f'(-1) = \lim_{h \to 0} \frac{(-1 + h)^2 - (-1)^2}{h}$$
$$= \lim_{h \to 0} \frac{1 - 2h + h^2 - 1}{h} \qquad = \lim_{h \to 0}(-2 + h) = -2,$$

$$f'(0) = \lim_{h \to 0} \frac{(0 + h)^2 - (0)^2}{h}$$
$$= \lim_{h \to 0} \frac{h^2}{h} \qquad = \lim_{h \to 0} h = 0,$$

(4.24)

$$f'(1) = \lim_{h \to 0} \frac{(1 + h)^2 - (1)^2}{h}$$
$$= \lim_{h \to 0} \frac{1 + 2h + h^2 - 1}{h} \qquad = \lim_{h \to 0}(2 + h) = 2,$$

$$f'(2) = \lim_{h \to 0} \frac{(2 + h)^2 - (2)^2}{h}$$
$$= \lim_{h \to 0} \frac{4 + 4h + h^2 - 4}{h} \qquad = \lim_{h \to 0}(4 + h) = 4.$$

Notice that we did essentially the same calculation each time. Instead of repeating it for each value of x that we care about, we can use algebra to do the calculation for every value of x, once and for all, as in Figure 4.9:

(4.25)
$$f'(x) = \lim_{h \to 0} \frac{(x + h)^2 - (x)^2}{h}$$
$$= \lim_{h \to 0} \frac{x^2 + 2xh + h^2 - x^2}{h}$$
$$= \lim_{h \to 0}(2x + h)$$
$$= 2x.$$

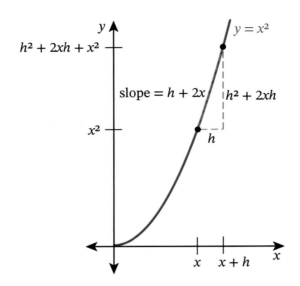

Figure 4.9. The calculation of $f'(x)$ is just like the calculation of $f'(1)$.

We have now explored the derivative function $f'(x)$ (of $f(x) = x^2$) from all four perspectives:

(1) As an algorithm. For each value of a, we can compute the slope $f'(a)$ of the graph $y = x^2$ using the methods of Section 4.4.

(2) As a table:

x	$f'(x)$
-2	-4
-1	-2
0	0
1	2
2	4

(3) As a graph (Figure 4.8).

(4) As a formula: $f'(x) = 2x$.

There are several different notations for the derivative function. The most common are

(4.26) $f'(x)$ and $\dfrac{df(x)}{dx}$, or $\dfrac{df}{dx}$ for short.

If there is a variable y, and $y = f(x)$, we sometimes write dy/dx instead of df/dx. The letter d stands for "difference", and the symbol Δ that we previously used for the change in a variable is just the Greek equivalent of d. When computing difference quotients we write $\Delta y/\Delta x$, but when we talk about the limit of that quotient we write dy/dx.

We also sometimes use a dot to mean a derivative with respect to time. That is, if $y = f(t)$, then

(4.27) $$\dot{y} = \frac{df}{dt} = f'(t).$$

The dot notation is especially useful in physics, where many quantities depend both on position (x) and time (t), and we use a dot to denote a derivative with respect to time. Finally, if $f(x)$ is given by a formula, say $f(x) = x^2 \sin(x)$, we might write $\left(x^2 \sin(x)\right)'$ instead of $f'(x)$ and $d(x^2 \sin(x))/dx$ instead of df/dx.

So why do we have all these different notations? One reason is historical (and stupid). The two primary inventors of calculus were Isaac Newton and Gottfried Leibniz. They didn't get along, and there was a big controversy over who should get credit for what. Newton used f' and $\dot{f}$, so generations of British scientists followed his example. Leibniz used dy/dx and df/dx and df/dt, so generations of continental European scientists followed *his* example. There was a big rivalry between British and continental scientists, neither side would budge, and both notations stuck.

A better reason is that there are real advantages to each notation. We'll see that when we get to the chain rule in Chapter 5. Leibniz's notation gets to the heart of what the chain rule means and is much better for doing related rates problems, while Newton's notation is much better for computing the derivative of $\sin(x^2 - 7)$. It's worth your while to learn both of these notations.

We close this section with a worked problem:

Problem: Use a linear approximation to estimate 9.975^2 and 10.05^2.

Solution: We are studying the function $f(x) = x^2$. Because $f'(x) = 2x$, $f'(10) = 20$. The microscope equation tells us that, for x close to 10,

(4.28)
$$x^2 = f(x) \approx f(10) + f'(10)(x - 10)$$
$$= 100 + 20(x - 10).$$

Thus

(4.29)
$$9.975^2 \approx 100 + 20(-0.025) \quad = 99.5, \text{ and}$$
$$10.05^2 \approx 100 + 20(0.05) \qquad = 101.$$

The exact values turn out to be $9.975^2 = 99.500625$ and $10.05^2 = 101.0025$. Our linear approximation is both simple and accurate.

4.6. Chapter Summary

The Main Ideas.

- We can estimate rates of change from data using forward, backward, or centered difference quotients. If $h > 0$, those are

(4.30) $$\frac{f(x + h) - f(x)}{h}, \quad \frac{f(x) - f(x - h)}{h}, \quad \text{and} \quad \frac{f(x + h) - f(x - h)}{2h},$$

respectively. Sometimes there is only enough data to do it one of the three ways, but when we have a choice we should generally use centered differences, which are more accurate.

- Rates of change aren't just for things that vary in time. If a factory makes x items, then the rate of change with respect to x of the cost, revenue and profit are called the **marginal cost**, **marginal revenue** and **marginal profit**. Decisions on whether to increase or decrease production should be based on whether the marginal profit is positive or negative, or equivalently on whether the marginal revenue is bigger or less than the marginal cost.

- All smooth functions look linear when you zoom in enough, so it makes sense to look for linear approximations.

- If a function $f(x)$ changes at rate $f'(a)$ when $x = a$, then the **linear approximation**

$$(4.31) \qquad f(x) \approx f(a) + f'(a)(x - a),$$

also called the **microscope equation**, is very accurate whenever x is sufficiently close to a.

- The **derivative** $f'(a)$ of the function $f(x)$ at $x = a$ can be understood as
  - the rate of change of the function when $x = a$,
  - the slope of the line tangent to the graph at $x = a$,
  - the multiplier in the best linear approximation,
  - the limit $\lim_{x \to a} \frac{f(x)-f(a)}{x-a}$, and
  - the limit $\lim_{h \to 0} \frac{f(a+h)-f(a)}{h}$.

- We don't need tables of values to compute derivatives. Most of the time, it's easier to use algebra.

- The derivative **function** is the function $f'(x)$ whose value at each point is the derivative of $f(x)$ at that point.

- The **value** of $f'(x)$ gives the **rate of change** of $f(x)$, but does not say anything about the value of $f(x)$. Equivalently, the **height** of the graph of $f'(x)$ gives the **slope** of the graph of $f(x)$, but does not say anything about the height of the graph of $f(x)$.

- The sign of $f'(x)$ tells whether $f(x)$ is increasing.
  - When $f'(x) > 0$, $f(x)$ is **increasing**.
  - When $f'(x) < 0$, $f(x)$ is **decreasing**.
  - When $f(x)$ hits a maximum or a minimum, then $f'(x) = 0$ (or does not exist).

- It is often possible to find a formula for $f'(x)$ by using algebra to compute the limit

$$(4.32) \qquad f'(x) = \lim_{h \to 0} \frac{f(x + h) - f(x)}{h}.$$

Expectations. You should be able to:

- Use tables of data to compute rates of change.
- Use rates of change to make linear approximations (equation (4.31), the microscope equation).
- Explain what marginal quantities mean and how they are used.
- Explain how the different definitions of $f'(a)$ are related.
- Compute the derivatives of simple functions such as $f(x) = x^2$ and $f(x) = 1/x$ using limits and algebra.
- Relate the derivative of a function to the shape of its graph.
- Relate the shape of the graph of a function to the graph of its derivative.

4.7. Exercises

MATLAB tip: A number of the following exercises involve graphing functions with MATLAB and zooming in and out of the picture. The graphs are most easily generated using the "plot" or "fplot" commands. The drop-down menus for the figure itself contain commands for zooming in and out. You can generate multiple graphs on the same axes directly using the "plot" command (type "help plot" at the MATLAB prompt for details), or by adding new graphs one at a time with the help of the "hold on" command.

Population Growth. The following table gives the populations of Texas and Puerto Rico (in millions, to 2 decimal places) between 2001 and 2020. Use this data for Exercises 4.1–4.3.

Year	TX	PR	Year	TX	PR
2001	21.33	3.67	2011	25.67	3.56
2002	21.72	3.66	2012	26.06	3.54
2003	22.10	3.65	2013	26.45	3.50
2004	22.49	3.64	2014	26.96	3.45
2005	22.93	3.63	2015	27.47	3.38
2006	23.51	3.62	2016	27.86	3.28
2007	23.90	3.62	2017	28.30	3.16
2008	24.33	3.61	2018	28.63	3.04
2009	24.78	3.60	2019	29.00	2.93
2010	25.25	3.58	2020	29.90	2.86

4.1. We are going to compare forward, backward and centered differences for understanding how fast Texas was growing in 2006. Let $TX(t)$ be the population of Texas in year t.
 (a) Using a forward difference with $\Delta t = 5$ years, estimate the rate of change $TX'(2006)$. Repeat with $\Delta t = 2$. Repeat again with $\Delta t = 1$.
 (b) Now use backward differences with $\Delta t = 5$ years, 2 years and 1 year to estimate $TX'(2006)$.
 (c) Finally, use centered differences with $\Delta t = 5$ years, 2 years and 1 year to estimate $TX'(2006)$.
 (d) Which of these nine estimates do you think is the most accurate? Why?

4.2. Now that we know the population of Texas in 2006 and how fast it was changing, let's see what that says about the population of Texas in other years.
 (a) Using your best estimated value of $TX'(2006)$ from Exercise 4.1, write the microscope equation for $TX(t)$.
 (b) Use the linear approximation from part (a) to estimate the population of Texas in 2010, 2015, and 2020.
 (c) Using MATLAB, generate a graph that shows both the actual population and the linear approximation from $t = 2001$ through $t = 2020$ on the same set of axes. Would you say that the linear approximation does a good job of describing population growth in Texas?

4.3. Now let's study the population $PR(t)$ of Puerto Rico in year t.
 (a) Using a forward difference with $\Delta t = 2$ years, estimate $PR'(2006)$. Repeat with $\Delta t = 1$.
 (b) Now use backward differences with $\Delta t = 2$ years and 1 year to estimate $PR'(2006)$.
 (c) Finally, use centered differences with $\Delta t = 2$ years and 1 year to estimate $PR'(2006)$.
 (d) Which of these six estimates do you think is the most accurate? Why?

4.4. The problem with using a small Δt to estimate $PR'(2006)$ is that, after rounding to the nearest 10,000, we don't see any change at all between 2006 and 2007! To get accurate results we have to make sure that the change in our function is bigger than our round-off errors, or our actual errors in measurement.
 (a) Here is Puerto Rico's population (in millions) to three decimal places instead of two:

Year	PR
2005	3.631
2006	3.623
2007	3.616
2008	3.608

  Using this more accurate data set, estimate $PR'(2006)$.
 (b) Explain why we needed more accurate data to estimate $PR'(2006)$, but we didn't need more accurate data to estimate $TX'(2006)$.

Food for Thought: In general, using the smallest possible Δt is a good idea when our data is exact, as with a mathematical function like x^2 or e^x. However, when there are errors in the data, it's sometimes necessary to use a bigger baseline to average out random fluctuations and rounding errors.

Other Quantities That Change

4.5. The number of aphids on a sorghum plant is being monitored. The graph below shows the estimates for the number of aphids per leaf t days after the study started.

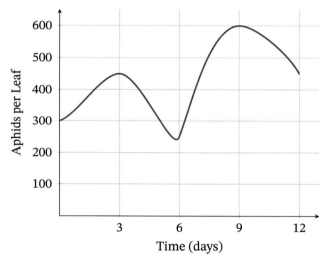

(a) Find the average rate of change in the number of aphids per leaf on the following time intervals.
  (i) $[0, 3]$
  (ii) $[3, 6]$
  (iii) $[6, 9]$
  (iv) $[9, 12]$
(b) On which of the time interval(s) in part (a) is the number of aphids per leaf increasing?
(c) On which of the time interval(s) in part (a) is the number of aphids per leaf decreasing?
(d) Discuss the relationship between your results in parts (a), (b), and (c).
(e) What is another interpretation of the phrase "average rate of change"?

4.6. The year 2011 was hot and dry in the southern United States. This caused water levels in Lake Travis, Austin's main water reservoir, to drop far below capacity. The following table shows the minimum water level, in feet above sea level, for each month in 2011 and 2012. Lake levels stayed low until the 2015 hurricane season brought an end to the drought. (For reference, Lake Travis is considered full when the water level is at 681 feet. The dam that controls the lake can handle water levels as high as 714 feet.)

Month	2011 level	2012 level
January	666.88	626.09
February	665.86	626.75
March	663.65	631.06
April	657.39	638.55
May	650.28	638.01
June	646.16	640.18
July	639.94	639.13
August	633.92	635.49
September	629.01	634.01
October	627.09	633.64
November	626.09	632.15
December	626.26	630.95

(a) Using MATLAB, graph the height of the lake as a function of time from the beginning of 2011 through the end of 2012.

(b) This graph is not at all close to a straight line, so it doesn't make sense to approximate the *entire* curve with a single linear function. However, there are shorter periods of time when the water level shrank or grew at a more or less constant rate, or stayed steady. Identify at least two such time periods.

(c) For one of the time periods you identified in part (b), draw a straight line (either electronically or with a pencil) that comes close to tracking the curve. What is the slope of that line? (Be sure to express that slope in appropriate units.) Does that correspond to the actual difference quotient $\Delta f / \Delta t$ for an interval?

Functions That Repeat

4.7.   For this exercise, look at the graph of the function $f(x) = \sin(x)$ between $x = -1$ and $x = 12$. While any graph will do, a graph generated with MATLAB is best, because it will allow you to zoom in on different portions of the graph.

(a) What are the x-intercepts of $\sin(x)$? That is, where is the function zero?

(b) On what interval(s) is the function $f(x)$ increasing? Decreasing?

(c) At what x value(s) is $f(x)$ increasing most rapidly? Approximately how fast is $f(x)$ increasing at this value of x?

(d) What is the largest value of $f(x)$ on the interval $-1 \le x \le 12$?

(e) Zoom in on the portion of the graph near $x = 0$ until it looks like a straight line. What is the slope of this line?

(f) Use the slope you found in part (e) and the point $(0, f(0))$ to find a linear approximation to $\sin(x)$ for x close to 0.

4.8.   For this exercise, look at the graphs of $h(x) = \cos(x)$ and $g(x) = \cos(2x)$ between $x = -2\pi$ and $x = 2\pi$.
  (a) How far apart are the x-intercepts of $h(x)$? How far apart are the x-intercepts of $g(x)$?
  (b) The graph of $g(x)$ has a repeating pattern. How often does it repeat? The graph of $h(x)$ also has a repeating pattern. How often does it repeat?
  (c) Compare the graphs of $h(x)$ and $g(x)$. Can you get one of them by compressing or stretching the other vertically? If so, how? Can you get one of them by compressing or stretching the other horizontally? If so, how?
  (d) How fast is $h(x)$ increasing at its steepest point? How fast is $g(x)$ increasing at its steepest point?

4.9.   In Exercises 4.7 and 4.8, the functions involved had a repeating pattern. Give three examples of applications where a function that repeats would be useful for modeling a behavior.

Approximate Derivatives and Microscopes

4.10.   Approximate the derivative of each function, at the given value $x = a$, in two different ways.
  • Use a computer microscope (e.g., MATLAB) to view the graph of f near $x = a$. Zoom in until the graph looks straight and find its slope.
  • Use a calculator to find the value of the centered difference quotient

  $$\frac{f(a + h) - f(a - h)}{2h}$$

  for $h = 0.1, 0.01, 0.001$, and 0.0001. Based on these quotients, give your best estimate for $f'(a)$.
  (a) $f(x) = 1/x^2$ at $x = 1$
  (b) $f(x) = x^3$ at $x = 2$

4.11.   Write the microscope equation for each function at the indicated point. You will want to use your results from Exercise 4.10.
  (a) $f(x) = 1/x^2$ at $x = 1$
  (b) $f(x) = x^3$ at $x = 2$

4.12.   This exercise uses the microscope equation for $f(x) = x^3$ at $x = 2$ that you constructed in Exercise 4.11b.
  (a) What is Δx when $x = 2.05$? What estimate does the microscope equation give for $f(2.05)$? How does that compare to the exact value of $f(2.05)$, namely 8.615125?
  (b) What estimate does the microscope equation give for $f(2.02)$? How far is that from the exact value of 8.242408?
  (c) What estimate does the microscope equation give for $f(1.99)$? How far is this from the exact value of 7.880599?
  (d) The error in parts (b)–(d) is roughly proportional to what power of Δx?

4.13.   Suppose $h(x)$ is a function for which $h(3) = 10$ and $h'(3) = 1.5$.
  (a) Write the microscope equation for h at $x = 3$.

(b) Draw the graph of what you would see in a microscope. Do you need a formula for $h(x)$ in order to do this?

(c) If $x = 3.2$, what is Δx in the microscope equation? What estimate does the microscope equation give for Δh? What estimate does the microscope equation then give for $h(3.2)$?

(d) What estimates does the microscope equation give for $h(3.25)$, $h(2.9)$, and $h(5)$? Do you think these estimates are equally reliable? Why or why not?

4.14. Suppose $f(t)$ is a function for which $f(-2) = 5$ and $f'(-2) = 2$.
(a) Write the microscope equation for $f(t)$ near $t = -2$.
(b) Draw the graph of what you see in the microscope.
(c) Estimate $f(-2.5)$ and $f(-2.75)$.
(d) For what value of t near -2 would you estimate that $f(t) = 6$? For what value of t would you estimate $f(t) = 5.5$?

4.15. Suppose that $f(a) = b$, $f'(a) = -2$ and k is a small number. Which of the following is the estimate for $f(a+k)$ that comes from the microscope equation? Explain.

$$a + 2k \qquad a - 2k \qquad b + 2k \qquad b - 2k$$

$$2a + 2b \qquad 2a - 2b \qquad 2b - k \qquad f'(a - 2k)$$

4.16. Suppose that f is differentiable at a, and that h is a very small number. Explain why each expression either is or isn't a reasonable approximation of $f'(a)$.

(a) $\dfrac{f(a - h) - f(a)}{h}$ (b) $\dfrac{f(a + 2h) - f(a - 2h)}{4h}$

(c) $\dfrac{f(a + h) - f(h)}{h}$ (d) $\dfrac{f(a + h) - f(a - h)}{h}$

Functions and Their Derivatives

4.17. If a and b are any two numbers, then

$$(a + b)^3 = a^3 + 3a^2b + 3ab^2 + b^3.$$

Using this identity, and the formula $f'(1) = \lim\limits_{h \to 0} \dfrac{f(1 + h) - f(1)}{h}$, compute the derivative of $f(x) = x^3$ at $x = 1$.

4.18. As in Exercise 4.17, let $f(x) = x^3$. Using the formula

$$f'(a) = \lim\limits_{h \to 0} \dfrac{f(a + h) - f(a)}{h}$$

for the derivative of $f(x)$ at an arbitrary point $x = a$, compute the following derivatives.
(a) $f'(-1)$ (b) $f'(0)$ (c) $f'(2)$
(d) Repeating the calculation, only with x instead of -1, 0, or 2, find the formula for $f'(x)$.

4.19. If a and b are any two numbers, and if $a \neq b$, then

$$\frac{b^3 - a^3}{b - a} = b^2 + ab + a^2.$$

Let $f(x) = x^3$. Using the identity above and the formula

$$f'(a) = \lim_{x \to a} \frac{f(x) - f(a)}{x - a},$$

compute the following derivatives.
(a) $f'(0)$ (b) $f'(1)$ (c) $f'(2)$
(d) What is the problem with using this approach to find the formula for $f'(x)$?

4.20. Consider the graph of a mystery function $f(x)$ between $x = 0$ and $x = 3$. We know that $f(x)$ is differentiable and that $f(0) = 1$. We are also told that

$$\begin{aligned}
0.25 \leq f'(x) \leq 1 &\qquad \text{when } 0 \leq x \leq 1, \\
0 \leq f'(x) \leq 2 &\qquad \text{when } 1 \leq x \leq 2, \\
-0.5 \leq f'(x) \leq 0 &\qquad \text{when } 2 \leq x \leq 3.
\end{aligned}$$

(a) First consider $f(x)$ for x between 0 and 1. Write an expression (in terms of x) for the largest possible value of $f(x)$. Write an expression for the smallest possible value of $f(x)$. What are the possible values of $f(1)$?
(b) Repeat part (a) for x between 1 and 2, using what you just learned about $f(1)$. What are the possible values of $f(2)$?
(c) Repeat part (a) for x between 2 and 3.
(d) Draw and shade the smallest region in the x-y plane in which the graph of $f(x)$ must lie.

4.21. Suppose that for a function $g(x)$, $g(0) = 0$, and that $0 \leq g'(x) \leq 2$ for all values of x between 0 and 4.
(a) Explain why the graph of g must lie entirely in the shaded region below:

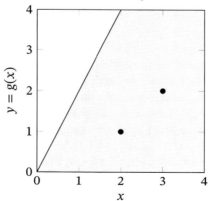

(b) Suppose that $g(2) = 1$ and $g(3) = 2$, as indicated in the figure. Draw the smallest region in which the graph of g must lie.
(*Hint*: How do the conditions on $g'(x)$ constrain the values of $g(x)$ to the right *and to the left* of the indicated points?)

4.22.   For each function graphed below, sketch the graph of its derivative.

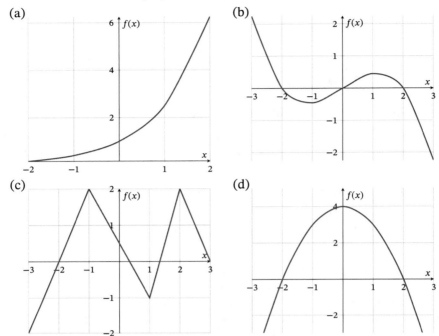

4.23.   **Metabolic Cost of Predation.** Consider an omnivore that feeds on a wide
variety of prey. Let x denote the density of a specific prey item and let c be
the **metabolic cost** to the predator of searching for and capturing this specific
prey. That is, the predator has to use up an amount of energy c (measured in
calories) to catch each prey animal. We would expect the density (prey per
area) of a specific prey item to be a function of that metabolic cost. That is,
we would have $x = F(c)$ for some function F. For example, if the metabolic
cost of capturing the prey is low, then we would expect a high rate of predation,
causing the density of prey to drop. However, if the metabolic cost of predation
is high, then the predator might choose to leave the prey alone and look for a
different kind of food, allowing the density of prey to increase.

(a) What strategies might prey employ in order to increase the metabolic cost
to predators?

(b) Carefully draw by hand a graph that describes a plausible relationship between x and c. (There is more than one right answer.) Be sure to label the
axes appropriately and to explain why you chose the particular shape of
your graph.

(c) Carefully draw by hand the graph of the derivative of the function that you
graphed in part (b).

Chapter 5

Computing and Using Derivatives (What Goes Up Has to Stop Before It Comes Down)

There are three questions you should ask about every mathematical concept that you encounter:

- What is it?

- How do you compute it?

- What is it good for?

For derivatives, Chapter 4 was all about the first question. We learned that derivatives are limits of difference quotients and are closely connected to linear approximations. In this chapter, we'll tackle the second and third questions.

In the first half of the chapter, we'll develop some rules and formulas for computing derivatives. In Chapter 4 we saw that the derivative of $f(x) = x^2$ is $f'(x) = 2x$, but how can we find the derivative of $g(x) = \dfrac{5x^3 - 3x + \sin(x^2)}{\cos(2^x + 17)}$? Working that out directly from the definition $g'(x) = \lim_{h \to 0} \dfrac{g(x + h) - g(x)}{h}$ is ridiculously hard, but with the methods we are going to develop, computing derivatives of even the ugliest functions will become easy.

In the second half of the chapter, we'll see what derivatives are good for. The biggest application is finding the values of a variable x that maximize or minimize a function, for instance figuring out what heart rate delivers the most oxygen to our muscles. Life is full of trade-offs, and evolution works to **optimize** the way that our bodies (and the bodies of our enemies, like bacteria and viruses) work.

Another important application is to use properties of $f'(x)$ to sketch the graph of a function $f(x)$, thereby visualizing how the function behaves. "A picture is worth a thousand words!" We'll take the derivative of $f'(x)$ (also called the **second derivative** of $f(x)$, denoted $f''(x)$ or d^2f/dx^2) and use it to tell us even more about the graph of $f(x)$. Finally, we will see how derivatives can help us solve complicated equations.

5.1. Building Blocks

Most functions that you'll see are built by combining several very basic functions:

- *Powers of x.* That is, $f(x) = x^n$. This includes not only functions like x^2 and x^3, but also functions like $\frac{1}{x} = x^{-1}$, $\sqrt{x} = x^{1/2}$ and $1 = x^0$.
- *The trigonometric functions* $\sin(x)$ *and* $\cos(x)$. Later we'll add $\tan(x)$, $\cot(x)$, $\sec(x)$, and $\csc(x)$ to the list, but those can be built from $\sin(x)$ and $\cos(x)$.
- *The exponential function* $f(x) = a^x$, *where a is a fixed positive number.* Eventually, we'll add logarithms, which are inverse exponentials, to the list.

So let's figure out the derivatives of these functions, starting with simple powers of x.

The Power Law (Newton's Hammer). We already saw that the derivative of x^2 is $2x$. The derivative of x^3 is

$$(5.1) \qquad (x^3)' \;=\; \lim_{h\to 0} \frac{(x+h)^3 - x^3}{h}$$

$$= \lim_{h\to 0} \frac{(x^3 + 3x^2h + 3xh^2 + h^3) - x^3}{h}$$

$$= \lim_{h\to 0} \frac{3x^2h + 3xh^2 + h^3}{h}$$

$$= \lim_{h\to 0}(3x^2 + 3xh + h^2)$$

$$= 3x^2.$$

The details of the expansion of $(x+h)^3$ aren't very important. What's important is that

$$(5.2) \qquad (x+h)^3 = x^3 + 3x^2h + \text{terms with 2 or more powers of } h,$$

meaning that the extra terms involve h^2 or h^3, but not h to a power that is less than 2. Subtracting x^3 and dividing by h, we get that

$$(5.3) \qquad \frac{(x+h)^3 - x^3}{h} = 3x^2 + \text{terms with 1 or more powers of } h.$$

However, all terms with 1 or more power of h go to zero as $h \to 0$, so

$$(5.4) \quad \lim_{h\to 0}\frac{(x+h)^3 - x^3}{h} \;=\; \lim_{h\to 0}\frac{3x^2h + \text{terms with 2 or more powers of } h}{h}$$

$$= \lim_{h\to 0}(3x^2 + \text{terms with 1 or more power of } h)$$

$$= 3x^2.$$

We will use the notation $O(h^2)$ to mean "terms with 2 or more powers of h", and $O(h)$ to mean "terms with 1 or more power of h". This simplifies our algebra, as when we add two $O(h^2)$ terms, or when we multiply an $O(h^2)$ term by x, we still get $O(h^2)$. The first few powers of $x + h$ are then as follows.

$$(5.5) \quad (x+h)^2 = \qquad x^2 + 2xh + h^2 \qquad = x^2 + 2xh + O(h^2);$$

$$\begin{aligned} (x+h)^3 &= \qquad (x+h)^2(x+h) \\ &= \qquad (x^2 + 2xh + O(h^2))(x+h) \\ &= \quad x^3 + x^2h + 2x^2h + 2xh^2 + O(h^2) \quad = x^3 + 3x^2h + O(h^2); \end{aligned}$$

$$\begin{aligned} (x+h)^4 &= \qquad (x+h)^3(x+h) \\ &= \qquad (x^3 + 3x^2h + O(h^2))(x+h) \qquad = x^4 + 4x^3h + O(h^2). \end{aligned}$$

Do you see the pattern? No matter how many powers of $(x + h)$ we take, we always have

$$(5.6) \qquad (x+h)^n = x^n + nx^{n-1}h + O(h^2).$$

To make absolutely sure that equation (5.6) works for all powers n, we apply a recursive argument called **mathematical induction**. Once we know that equation (5.6) works for a particular value of n (say, $n = k$), we can check it for the next value of n:

$$(5.7) \qquad \begin{aligned} (x+h)^{k+1} &= \quad (x+h)^k(x+h) \\ &= \quad \left(x^k + kx^{k-1}h + O(h^2)\right)(x+h) \\ &= \quad x^{k+1} + kx^kh + x^kh + O(h^2) \\ &= \quad x^{k+1} + (k+1)x^kh + O(h^2), \end{aligned}$$

which is equation (5.6) with $n = k + 1$. As we already know that equation (5.6) works for $n = 1$, the recursion tells us that it must also work for $n = 2$, so it must also work for $n = 3$, so it must also work for $n = 4$, and so on forever.

Equation (5.6) tells us that

$$(5.8) \qquad \begin{aligned} (x^n)' = \frac{dx^n}{dt} &= \quad \lim_{h \to 0} \frac{(x+h)^n - x^n}{h} \\ &= \quad \lim_{h \to 0} \frac{x^n + nx^{n-1}h + O(h^2) - x^n}{h} \\ &= \quad \lim_{h \to 0} \left(nx^{n-1} + O(h)\right) \\ &= \quad nx^{n-1}. \end{aligned}$$

That's it! We have just figured out the derivative of infinitely many powers of x.

The **power law** formula

$$(5.9) \qquad\qquad\qquad (x^n)' = nx^{n-1}$$

is so useful that it even has a nickname: **Newton's hammer**. Strictly speaking, we have only proven Newton's hammer when n is a whole number, but the same formula also works when n is a fraction or is negative. Once we develop the product, quotient and chain rules, we'll see why.

There is an important special case of this formula. When $n = 0$, then $x^n = x^0 = 1$. By definition, constants don't change, so the derivative of a constant is zero. This agrees with our formula, because when $n = 0$, $nx^{n-1} = 0x^{-1} = 0$.

Sines and Cosines. Next we turn to the trigonometric functions $\sin(x)$ and $\cos(x)$. We'll go into these in much more detail in the next chapter, but for now just look at their graphs. The graph of $\sin(x)$, shown in Figure 5.1, shows that $\sin(x)$ is increasing whenever $\cos(x)$ is positive, and is decreasing whenever $\cos(x)$ is negative. This suggests that the derivative of $\sin(x)$ has something to do with $\cos(x)$. Similarly, $\cos(x)$ is increasing whenever $\sin(x)$ is *negative* and is decreasing whenever $\sin(x)$ is *positive*, so the derivative of $\cos(x)$ has something to do with $-\sin(x)$.

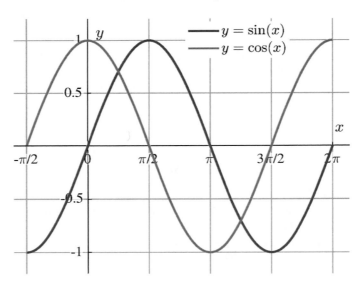

Figure 5.1. The graphs of $\sin(x)$ and $\cos(x)$

In fact, as long as we work in radians,[1] the derivatives of $\sin(x)$ and $\cos(x)$ are simply

$$(5.10) \qquad \frac{d\sin(x)}{dx} = \cos(x) \qquad \text{and} \qquad \frac{d\cos(x)}{dx} = -\sin(x).$$

Equation (5.10) *looks* right, but how do we know that it's really true? To actually compute the derivatives of $\sin(x)$ and $\cos(x)$, we need the addition of angle formulas from trigonometry. (See the supplemental materials at the end of Chapter 6 for a

---

[1]If we work in degrees, then the derivative of $\sin(x)$ is $\frac{\pi}{180}\cos(x)$, which is ugly. When doing calculus, we always use radians.

derivation of these formulas, as well as for a general review of trigonometry.) If A and B are any angles, then

(5.11)
$$\begin{aligned}
\sin(A+B) &= \sin(A)\cos(B) + \cos(A)\sin(B), \\
\cos(A+B) &= \cos(A)\cos(B) - \sin(A)\sin(B).
\end{aligned}$$

This means that

(5.12)
$$\begin{aligned}
\frac{\sin(x+h) - \sin(x)}{h} &= \frac{\sin(x)\cos(h) + \cos(x)\sin(h) - \sin(x)}{h} \\
&= \sin(x)\left(\frac{\cos(h)-1}{h}\right) + \cos(x)\frac{\sin(h)}{h}, \\
\frac{\cos(x+h) - \cos(x)}{h} &= \frac{\cos(x)\cos(h) - \sin(x)\sin(h) - \cos(x)}{h} \\
&= \cos(x)\left(\frac{\cos(h)-1}{h}\right) - \sin(x)\frac{\sin(h)}{h}.
\end{aligned}$$

Table 5.1. Behavior of $\frac{\sin(h)}{h}$ and $\frac{\cos(h)-1}{h}$ as $h \to 0$

h	$\sin(h)$	$\frac{\sin(h)}{h}$	$\cos(h) - 1$	$\frac{\cos(h)-1}{h}$
-1.00	-0.8414700	0.84147	-0.459700	0.45970
-0.10	-0.0998330	0.99833	-0.004996	0.04996
-0.01	-0.0099998	0.99998	-0.000050	0.00500
0.01	0.0099998	0.99998	-0.000050	-0.00500
0.10	0.0998330	0.99833	-0.004996	-0.04996
1.00	0.8414700	0.84147	-0.459700	-0.45970

To understand these limits as $h \to 0$, we have to understand how the quotients $\frac{\cos(h)-1}{h}$ and $\frac{\sin(h)}{h}$ behave as $h \to 0$. As you can see from Table 5.1, the closer h gets to 0, the closer $\frac{\sin(h)}{h}$ gets to 1 and the closer $\frac{\cos(h)-1}{h}$ gets to 0.[2] Thus

(5.13)
$$\begin{aligned}
\frac{d\sin(x)}{dx} &= \lim_{h\to 0} \frac{\sin(x+h) - \sin(x)}{h} \\
&= \lim_{h\to 0}\left(\sin(x)\left(\frac{\cos(h)-1}{h}\right) + \cos(x)\frac{\sin(h)}{h}\right) \\
&= 0\sin(x) + 1\cos(x) \\
&= \cos(x).
\end{aligned}$$

---

[2]If you don't find the table convincing, look at the supplemental materials in Section 5.8, where we provide rigorous computations of these limits.

The derivative of $\cos(x)$ is similar:

$$(5.14) \qquad \frac{d\cos(x)}{dx} = \lim_{h\to 0} \frac{\cos(x+h) - \cos(x)}{h}$$

$$= \lim_{h\to 0} \left(\cos(x) \left(\frac{\cos(h) - 1}{h} \right) - \sin(x) \frac{\sin(h)}{h} \right)$$

$$= 0\cos(x) - 1\sin(x)$$

$$= -\sin(x).$$

Exponential Functions. Finally, we consider exponential functions. We will go over exponentials and logarithms systematically in Section 6.2, but for now we just need the basic fact that

$$(5.15) \qquad\qquad\qquad a^{b+c} = a^b a^c.$$

In particular, $a^{x+h} = a^x a^h$, so

$$(5.16) \qquad \frac{da^x}{dx} = \lim_{h\to 0} \frac{a^{x+h} - a^x}{h}$$

$$= \lim_{h\to 0} \frac{a^h a^x - a^x}{h}$$

$$= \lim_{h\to 0} \frac{a^h - 1}{h} a^x$$

$$= k_a a^x,$$

where

$$(5.17) \qquad\qquad\qquad k_a = \lim_{h\to 0} \frac{a^h - 1}{h}.$$

That is, the derivative of a^x is proportional to a^x, and the proportionality constant depends on a. From the data in Table 5.2, we see that k_2 is around 0.69, k_3 is around 1.10, and k_4 is around 1.38.

Table 5.2. Estimating k_2, k_3, and k_4 numerically

h	$\frac{2^h - 1}{h}$	$\frac{3^h - 1}{h}$	$\frac{4^h - 1}{h}$
-1.000	0.50000	0.66667	0.75000
-0.100	0.66967	1.04042	1.29449
-0.010	0.69075	1.09260	1.37673
-0.001	0.69291	1.09801	1.38533
0.001	0.69339	1.09922	1.38726
0.010	0.69555	1.10467	1.39595
0.100	0.71773	1.16123	1.48698
1.000	1.00000	2.00000	3.00000

Because $k_2 < 1 < k_3$, there has to be a number, somewhere between 2 and 3, for which the constant is exactly 1. We call this number **Euler's number** or **Napier's constant** (but not *Euler's constant*, which is a different number), and denote it with the letter e. Numerically, $e \approx 2.71828$.

This makes $f(x) = e^x$ a very special function. Because $k_e = 1$, we have

(5.18)
$$\frac{de^x}{dx} = k_e e^x = e^x,$$

so e^x is its own derivative! We also have $e^0 = 1$, because $a^0 = 1$ for any value of a. This means that $y = e^x$ is the solution to the rate equation

(5.19)
$$y' = y$$

with initial condition $y(0) = 1$. We previously saw this rate equation (with an extra factor of r) in the context of compound interest. If you borrow a dollar from a loan shark at 100% annual interest (compounded continuously), you will owe him e^t dollars after t years. Be sure to pay him back quickly, as if you wait 10 years you'll wind up owing him over $22,000!

Summary. The results of this section are summarized in Table 5.3. Be sure to *memorize it!*

Table 5.3. Derivatives of basic functions

$f(x)$	$f'(x)$
x^n	nx^{n-1}
$\sin(x)$	$\cos(x)$
$\cos(x)$	$-\sin(x)$
a^x	$k_a a^x$
e^x	e^x

5.2. Adding, Subtracting, Multiplying, and Dividing Functions

So far we've figured out the derivatives of functions like x^3, $\sin(x)$, and e^x. But what about functions like $x^3 + 2\sin(x) + 17e^x$? Or $5x^3 \sin(x)$? Or $\frac{e^x}{x^3 + \sin(x)}$? It turns out that if we know the derivatives of two functions $f(x)$ and $g(x)$, then it's fairly easy to find the derivatives of $f(x) + g(x)$ and $f(x) - g(x)$ and $cf(x)$, where c is any constant. Finding the derivatives of the product $f(x)g(x)$ and the quotient $f(x)/g(x)$ is a little more complicated, requiring the **product rule** and the **quotient rule**, which we will learn in this section. With those rules, we'll be able to handle all of the examples listed at the beginning of this paragraph.

The Trivial Rules: Adding, Subtracting, and Multiplying by Constants.
Suppose that two different kinds of trout, cutthroat and rainbow, live in the same lake. At time t, the population of cutthroat trout is $T_c(t)$, while the population of rainbow trout is $T_r(t)$. Suppose that the cutthroat trout population is growing at a rate of 2000 trout/year, while the rainbow trout population is growing at 3000 trout/year. How fast is the total trout population growing?

The answer, of course, is 5000 trout/year. If there are 2000 extra cutthroat trout and 3000 extra rainbow trout, then there are 5000 extra trout. That is,

$$(5.20) \qquad \frac{d(T_c + T_r)}{dt} = \frac{dT_c}{dt} + \frac{dT_r}{dt}.$$

If there is a competition between fishermen who like cutthroat trout and those who like rainbow trout, then we might ask how fast the difference between the two populations is changing. The answer, of course, is

$$(5.21) \qquad \frac{d(T_c - T_r)}{dt} = \frac{dT_c}{dt} - \frac{dT_r}{dt}.$$

Suppose that an average cutthroat trout weighs about 5 pounds.[3] How fast is the *total weight* of the cutthroat trout population increasing? As the weight of 2000 extra trout is 10,000 pounds, the weight is growing at 10,000 pounds/year. That is, $(5T_c)' = 5(T_c')$.

In terms of functions $f(x)$ and $g(x)$ and a constant c, we have the following:

The basic rules:

$$(5.22) \qquad \begin{aligned} \frac{d}{dx}(f(x) + g(x)) &= \frac{df}{dx} + \frac{dg}{dx}, \\ \frac{d}{dx}(f(x) - g(x)) &= \frac{df}{dx} - \frac{dg}{dx}, \\ \frac{d}{dx}(cf(x)) &= c\frac{df}{dx}. \end{aligned}$$

These can be combined into a single rule. If a and b are constants, then

$$(5.23) \qquad \frac{d}{dx}(af(x) + bg(x)) = a\frac{df}{dx} + b\frac{dg}{dx},$$

or in Newton's notation,

$$(5.24) \qquad (af(x) + bg(x))' = af'(x) + bg'(x).$$

The same rule applies to combinations of three, four, or more functions. Thus

$$(5.25) \qquad \begin{aligned} (x^3 + 2\sin(x) + 17e^x)' &= (x^3)' + 2(\sin(x))' + 17(e^x)' \\ &= 3x^2 + 2\cos(x) + 17e^x. \end{aligned}$$

---

[3]This varies by species. Lake trout can be bigger, while coastal sea-run trout are smaller, and river trout are *much* smaller.

The combined rule (5.24) can be proven directly, as follows:

$$(5.26) \quad (af(x) + bg(x))' = \lim_{h \to 0} \frac{(af(x+h) + bg(x+h)) - (af(x) + bg(x))}{h}$$

$$= \lim_{h \to 0} a\left(\frac{f(x+h) - f(x)}{h}\right) + b\left(\frac{g(x+h) - g(x)}{h}\right).$$

As $h \to 0$, $\frac{f(x+h) - f(x)}{h}$ gets closer and closer to $f'(x)$, while $\frac{g(x+h) - g(x)}{h}$ gets closer and closer to $g'(x)$, so $a\frac{f(x+h) - f(x)}{h} + b\frac{g(x+h) - g(x)}{h}$ gets closer and closer to $af'(x) + bg'(x)$.

The Product Rule. If we know the derivative of $f(x)$ and the derivative of $g(x)$, then what is the derivative of $f(x)g(x)$? An obvious guess, which almost every calculus student has believed at one time or another, is

$$(5.27) \qquad (f(x)g(x))' = f'(x)g'(x). \qquad \text{(Wrong!)}$$

Let's test whether it works using functions that we already understand.

If $f(x) = x^2$ and $g(x) = x^3$, then $f(x)g(x) = x^5$. The derivative of x^2 is $2x$, the derivative of x^3 is $3x^2$, and $f'(x)g'(x) = (2x)(3x^2) = 6x^3$. However, the derivative of x^5 is $5x^4$, not $6x^3$. The naive rule (5.27) *does not work*.

Let's try another example. If $f(x) = 1$ and $g(x) = e^x$, then $f'(x) = 0$, so $f'(x)g'(x) = 0$. But $f(x)g(x) = e^x$, and the derivative of e^x is not 0. Again, rule (5.27) fails.

A good way to check whether a formula is plausible is to see if the units line up. Suppose we have a rectangle of length $L(t)$ and width $W(t)$, and both the length and the width are changing with time. Then $L(t)$ has units of length (say, meters), as does $W(t)$, t has units of time (say, seconds), and both $L'(t)$ and $W'(t)$ have units of meters/second. Their product has units of square meters/square second. But the area $A(t) = L(t)W(t)$ has units of square meters, so A' has units of square meters/second. Something with units of m^2/s can't possibly be equal to something with units of m^2/s^2! Formula (5.27) isn't just wrong, it's nonsensical.

So if that isn't the right formula, what is? We'll work that out with an example about an oil well.

Suppose that an oil well produces oil at a rate Q, and that oil sells for a price p. Right now, $Q = 200$ barrels/year, and $p = \$70$/barrel. So the revenue $R = pQ$ generated by the well is \$14,000/year. What happens if the production increases to 205 barrels/year? What happens if the price increases to \$73/barrel? What happens if the production increases to 205 barrels/year *and* the price increases to \$73/barrel?

If the production goes up and the price stays the same, then we can treat p as a constant. We have $R = 70Q$, so the change in R is 70 times the change in Q. That is,

$$(5.28) \qquad \Delta R = 70\Delta Q = 70(205 - 200) = 350.$$

Five more barrels per year at \$70/barrel is worth an extra \$350/year.

If the price goes up and the production stays the same, we can treat Q as a constant and write

$$(5.29) \qquad \Delta R = 200\Delta p = 200(73 - 70) = 600.$$

Three more dollars for each of the 200 barrels gives us an extra \$600/year.

If both the production and price go up, then our new revenue (in dollars/year)is

(5.30) $(70 + 3)(200 + 5) \;\;=\;\; 70(200) + 70(5) + 3(200) + 3(5)$

$=\;\; 14{,}000 + 350 + 600 + 15.$

The first term is our old revenue, the second is the effect of raising production, and the third is the effect of raising the price. The fourth term, involving both the change in production and the change in price, is much smaller than the others. To find the *change* in revenue, we subtract our original revenue of 14,000 from this sum. More generally, if Q changes to $Q + \Delta Q$ and p changes to $p + \Delta p$, then

(5.31) $\Delta R \;\;=\;\; (p + \Delta p)(Q + \Delta Q) - pQ$

$=\;\; pQ + p\Delta Q + Q\Delta p + \Delta p\Delta Q - pQ$

$=\;\; p\Delta Q + Q\Delta p + \Delta p\Delta Q.$

On the last line, the first term is the effect of just changing Q and the second term is the effect of just changing p. If we're talking about small changes, then the cross-term $\Delta p\Delta Q$ is a lot smaller than the two main terms.

The same reasoning applies to functions in general. Suppose that $f(x)$ and $g(x)$ are functions and we change x by a small amount h, resulting in changes of size $\Delta f = f(x + h) - f(x)$ and $\Delta g = g(x + h) - g(x)$ to f and g, and a resultant change $\Delta(fg) = f(x + h)g(x + h) - f(x)g(x)$ to fg. Then

(5.32) $\Delta(fg) = f(x)\Delta g + g(x)\Delta f + \Delta f \Delta g,$

so

(5.33) $\dfrac{\Delta(fg)}{h} = f(x)\dfrac{\Delta g}{h} + g(x)\dfrac{\Delta f}{h} + h\dfrac{\Delta f}{h}\dfrac{\Delta g}{h}.$

As $h \to 0$, $\frac{\Delta f}{h}$ approaches $f'(x)$ (as that's the definition of $f'(x)$) and $\frac{\Delta g}{h}$ approaches $g'(x)$. The first term on the right side approaches $f(x)g'(x)$, the second approaches $g(x)f'(x)$, and the third goes to $0f'(x)g'(x) = 0$. Thus, in both Newton's notation and Leibniz's, we have the following:

The product rule:

(5.34) $(f(x)g(x))' \;\;=\;\; f(x)g'(x) + f'(x)g(x),$

$\dfrac{d(f(x)g(x))}{dx} \;\;=\;\; f(x)\dfrac{dg(x)}{dx} + \dfrac{df(x)}{dx}g(x).$

In both versions, the first term is the effect of g changing and the second term is the effect of f changing.

Let's see how that works with some examples, including the ones that (5.27) failed at:

- If $f(x) = x^2$ and $g(x) = x^3$, then

(5.35) $f(x)g'(x) + f'(x)g(x) = x^2(3x^2) + 2x(x^3) = 5x^4,$

which is the derivative of $f(x)g(x) = x^5$.

- If $f(x) = 1$ and $g(x) = e^x$, then $f'(x) = 0$ and $g'(x) = e^x$, so

(5.36)
$$f(x)g'(x) + f'(x)g(x) = 1(e^x) + 0(e^x) = e^x,$$

which is the derivative of $f(x)g(x) = e^x$.

- If $f(x) = 5x^3$ and $g(x) = \sin(x)$, then $f'(x) = 15x^2$ and $g'(x) = \cos(x)$, so

(5.37)
$$\left(5x^3 \sin(x)\right)' = 15x^2 \sin(x) + 5x^3 \cos(x).$$

- If $f(x) = g(x) = \sqrt{x}$, then we don't need the product rule to take the derivative of $f(x)g(x) = x$; instead, we use it to figure out the derivative of $\sqrt{x}$. Because the derivative of x is $x' = 1$, we have

(5.38)
$$x = \sqrt{x}\sqrt{x},$$
$$(x)' = \sqrt{x}(\sqrt{x})' + (\sqrt{x})'\sqrt{x},$$
$$1 = 2\sqrt{x}(\sqrt{x})',$$
$$(\sqrt{x})' = \frac{1}{2\sqrt{x}},$$

in agreement with Newton's hammer: $(x^{1/2})' = \frac{1}{2}x^{-1/2}$.

The Quotient Rule. Next we turn to the ratio of two functions. Suppose that $h(x) = f(x)/g(x)$. How can we write the derivative of $h(x)$ in terms of $f'(x)$ and $g'(x)$? Because $f(x) = g(x)h(x)$, we already have a rule that relates $f'(x)$, $g'(x)$, and $h'(x)$. We just have to apply the product rule to $f = gh$ and solve for h':

(5.39)
$$f(x) = g(x)h(x),$$
$$f'(x) = g(x)h'(x) + g'(x)h(x),$$
$$g(x)h'(x) = f'(x) - g'(x)h(x),$$
$$h'(x) = \frac{f'(x) - g'(x)(f(x)/g(x))}{g(x)}$$
$$= \frac{f'(x)g(x) - g'(x)f(x)}{(g(x))^2}.$$

In both Newton's and Leibniz's notation, we have the following:

The quotient rule:

(5.40)
$$\left(\frac{f}{g}\right)' = \frac{f'g - g'f}{g^2},$$
$$\frac{d}{dx}\left(\frac{f}{g}\right) = \frac{\frac{df}{dx}g - \frac{dg}{dx}f}{g^2}.$$

The hardest thing about remembering the quotient rule is keeping straight which term is positive and which is negative. Think of it this way: if we have a ratio $f(x)/g(x)$ of positive functions, then increasing the numerator $f(x)$ makes the ratio *bigger*. This means the term involving $f'(x)$ must be *positive*. On the other hand, increasing the denominator $g(x)$ makes the ratio *smaller*, so the term involving $g'(x)$ must be *negative*.

Here are some examples involving the quotient rule.

- If $f(x) = 1$ and $g(x) = x^n$, then $f'(x) = 0$ and $g'(x) = nx^{n-1}$, so

$$(5.41) \qquad (x^{-n})' = \left(\frac{1}{x^n}\right)'$$
$$= \frac{(1)'x^n - (x^n)'1}{x^{2n}} = \frac{-nx^{n-1}}{x^{2n}} = -nx^{-n-1}.$$

This shows that Newton's hammer works for negative powers as well as positive powers of x.

- If $f(x) = e^x$ and $g(x) = x^3 + \sin(x)$, then $f'(x) = e^x$ and $g'(x) = 3x^2 + \cos(x)$, so

$$(5.42) \qquad \left(\frac{e^x}{x^3 + \sin(x)}\right)' = \frac{(x^3 + \sin(x))e^x - (3x^2 + \cos(x))e^x}{(x^3 + \sin(x))^2}$$
$$= \frac{(x^3 - 3x^2 + \sin(x) - \cos(x))e^x}{(x^3 + \sin(x))^2}.$$

- Besides $\sin(x)$ and $\cos(x)$, we also have the trig functions $\tan(x) = \sin(x)/\cos(x)$, $\sec(x) = 1/\cos(x)$, $\cot(x) = \cos(x)/\sin(x)$ and $\csc(x) = 1/\sin(x)$. The derivative of $\tan(x)$ is:

$$(5.43) \qquad \left(\frac{\sin(x)}{\cos(x)}\right)' = \frac{\cos(x)(\sin(x))' - \sin(x)(\cos(x))'}{\cos^2(x)}$$
$$= \frac{\cos(x)\cos(x) - \sin(x)(-\sin(x))}{\cos^2(x)}$$
$$= \frac{\cos^2(x) + \sin^2(x)}{\cos^2(x)}$$
$$= \frac{1}{\cos^2(x)}$$
$$= \sec^2(x),$$

as $\sin^2(x) + \cos^2(x) = 1$.

- Similarly, you can (should!) use the quotient rule to compute the derivatives of $\sec(x)$, $\cot(x)$, and $\csc(x)$. I mean *you*, not *we*! The answers appear in Table 5.4, but don't peek until you've done the calculations yourself.

Table 5.4. More results about derivatives

Function	Derivative	Comments
$af(x)+bg(x)$	$af'(x) + bg'(x)$	Obvious
$f(x)g(x)$	$f'(x)g(x)+g'(x)f(x)$	The product rule is **not** $f'(x)g'(x)$.
$\dfrac{f(x)}{g(x)}$	$\dfrac{f'(x)g(x)-g'(x)f(x)}{g(x)^2}$	Quotient rule. Watch the signs.
x^{-n}	$-nx^{-n-1}$	Newton's hammer still works.
$\tan(x)$	$\sec^2(x)$	which equals $1/\cos^2(x)$
$\sec(x)$	$\sec(x)\tan(x)$	which equals $\sin(x)/\cos^2(x)$
$\cot(x)$	$-\csc^2(x)$	which equals $-1/\sin^2(x)$
$\csc(x)$	$-\csc(x)\cot(x)$	which equals $-\cos(x)/\sin^2(x)$

5.3. The Chain Rule

A tablespoon of commercial butter contains about 11 grams of fat, as well as some water and a negligible amount of protein and sugar. Each gram of fat has about 9 calories[4] of energy. So how many calories are in a tablespoon of butter?

This an easy problem:

$$\begin{aligned}
\text{1 tablespoon of butter} &= \text{11 grams of fat,} \\
\text{1 gram of fat} &= \text{9 calories,} \\
\text{1 tablespoon of butter} &= \text{11(9 calories)} \\
&= \text{99 calories.}
\end{aligned} \tag{5.44}$$

In a two-step chain, where tablespoons go to grams and grams go to calories, the **conversion factor** from tablespoons to calories equals the product of the conversion factors from tablespoons to grams and from grams to calories.

Now suppose that you have a stick of butter that is more than one tablespoon. Let x be the number of tablespoons of butter, y the number of grams of fat, and z the number of calories. We have

$$\begin{aligned}
y &= 11x &\text{so} \quad \frac{dy}{dx} &= 11, \\
z &= 9y &\text{so} \quad \frac{dz}{dy} &= 9, \\
z &= 99x &\text{so} \quad \frac{dz}{dx} &= 99.
\end{aligned} \tag{5.45}$$

That is,

The chain rule (Leibniz's notation):

$$\frac{dz}{dx} = \frac{dz}{dy}\frac{dy}{dx}. \tag{5.46}$$

---

[4] Actually, about 9 *kilocalories* of energy, or enough to heat a liter of water by nine degrees Celsius. When talking about food, we usually drop the prefix "kilo".

We derived the chain rule in this simple example, where x, y, and z were all proportional to one another. However, it applies much more generally.

Suppose that x, y, and z are variables, that y is a function of x, and that z is a function of y. For instance, we might have $y = x^2$ and $z = \sin(y) = \sin(x^2)$. More generally, suppose that $y = g(x)$ and $z = f(y) = f(g(x))$. As long as f and g are differentiable, the microscope equation tells us that small changes in x result in proportional changes in y, and that small changes in y result in proportional changes in z. There are conversion factors m_1, m_2, and m_3 such that

$$(5.47) \qquad\qquad \begin{aligned} \Delta y &\approx m_1 \Delta x, \\ \Delta z &\approx m_2 \Delta y, \qquad \text{and} \\ \Delta z &\approx m_3 \Delta x, \end{aligned}$$

where $m_1 = \frac{dy}{dx}$, $m_2 = \frac{dz}{dy}$, and $m_3 = \frac{dz}{dx}$. As m_3 is obviously equal to $m_1 m_2$, we conclude that $\dfrac{dz}{dx} = \dfrac{dz}{dy}\dfrac{dy}{dx}$.

For instance, suppose that $y = x^2 - 1$ and $z = 3e^y = 3e^{x^2-1}$. When $x = 1$, $y = 1^2 - 1 = 0$, so $z = 3e^0 = 3$. Because $dy/dx = 2x = 2$, every small change in x results in twice as big a change in y. If x becomes $1 + h$, then y will become about $0 + 2h = 2h$. Because $dz/dy = 3e^y = 3$, every change in y results in three times as big a change in z, so if y becomes $2h$, z must increase by $6h$ and become approximately $3 + 6h$. Restating this example in equations, we have

$$(5.48) \qquad\qquad \begin{aligned} y &= x^2 - 1 &&\text{so} & \frac{dy}{dx} &= 2x, \\ z &= 3e^y &&\text{so} & \frac{dz}{dy} &= 3e^y = 3e^{x^2-1}, \\ \frac{dz}{dx} &= \frac{dz}{dy}\frac{dy}{dx} &&\text{so} & \frac{dz}{dx} &= 6xe^{x^2-1}. \end{aligned}$$

Plugging in $x = 1$ gives us $\frac{dy}{dx} = 2$, $\frac{dz}{dy} = 3$ and $\frac{dz}{dx} = 6$, but the formulas also apply at other values of x. The derivative of $3e^{x^2-1}$ with respect to x equals $6xe^{x^2-1}$ everywhere.

The chain rule in this form (5.46) is very useful for **related rates** problems. For instance, suppose that we are pumping air into a large balloon at a rate of 3.7 cubic feet/minute. How fast is the radius of the balloon growing when the radius is 1 foot? When the radius is 5 feet?

In this problem, our three variables are the radius r, the volume V, and the time t. We imagine a chain going from time to radius to volume. We know how volume is related to radius:

$$(5.49) \qquad\qquad V = \frac{4}{3}\pi r^3, \quad \text{so} \quad \frac{dV}{dr} = 4\pi r^2.$$

We know that $dV/dt = 3.7$ cubic feet/minute. We just need to write down the chain rule and solve for dr/dt:

(5.50)
$$\frac{dV}{dt} = \frac{dV}{dr}\frac{dr}{dt},$$

$$3.7 \text{ ft}^3/\text{min} = 4\pi r^2 \frac{dr}{dt},$$

$$\frac{dr}{dt} = \frac{3.7 \text{ ft}^3/\text{min}}{4\pi r^2}$$

$$= \frac{0.2944 \text{ ft}^3/\text{min}}{r^2}.$$

When $r = 1$ foot, this is 0.2944 ft/min. (r^2 has units of square feet, and cubic feet divided by square feet make feet.) When $r = 5$ feet, this is 0.01178 ft/min. We are pumping air into the balloon at a constant rate, but the radius grows much more slowly when $r = 5$ feet than when $r = 1$ foot. You may have noticed a similar effect when blowing up party balloons. At first they seem the grow quickly, but then you get out of breath and they hardly seem to grow at all.

So far we have studied the chain rule in Leibniz's notation, which is great for understanding why it works, and great for related rates problems. Now we're going to switch to Newton's notation, which is better for solving problems like "find the derivative of $\sin(e^x + x^3)$".

Instead of talking about three variables x, y, and z, we'll talk about an input variable x, two functions f and g, and the compound function $f \circ g(x) = f(g(x))$. It's sometimes useful to let $y = g(x)$ and $z = f(y) = f(g(x))$, but that isn't always necessary. Translating from Leibniz to Newton,

(5.51)
$$\frac{dy}{dx} \quad \text{becomes} \quad g'(x),$$

$$\frac{dz}{dy} \quad \text{becomes} \quad f'(y)$$

$$\text{which equals} \quad f'(g(x)),$$

$$\frac{dz}{dx} \quad \text{becomes} \quad (f(g(x)))',$$

and we get

> **The chain rule (Newton's notation):**
>
> (5.52) $\qquad (f(g(x)))' = f'(g(x))\,g'(x).$

Note that the derivative of f is evaluated at y, which is to say at $g(x)$, and not at x. The derivative of $\sin(x^2)$ is $2x\cos(x^2)$, not $2x\cos(x)$.

There is also a hybrid form of the chain rule, where we write u in place of $g(x)$:

(5.53)
$$\frac{df(u)}{dx} = f'(u)\frac{du}{dx}.$$

In particular, if u is any function of x (such as $x^2 - 1$ or $\cos(x)$ or $e^x + 7$ or whatever you want), we have

(5.54)
$$\frac{d(u^n)}{dx} = nu^{n-1}\frac{du}{dx},$$
$$\frac{d\sin(u)}{dx} = \cos(u)\frac{du}{dx},$$
$$\frac{d\cos(u)}{dx} = -\sin(u)\frac{du}{dx},$$
$$\frac{de^u}{dx} = e^u\frac{du}{dx}.$$

Here are some examples:

- The derivative of $\sin(x^2)$ with respect to x is $2x\cos(x^2)$. This is the second equation of (5.54) with $u = x^2$.
- The derivative of $\sin^2(x)$ is $2\sin(x)\cos(x)$. This is the first equation of (5.54) with $u = \sin(x)$.
- If k is a constant, then the derivative of e^{kx} is ke^{kx}. This will come up a lot when we study growth models.
- The chain rule tells us how to combine two links of a chain. If our chain is longer than two links, we can apply the rule multiple times. For instance, $\sin(e^{5x})$ is a three-step chain: x becomes $5x$, which becomes e^{5x}, which becomes $\sin(e^{5x})$. Its derivative is:

(5.55)
$$\begin{aligned}
\left(\sin\left(e^{5x}\right)\right)' &= \cos\left(e^{5x}\right) \times \left(e^{5x}\right)' \\
&= \cos\left(e^{5x}\right) \times e^{5x} \times (5x)' \\
&= 5e^{5x}\cos\left(e^{5x}\right)
\end{aligned}$$

- The chain rule can be combined with the product rule and the quotient rule. For instance, if we want to compute the derivative of the function $g(x) = \frac{5x^3 - 3x + \sin(x^2)}{\cos(2^x + 17)}$ that we saw at the beginning of the chapter, we first apply the quotient rule to get that $g'(x)$ equals

(5.56)
$$\frac{\cos(2^x + 17)\left(5x^3 - 3x + \sin(x^2)\right)' - \left(\cos(2^x + 17)\right)'\left(5x^3 - 3x + \sin(x^2)\right)}{\cos^2(2^x + 17)}.$$

Next we use the sum rule and the chain rule to evaluate

(5.57)
$$\left(5x^3 - 3x + \sin(x^2)\right)' = 15x^2 - 3 + 2x\cos(x^2).$$

Remembering that the derivative of 2^x is $k_2 2^x$, where $k_2 \approx 0.69$, we use the chain rule to get

(5.58)
$$\left(\cos(2^x + 17)\right)' = -\sin(2^x + 17)(k_2 2^x).$$

Combining everything, we get that the derivative of $g(x)$ is

(5.59)
$$\frac{\cos(2^x + 17)\left(15x^2 - 3 + 2x\cos(x^2)\right) + k_2 2^x \sin(2^x + 17)\left(5x^3 - 3x + \sin(x^2)\right)}{\cos^2(2^x + 17)}.$$

The derivatives of even the ugliest functions (and this one was pretty darn ugly) can be evaluated patiently, *one step at a time*. We don't have to solve the whole problem by applying a single rule. We just have to turn it into a slightly easier problem, and then repeat until we're done.

- So far we have been using the chain rule to get the derivative of $f \circ g$ from the derivatives of f and g. However, it can also be used to get the derivative of f from the derivatives of $f \circ g$ and g, or the derivative of g from the derivatives of $f \circ g$ and f.

  For instance, we originally obtained Newton's hammer for positive integer powers of x, and then we extended it to negative integer powers with the quotient rule. What about fractional powers? If $x^{p/q}$ is a fractional power of x (where p and q are integers and $q \neq 0$), we can write $x^p = u^q$, where $u = g(x) = x^{p/q}$. Using the first equation of (5.54),

(5.60)
$$\begin{aligned}
(x^p)' &= (u^q)', \\
px^{p-1} &= qu^{q-1}u', \\
px^{p-1} &= qx^{p-\frac{p}{q}}(x^{p/q})', \\
(x^{p/q})' &= \frac{p}{q}x^{\frac{p}{q}-1}.
\end{aligned}$$

Newton's hammer works for fractional powers, too!

5.4. Optimization

In this section we'll take a look at two examples of functions that have a control parameter and see how we can adjust the control parameter to maximize the function. Our method can be summed up in the 3rd Pillar of Calculus:

> *What goes up has to stop before it comes down.*

Blood flow and heart rate. The amount of blood that a heart pumps is a function of how fast it beats. Let p (for "pulse") be the heart rate in beats per minute and let $H(p)$ be the blood flow, in milliliters per minute. You may not think of your pulse as a control parameter because you don't consciously control it, but your brain does adjust your pulse to get the right amount of blood into your tissues. The function $H(p)$ varies from person to person and depends on age, sex, size, health, and general level of fitness. However, the function for an average 40-year-old person in good health is approximately[5]

(5.61)
$$H(p) = 5p\sqrt{270 - p}.$$

How much blood can the heart actually pump and how fast does it have to beat in order to pump that much blood?

Our first step is to translate this medical question into a mathematical question: *What is the maximum value of $H(p)$, and for what value of p do we achieve that maximum?*

---

[5]This formula is *not* taken from the medical literature. However, the overall shape is remarkably accurate.

Next we look at the derivative $H'(p)$, which we compute with the product rule and the chain rule:

$$
\begin{aligned}
(5.62) \qquad H'(p) &= (5p\sqrt{270 - p})' \\
&= 5\sqrt{270 - p} + 5p(\sqrt{270 - p})' \\
&= 5\sqrt{270 - p} + \frac{5p}{2\sqrt{270 - p}}(270 - p)' \\
&= 5\sqrt{270 - p} - \frac{5p}{2\sqrt{270 - p}} \\
&= \frac{10(270 - p) - 5p}{2\sqrt{270 - p}} \\
&= \frac{5(540 - 3p)}{2\sqrt{270 - p}}.
\end{aligned}
$$

Note that $H'(p)$ is positive when $p < 180$, is zero when $p = 180$, and is negative when $p > 180$. For instance, $H'(60) = 900/\sqrt{210} \approx 62.1$, so at 60 beats per minute we can increase our blood flow by increasing the heart rate; each additional beat per minute pumps 62.1 more milliliters of blood per minute. Meanwhile, $H'(220) = -300/\sqrt{50} \approx -42.5$, so at 220 beats per minute we can increase our blood flow by decreasing the heart rate, as each additional beat per minute pumps 42.5 fewer milliliters of blood per minute.

Figure 5.2. Blood flow as a function of heart rate

The situation is shown in Figure 5.2. When $p < 180$, we can get more blood flow by speeding up the heart rate. When $p > 180$, we can get more blood flow by slowing down the heart rate. The maximum flow is achieved when $p = 180$ and is $H(180) \approx 8538$ milliliters per minute.

> 180 beats per minute is the **maximum heart rate** (MHR or HR_{MAX}) of this average individual. If this person's pulse is close to 180, say while running the Boston Marathon, then his blood flow is close to its maximum value of 8538 milliliters per minute. If he does something that requires more than 8538 milliliters per minute of blood, like running up Heartbreak Hill, then his heart won't be able to keep up. The result is likely to be collapse or heart failure.
>
> Fortunately, the human body sends us very strong signals to stop, such as extreme heat and nausea, when we get within 10 or 20 points of MHR. To stay safe, we need to listen to those signals. We also need to know our personal limits. A common rule of thumb is MHR = 220 − age, but there is a lot of variation from person to person. Before engaging in super-strenuous exercise, please have your MHR tested in a doctor's office or sports lab.

To recap, the sign of $H'(p)$ tells us what happens as we make changes. When $H'(p) > 0$, $H(p)$ is an increasing function so we can get more flow by increasing p. When $H'(p) < 0$, $H(p)$ is a decreasing function so we can get more flow by decreasing p. The **maximum** of $H(p)$ occurs when $H(p)$ stops increasing and is about to start decreasing. That's the point where $H'(p)$ goes from positive to negative, and in particular equals zero. (If we were trying to minimize a function, we would look for where the derivative stops being negative and starts being positive.)

> *The standard method for finding both maxima and minima is to take the derivative of the function we are studying, set it equal to zero, and solve for our input variable.*

Widgets and profit. In Section 4.2 we encountered a struggling widget factory. Producing x widgets in a month costs

(5.63) $$C(x) = 22{,}000 + 7x$$

dollars. The marketing department determines that, if the factory produces x widgets per month, it can sell them for a price of

(5.64) $$p(x) = \left(10 - \frac{x}{10{,}000}\right) \text{ dollars/widget,}$$

giving them a monthly revenue of

(5.65) $$R(x) = xp(x) = \left(10x - \frac{x^2}{10{,}000}\right) \text{ dollars.}$$

The monthly profit,[6] in dollars, is then

(5.66) $$P(x) = R(x) - C(x) = 3x - \frac{x^2}{10{,}000} - 22{,}000.$$

The factory is currently producing 10,000 widgets/month, selling them for $9 each, for a revenue of $90,000/month, which is less than the monthly cost of $C(10{,}000) = \$92{,}000$. Because the factory is losing money, corporate headquarters wants to close it. Is it possible to turn things around and save the factory?

---

[6]It's unfortunate that price and profit start with the same letter, but that's the way it goes. We use a lower case p to denote price and a capital P to denote profit.

To decide this question, we compute the marginal cost, revenue, and profit using Newton's hammer:

(5.67)
$$C'(x) = 7,$$
$$R'(x) = 10 - \frac{x}{5,000}, \quad \text{and}$$
$$P'(x) = 3 - \frac{x}{5,000}.$$

See Figure 5.3. As long as $x < 15,000$, the marginal profit is positive, so the factory can make more money by increasing production. Once x exceeds 15,000, the marginal profit is negative, so it's better to decrease production. The ideal production level is therefore $x = 15,000$.

This calculation shows that the best production level is at $x = 15,000$, but it doesn't tell us anything about how much money the factory is making or losing. For that, we have to plug $x = 15,000$ into our formula for $P(x)$:

(5.68)
$$P(15,000) = 3(15,000) - \frac{(15,000)^2}{10,000} - 22,000 = 500.$$

At the new production level the factory turns a (small) profit. The factory stays open and the workers' jobs are saved.

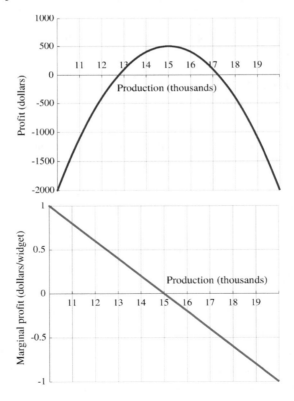

Figure 5.3. The profit $P(x)$ and marginal profit $P'(x)$ as functions of x

5.5. The Shape of a Graph

We have seen that the sign of the derivative $f'(x)$ of $f(x)$ tells us whether $f(x)$ is increasing, and that the places where $f'(x) = 0$ are candidates for being maxima or minima. In this section we'll elaborate on these ideas. We'll see how studying the sign of $f(x)$, the sign of $f'(x)$, and the sign of the derivative of $f'(x)$—written $f''(x)$ and called the **second derivative** of $f(x)$—tells us a lot about the graph of $f(x)$. We will use two polynomials for our case studies: $f(x) = x^3 - 3x$ and $g(x) = x^4 - 6x^2$.

The Graph of $f(x) = x^3 - 3x$**.** The first thing to ask about any function is

Where is it positive, and where is it negative?

We don't actually have to check every point. Once we know that $f(-10) = -970$, we have a pretty good idea that $f(-10.01)$ and $f(-9.99)$ will also be negative. In fact, there are only two ways that a function can go from negative to positive or vice versa:

1. it can cross through zero, or

2. it can jump.

In this case, the function $f(x)$ is a continuous polynomial (actually, *all* polynomials are continuous), and so it can't jump. The only places where $f(x)$ can change sign are at the places where $f(x) = 0$. These are called the **roots** of $f(x)$. In this case, finding the roots is easy:

$$(5.69) \qquad \begin{aligned} x^3 - 3x &= 0, \\ x(x^2 - 3) &= 0, \\ x = 0 \quad &\text{or} \quad x^2 = 3, \\ x = 0 \quad &\text{or} \quad x = \pm\sqrt{3}. \end{aligned}$$

We mark these on a number line as in Figure 5.4.

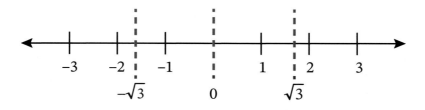

Figure 5.4. The roots of $f(x) = x^3 - 3x$

This divides the number line into four intervals: $(-\infty, -\sqrt{3})$, $(-\sqrt{3}, 0)$, $(0, \sqrt{3})$, and $(\sqrt{3}, \infty)$. The sign of $f(x)$ can't change within an interval; on each interval, the function is either completely positive or completely negative. To figure out which, we just pick a point in each interval and test it:

(5.70) $f(-10) = -970 < 0$ so $f(x)$ is negative on $(-\infty, -\sqrt{3})$,

$f(-1) = 2 > 0$ so $f(x)$ is positive on $(-\sqrt{3}, 0)$,

$f(1) = -2 < 0$ so $f(x)$ is negative on $(0, \sqrt{3})$,

$f(10) = 970 > 0$ so $f(x)$ is positive on $(\sqrt{3}, \infty)$.

There isn't anything special about the numbers -10, -1, 1, and 10. We could just as well have picked -5π, -1.27, 0.3, and 1776. But it's easier to evaluate our function at ± 10 and ± 1.

We update our number line with this new information, producing Figure 5.5, and call it a **sign chart** for $f(x)$.

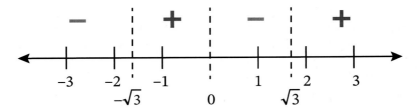

Figure 5.5. The sign chart of $f(x) = x^3 - 3x$

If we didn't know anything about the size of $f(x)$ and only had the information in the sign chart, we could still take a first stab at graphing $f(x)$, and we'd get something that looks like Figure 5.6.

Of course, this isn't a very *good* graph! On each interval we've drawn our function as a horizontal line because we don't know whether it's increasing or decreasing. To understand that, we need to look at $f'(x)$.

The next step is to make a sign chart for $f'(x) = 3x^2 - 3$. We repeat what we did before, only for $f'(x)$ instead of for $f(x)$. We first look for the places where $f'(x) = 0$ or

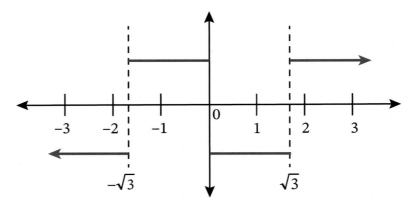

Figure 5.6. A schematic graph of $f(x) = x^3 - 3x$

$f'(x)$ doesn't exist. These are called the **critical points** of $f(x)$. As $f'(x) = 3x^2 - 3 = 3(x^2 - 1)$, these occur at $x = \pm 1$. Proceeding as before, we break our number line into three intervals, $(-\infty, -1)$, $(-1, 1)$, and $(1, \infty)$. We evaluate $f'(x)$ at three test points, say -2, 0, and 2.

$$(5.71) \qquad f'(-2) = 9 > 0 \qquad \text{so} \qquad f'(x) \text{ is positive on } (-\infty, -1),$$
$$\text{which means} \qquad f(x) \text{ is increasing on } (-\infty, -1).$$

$$f'(0) = -3 < 0 \qquad \text{so} \qquad f'(x) \text{ is negative on } (-1, 1),$$
$$\text{which means} \qquad f(x) \text{ is decreasing on } (-1, 1).$$

$$f'(2) = 9 > 0 \qquad \text{so} \qquad f'(x) \text{ is positive on } (1, \infty),$$
$$\text{which means} \qquad f(x) \text{ is increasing on } (1, \infty).$$

Figure 5.7. The sign chart of $f'(x) = 3x^2 - 3$

We can combine the sign chart for $f(x)$ (Figure 5.5) and the sign chart for $f'(x)$ (Figure 5.7) into one chart (Figure 5.8).

At $x = -1$, the function stops increasing and starts decreasing. In other words, $f(-1) = 2$ must be higher than the values of $f(x)$ at any nearby points. It isn't necessarily the highest point on the whole graph (in fact, $f(3) = 18$, which is much bigger than 2), but it's higher than all of its neighbors. We call a point like $(-1, 2)$ a **local maximum**. Similarly, when $x = 1$ the function stops decreasing and starts increasing, so that must be the bottom of a valley, and there is a **local minimum** at $(1, -2)$.

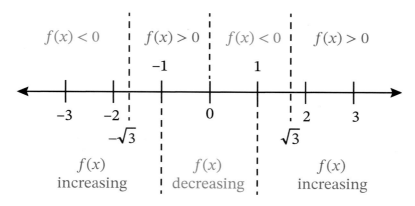

Figure 5.8. A combined sign chart for $f(x)$ and its derivative. The sign of the derivative tells us whether $f(x)$ is increasing or decreasing.

The sign chart for $f'(x)$ tells us everything we need to know to find the local maxima and minima. We summarize these observations:

> **First Derivative Test:** The local maxima of a differentiable function occur where the derivative changes from positive to negative. The local minima occur where the derivative changes from negative to positive.

Now we're ready to make a better sketch, drawing an upward-sloping line wherever $f'(x) > 0$ and a downward sloping line wherever $f'(x) < 0$, making sure that the sign of the function agrees with our sign chart for $f(x)$. That is, there are four possible local pictures of our function, shown in Figure 5.9.

Applying this technique to our number line, with the six intervals $(-\infty, -\sqrt{3})$, $(-\sqrt{3}, -1)$, $(-1, 0)$, $(0, 1)$, $(1, \sqrt{3})$, and $(\sqrt{3}, \infty)$, gives the improved sketch shown in Figure 5.10.

Finally, we want to consider which way the graph of $f(x)$ is curving. In a $\cup$-shaped graph like that of x^2, the derivative is always increasing, going from very negative to slightly negative to 0 to slightly positive to very positive. Curves with this shape are

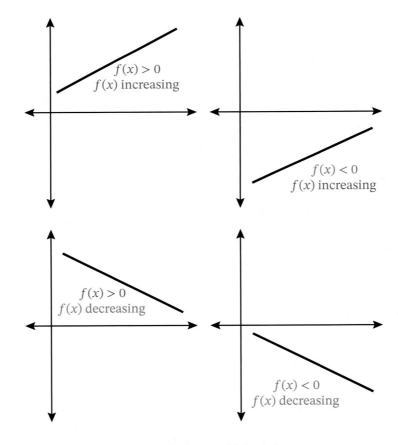

Figure 5.9. The four possible local pictures

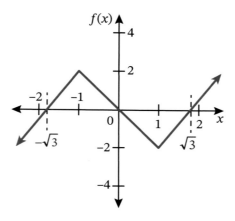

Figure 5.10. An improved sketch of the graph of $f(x) = x^3 - 3x$

said to be **concave up**. In a ∩-shaped graph like that of $-x^2$, the derivative is always decreasing, going from very positive to slightly positive to 0 to slightly negative to very negative. See Figure 5.11. Curves with this shape are said to be **concave down**. The **concavity** of a graph depends on whether $f'(x)$ is increasing or decreasing, which depends on whether the derivative of $f'(x)$, written $f''(x)$, is positive or negative. That is:

The graph of $f(x)$ is concave up $\Leftrightarrow$ $f'(x)$ is increasing $\Leftrightarrow$ $f''(x) > 0$.
The graph of $f(x)$ is concave down $\Leftrightarrow$ $f'(x)$ is decreasing $\Leftrightarrow$ $f''(x) < 0$.

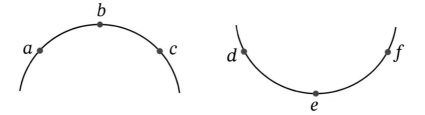

Figure 5.11. When the second derivative is negative, the first derivative can be positive (at a), zero (at b), or negative (at c). When the second derivative is positive, the first derivative can be negative (at d), zero (at e), or positive (at f).

When $f'(x) = 0$ and $f''(x) > 0$, the first derivative must be increasing, so it must be changing from negative to positive. Likewise, when $f'(x) = 0$ and $f''(x) < 0$, the first derivative must be decreasing, so it must be changing from positive to negative. Combined with the first derivative test, this gives us the following.

> **Second Derivative Test**: Points where $f'(x) = 0$ and $f''(x) < 0$ are local maxima. Points where $f'(x) = 0$ and $f''(x) > 0$ are local minima. If $f'(x) = 0$ and $f''(x) = 0$, then the second derivative test does not work, and we need to use the first derivative test instead.

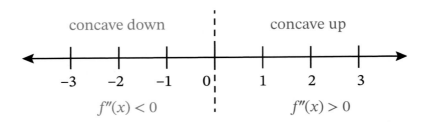

Figure 5.12. The sign chart for $f''(x) = 6x$

In our example, $f'(x) = 3x^2 - 3$, so $f''(x) = (f'(x))' = (3x^2 - 3)' = 6x$, so our sign chart for $f''(x)$ is very simple; see Figure 5.12. The places where $f''(x) = 0$ are called **subcritical points**, and the places where $f''(x)$ changes sign are called **points of inflection**.

Because $f'(-1) = 0$ and $f''(-1) < 0$, there is a local maximum at $x = -1$, and because $f'(1) = 0$ and $f''(1) > 0$ there is a local minimum at $x = 1$. In this example, we already knew about the local maximum and local minimum from the sign chart of $f'(x)$, but sometimes it's easier to apply the second derivative test than the first derivative test.

Combining everything, we can create Table 5.5 showing what is happening to $f(x)$ in each interval and at the points where the intervals meet.

Table 5.5. Behavior of $f(x) = x^3 - 3x$

Interval	Sign of f	Direction	Concavity	Comment
$x < -\sqrt{3}$	negative	up	down	
$x = \sqrt{3}$	0	up	down	root
$-\sqrt{3} < x < -1$	positive	up	down	
$x = -1$	positive	flat	down	local max
$-1 < x < 0$	positive	down	down	
$x = 0$	0	down	0	inflection pt
$0 < x < 1$	negative	down	up	
$x = 1$	negative	flat	up	local min
$1 < x < \sqrt{3}$	negative	up	up	
$x = \sqrt{3}$	0	up	up	root
$x > \sqrt{3}$	positive	up	up	

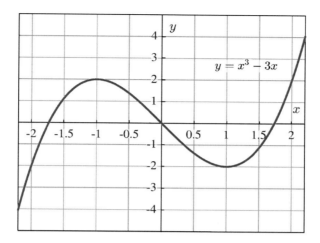

Figure 5.13. The actual graph of $f(x) = x^3 - 3x$

With this data we can adjust our sketch (Figure 5.10), smoothing the rough corners and making the graph concave down when $x < 0$ and concave up when $x > 0$. The resulting sketch probably wouldn't look *exactly* like the actual graph of $f(x)$, but it would have all the right qualitative features. The actual graph is shown in Figure 5.13.

The Graph of $x^4 - 6x^2$. As a second example, let's look at the function $g(x) = x^4 - 6x^2$. First we find the roots of $g(x)$:

$$(5.72) \qquad \begin{aligned} x^4 - 6x^2 &= 0, \\ x^2(x^2 - 6) &= 0, \\ x = 0 \quad \text{or} \quad x &= \pm\sqrt{6} \approx \pm 2.45. \end{aligned}$$

We pick test points at -3, -1, 1, and 3, and we compute

$$(5.73) \qquad \begin{aligned} f(-3) = 27 > 0 &\quad \text{so} \quad f(x) \text{ is positive on } (-\infty, -\sqrt{6}), \\ f(-1) = -5 < 0 &\quad \text{so} \quad f(x) \text{ is negative on } (-\sqrt{6}, 0), \\ f(1) = -5 < 0 &\quad \text{so} \quad f(x) \text{ is negative on } (0, -\sqrt{6}), \\ f(3) = 27 > 0 &\quad \text{so} \quad f(x) \text{ is positive on } (\sqrt{6}, \infty), \end{aligned}$$

as in the sign chart shown in Figure 5.14.

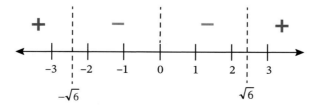

Figure 5.14. The sign chart for $g(x) = x^4 - 6x^2$

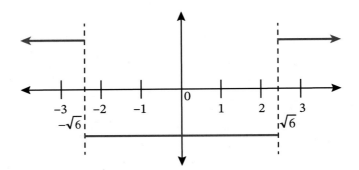

Figure 5.15. A schematic graph of $g(x) = x^4 - 6x^2$

Using just the sign chart for $g(x)$, we can get the (very!) rough sketch shown in Figure 5.15.

Note that the signs do not simply alternate, as in the previous example. A continuous function can only change sign by crossing through zero, but it doesn't have to change sign every time it hits zero. (Think about the functions x^2 and $-x^2$, which hit zero at $x = 0$ but never change sign.) Not every root is an axis crossing. Not every critical point is a local maximum or local minimum. Not every subcritical point is a point of inflection. Most are, but some aren't.

Next we compute

$$(5.74) \qquad g'(x) = 4x^3 - 12x, \qquad g''(x) = 12x^2 - 12.$$

Note how $g(x)$ is related to our previous example. As $g'(x) = 4f(x)$ and $g''(x) = 4f'(x)$, we already have our sign charts for $g'(x)$ (Figure 5.5) and $g''(x)$ (Figure 5.7), which we repeat in Figure 5.16 for convenience. $g'(x)$ is negative for $x < -\sqrt{3}$, positive for $-\sqrt{3} < x < 0$, negative for $0 < x < \sqrt{3}$, and positive for $x > \sqrt{3}$. This means that there must be local minima at $x = \pm\sqrt{3}$ and a local maximum at $x = 0$. At these extreme points, the values of g are $g(\pm\sqrt{3}) = -9$ and $g(0) = 0$.

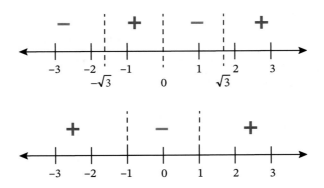

Figure 5.16. The sign charts of $g'(x)$ (top) and $g''(x)$ (bottom)

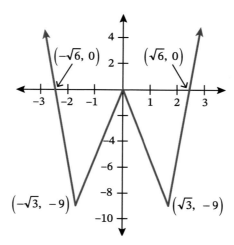

Figure 5.17. A better sketch of the graph of $g(x) = x^4 - 6x^2$

Using the information from the sign chart for $g'(x)$, we can make a better sketch (Figure 5.17) that shows the local maximum and local minima.

Meanwhile, $g''(x)$ is positive when $x < -1$ or $x > 1$ and negative when $-1 < x < 1$. This means that the graph of $g(x)$ is concave up on $(-\infty, -1)$, concave down on $(-1, 1)$, and concave up on $(1, \infty)$. Using the information from the sign chart for $g''(x)$, we can make a better sketch that resembles the true graph of $g(x)$, shown in Figure 5.18.

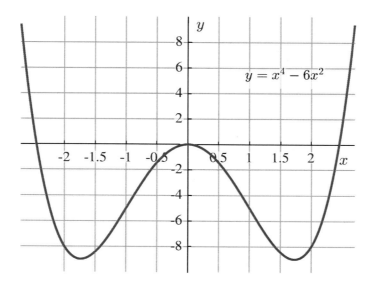

Figure 5.18. The actual graph of $g(x) = x^4 - 6x^2$

5.6. Newton's Method

Graphing More Complicated Functions. A key part of curve sketching was finding the roots, critical points, and subcritical points of $f(x)$. This involved solving the equations $f(x) = 0$, $f'(x) = 0$ and $f''(x) = 0$. In our examples, $f(x)$ and its derivatives were polynomials that were easily factored, which made producing our sign charts easy.

But what do we do when we have to deal with a more complicated function like $f(x) = e^x + \frac{x^2}{2} - 4x + 1$, with $f'(x) = e^x + x - 4$ and $f''(x) = e^x + 1$? In this example, $f''(x)$ is always positive, so $f'(x)$ is always increasing and $f(x)$ is always concave up. But what do the sign charts of f' and f look like?

Because $f'(x)$ is always increasing, there can be at most one critical point of $f(x)$, and because $f'(1) = e - 3 \approx -0.282$ is negative and $f'(2) = e^2 - 2 \approx 5.39$ is positive, that critical point (which is a local minimum by the first derivative test) is somewhere between 1 and 2. But where is it? We could try to home in on it with the bisection method of Section 3.7, but there is a better way, called **Newton's method**. We'll return to this example later in this section, but first let's develop the method in general.

Question 1: Suppose we have a linear function $L(x)$ with multiplier (a.k.a. slope) m, and that $L(a) = b$. Where does this function hit 0?

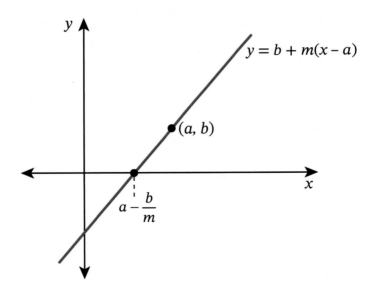

Figure 5.19. The x intercept of a line with slope m through (a, b) is $a - \frac{b}{m}$.

Answer: The equation of the linear function in point-slope form is

$$(5.75) \qquad\qquad L(x) = b + m(x - a).$$

See Figure 5.19. (Note that we are using point-slope form, not slope-intercept form, and that b is not the y-intercept!) Setting this equal to 0 gives

$$
\begin{align}
(5.76) \qquad b + m(x - a) &= 0, \\
m(x - a) &= -b, \\
x - a &= -\frac{b}{m}, \\
x &= a - \frac{b}{m}.
\end{align}
$$

Question 2: Suppose that we have a *nonlinear* function $f(x)$ such that $f(a) = b$ and $f'(a) = m$. Use a linear approximation to estimate where $f(x) = 0$.

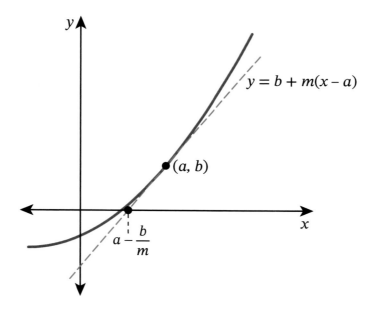

Figure 5.20. A root of a function with $f(a) = b$ and $f'(a) = m$ is approximately $a - \frac{b}{m}$.

Answer: This is the same problem; see Figure 5.20. The best linear approximation to our function near $x = a$ is

$$(5.77) \qquad f(x) \approx b + m(x - a)$$

which is zero at $x = a - \frac{b}{m}$.

Idea behind Newton's Method: If $x_0 = a$ is close to a root of $f(x)$, then

$$(5.78) \qquad a - \frac{f(a)}{f'(a)}$$

is very close to a root of $f(x)$.

The Actual Algorithm for Newton's Method. Newton's method uses this idea recursively to home in amazingly quickly on roots.

- Start with an initial point x_0 that is fairly close to a root. Unfortunately, there is no simple rule for how to pick x_0 or what "fairly close" means.
- Turn this into an improved guess, $x_1 = x_0 - \frac{f(x_0)}{f'(x_0)}$.
- Turn this into an improved guess, $x_2 = x_1 - \frac{f(x_1)}{f'(x_1)}$.
- Repeat as many times as necessary, always taking the new guess x_{n+1} to be

$$(5.79) \qquad\qquad x_{n+1} = x_n - \frac{f(x_n)}{f'(x_n)}.$$

- Stop when $f(x_n)$ is close enough to 0 (e.g., 0 to 15 decimal places).
- Convergence is typically extremely fast, with the number of decimal places doubling every step. For some functions like x^2 (whose derivative is 0 at the root), convergence is slower, but the method still works.
- If the method didn't seem to work, either there isn't a root where we thought there was, or our initial guess x_0 wasn't good enough. We may need to use the bisection method a few times to get a better x_0, and then switch to Newton's method.

Returning to our example, we want to graph the function $f(x) = e^x + \frac{x^2}{2} - 4x + 1$, which requires finding the roots of the function $g(x) = f'(x) = e^x + x - 4$. Here $g'(x) = e^x + 1$. Because e is close to 3, $g(1) = e - 3$ is fairly close to 0, so it is reasonable to pick $x_0 = 1$ as our initial guess. Table 5.6 shows our calculations.

Table 5.6. Solving $g(x) = 0$ with Newton's method

n	x_n	$g(x_n)$	$g'(x_n)$	$x_n - \frac{g(x_n)}{g'(x_n)}$
0	1	-0.281718172	3.71828183	1.07576569
1	1.07576569	0.0008003	3.93223722592	1.0737304832
2	1.0737304832	6.06861×10^{-6}	3.92627558542	1.07372893756
3	1.07372893756	1.37×10^{-11}	3.92627106245	1.07372893756

It only takes three steps to get from our initial guess $x_0 = 1$ to an estimate $x_3 = 1.07372893756$ that is accurate to 11 decimal places.

Now that we have found our root $x_3 = 1.07371893756$ of $g(x)$, which is a local minimum of $f(x)$, we evaluate $f(x_3) = 0.20780222789$. Because the value of $f(x)$ is positive at this minimum, $f(x)$ must be positive everywhere. More precisely, $f'(x)$ is negative for $x < x_3$ and is positive for $x > x_3$, and $f''(x)$ is positive everywhere. This means that the graph of $f(x)$ is roughly ∪-shaped, with a single minimum at $(1.07372893756, 0.20780222789)$, as shown in Figure 5.21.

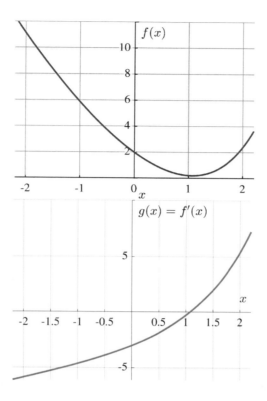

Figure 5.21. The graphs of $f(x) = e^x + \frac{x^2}{2} - 4x + 1$ and $g(x) = f'(x) = e^x + x - 4$

Return to Babylon. Next let's turn our attention to finding square roots. If a is a positive number, then the roots of $f(x) = x^2 - a$ are $\pm\sqrt{a}$. Here we have $f'(x) = 2x$, so the formula for Newton's method becomes

$$
\begin{aligned}
(5.80) \qquad x_{n+1} &= x_n - \frac{f(x_n)}{f'(x_n)} \\
&= x_n - \frac{x_n^2 - a}{2x_n} \\
&= \frac{x_n}{2} + \frac{a}{2x_n} \\
&= \frac{1}{2}\left(x_n + \frac{a}{x_n}\right).
\end{aligned}
$$

For $f(x) = x^2 - a$, Newton's method is *exactly* the same as the Babylonian method. That's why our approximation of $\sqrt{5}$ in Chapter 3 converged so quickly.

As with the Babylonian method, it's easy to program a computer to do all the calculations in Newton's method. Here is a sample MATLAB program that implements Newton's method for $g(x) = e^x + x - 4$ and generates the data in Table 5.6.

```
M=zeros(4,5);              % Initialize
x = 1;                     % Our initial guess
for j=1:4                  % Run the loop
        M(j,1)=j-1         % n runs from 0 to 3
        M(j,2)=x           % Write down xₙ
        M(j,3) = exp(x)+x-4    % That's g(xₙ)
        M(j,4) = exp(x)+1      % That's g'(xₙ)
        M(j,5) = M(j,2)-M(j,3)/M(j,4)   % xₙ – g(xₙ)/g'(xₙ)
        x = M(j,5)         % Set x to its new value
end;                       % Finish the loop
Answer=M                   % Print out the table.
```

When Newton's Method Fails. Newton's method can work brilliantly when three conditions are met:

(1) The function has a root.

(2) The function is differentiable near that root.

(3) Our initial guess is close to that root.

In such cases, our sequence of guesses converges quickly to that root, typically doubling the number of decimal places of accuracy at each step. However, things can fall apart when the conditions aren't met.

First, bad things can come from a bad initial guess, even if the function itself is well behaved. For instance, consider the function $f(x) = xe^{-x^2/2}$, shown in Figure 5.22.

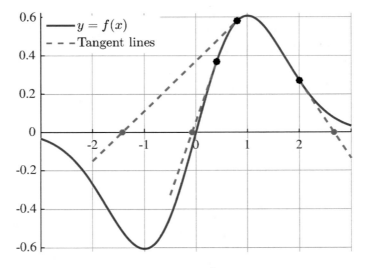

Figure 5.22. Newton's method for $f(x) = xe^{-x^2/2}$ works well starting from $x_0 = 0.4$ but not from $x_0 = 0.8$ or $x_0 = 2$.

This function has a single root at $x = 0$ and critical points at $x = \pm 1$. If we choose x_0 close enough to the origin (anything between $-\sqrt{2}/2$ and $\sqrt{2}/2$ will do), the method works and we quickly converge to 0. This region is called the **basin of attraction** for the root $x = 0$. If $x_0 > 1$, however, the tangent line leads us in the wrong direction. For instance, if $x_0 = 2$, then $x_1 = 8/3 \approx 2.67$. If x is between $\sqrt{2}/2$ and 1, or between -1 and $-\sqrt{2}/2$, then the tangent line takes us in the right direction, but overshoots. For instance, if $x_0 = 0.8$, then $x_1 \approx -1.42$.

Next, Newton's method will often fail if $f(x)$ is not differentiable at the root. For instance, the graph of the function $f(x) = x^{1/3}$ (see Figure 5.23) has a vertical tangent at $x = 0$, so there is no good linear approximation to the function there. In this case, we wind up overshooting the mark no matter where we start, and

$$(5.81) \qquad x_{n+1} = x_n - \frac{x_n^{1/3}}{\frac{1}{3}x^{-2/3}} = x_n - 3x_n = -2x_n.$$

If our first guess is $x_0 = 0.01$, then we'll get $x_1 = -0.02$, $x_3 = 0.04$, $x_4 = -0.08$, and so on.

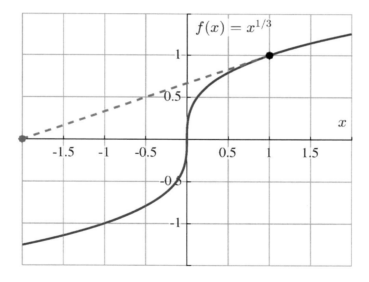

Figure 5.23. No matter where we start, Newton's method for $f(x) = x^{1/3}$ badly overshoots.

Finally, Newton's method can be deceived by points that are almost roots, but not quite. For instance, the graph of $f(x) = x^2 + 0.01$, shown in Figure 5.24, comes close to touching the x axis at $x = 0$. If we start at $x_0 = 1$, the first few estimates are $x_1 = 0.495$, $x_2 = 0.237399$, $x_3 = 0.97579$, and $x_4 = -0.0024510$, and it looks like we're approaching a root. However, at the fifth step we run away. The value of $f(x_4)$ is quite small, but the value of $f'(x_4)$ is even smaller, making $x_5 = x_4 - f(x_4)/f'(x_4) = 2.038758$ big.

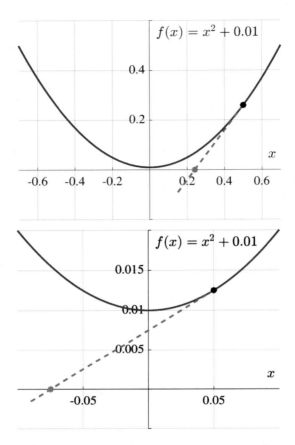

Figure 5.24. When doing one step of Newton's method on $f(x) = x^2 + 0.01$, the extra 0.01 doesn't really matter when x_0 is big, but can matter quite a bit when x_0 is small.

5.7. Chapter Summary

The Main Ideas.

- The derivatives of our basic building block functions are the following:

$f(x)$	$f'(x)$
x^n	nx^{n-1}
$\sin(x)$	$\cos(x)$
$\cos(x)$	$-\sin(x)$
$\tan(x)$	$\sec^2(x)$
a^x	$k_a a^x$
e^x	e^x

- Sums, differences, and scalar multiples of functions follow the obvious rule,

(5.82) $$(af(x) + bg(x))' = af'(x) + bg'(x).$$

- The **product rule** is

(5.83) $$(f(x)g(x))' = f(x)g'(x) + g(x)f'(x).$$

    The derivative is **not** $f'(x)g'(x)$.

- The **quotient rule** is

(5.84) $$\left(\frac{f(x)}{g(x)}\right)' = \frac{f'(x)g(x) - g'(x)f(x)}{(g(x))^2}.$$

- There are three forms of the chain rule. Learn all three, as there are places where each one is the most useful:

(5.85)
$$\frac{dz}{dx} = \frac{dz}{dy}\frac{dy}{dx} \qquad \text{Leibniz's chain rule;}$$

$$(f(g(x)))' = f'(g(x))\, g'(x) \qquad \text{Newton's chain rule;}$$

$$\frac{d}{dx}f(u) = f'(u)\frac{du}{dx} \qquad \text{the hybrid chain rule.}$$

- *What goes up has to stop before it comes down.* Local maxima and minima can only occur at critical points. Almost all optimization problems involve taking the derivative of something and setting it equal to 0.

- First derivative test:
  1. Local maxima occur where $f'(x)$ switches from positive to negative.
  2. Local minima occur where $f'(x)$ switches from negative to positive.

- Second derivative test:
  1. If $f'(x) = 0$ and $f''(x) < 0$, then there is a local maximum at x.
  2. If $f'(x) = 0$ and $f''(x) > 0$, then there is a local minimum at x.
  3. If $f'(x) = 0$ and $f''(x) = 0$, then we can't tell.

- The key to graphing is making sign charts of $f(x)$, $f'(x)$, and $f''(x)$.

- The sign of $f'(x)$ tells whether $f(x)$ is increasing or decreasing. The sign of $f''(x)$ tells whether the graph of $f(x)$ is concave up or concave down.

- Newton's method is the iteration $x_{n+1} = x_n - f(x_n)/f'(x_n)$. It homes in extremely quickly to a root of $f(x)$ if $f(x)$ is differentiable and our initial guess is close enough to that root.

Expectations. You should be able to:

- State from memory the derivatives of the basic functions, the product and quotient rules, and all three versions of the chain rule.

- Use the rules to analyze the derivative of a complicated function, breaking it down step by step until all that is left are derivatives of building blocks.

- Solve related rates problems using the chain rule.

- Optimize a function by finding its critical points and applying the first and second derivative tests.

- Use sign charts for $f(x)$, $f'(x)$ and $f''(x)$ to sketch the graph of a function, correctly indicating
  - regions where the function is positive,
  - regions where the function is negative,
  - regions where the function is increasing,
  - regions where the function is decreasing,
  - regions where the graph is concave up,
  - regions where the graph is concave down, and
  - the locations of roots, local maxima, local minima, and points of inflection.
- Use Newton's method to find roots of a complicated functions.
- Write (or adapt) a computer program to automate that computation.

5.8. Supplemental Material: Small Angle Approximations

In Section 5.1, we computed the limits

$$(5.86) \qquad \lim_{h \to 0} \frac{\sin(h)}{h} \quad \text{and} \quad \lim_{h \to 0} \frac{\cos(h) - 1}{h}$$

numerically, finding that the first limit is 1 and the second limit is 0. Here we'll derive these limits rigorously, and in the process we'll develop two very useful approximations,

$$(5.87) \qquad \sin(x) \approx x \quad \text{and} \quad \cos(x) \approx 1 - \frac{x^2}{2},$$

that apply whenever the angle x is small (up to half a radian or so, or just under 30 degrees). These are called **small angle approximations**.

Suppose that h is an angle between 0 and $\pi/2$. If we take a wedge W of angle h from the unit circle the area of W is a fraction $h/(2\pi)$ of the area of the entire circle and is therefore equal to $\frac{h}{2\pi}\pi = \frac{h}{2}$. We then consider a triangle T_1 inscribed in W and a triangle T_2 that contains W, as in Figure 5.25.

The triangle T_1 has a base of $\cos(h)$ and a height of $\sin(h)$, so it has an area of $\sin(h)\cos(h)/2$. The triangle T_2 has a base of 1 and a height of $\tan(h) = \frac{\sin(h)}{\cos(h)}$, so it has an area of $\frac{\sin(h)}{2\cos(h)}$. As T_1 is contained in W, and W is contained in T_2, we have

$$(5.88) \qquad \text{Area of } T_1 \quad < \quad \text{Area of } W \quad < \quad \text{Area of } T_2,$$
$$\frac{\sin(h)\cos(h)}{2} \quad < \quad \frac{h}{2} \quad < \quad \frac{\sin(h)}{2\cos(h)}.$$

Multiplying by $2/\sin(h)$ then gives

$$(5.89) \qquad \cos(h) < \frac{h}{\sin(h)} < \frac{1}{\cos(h)}.$$

Taking reciprocals switches the directions of the inequalities, giving us

$$(5.90) \qquad \frac{1}{\cos(h)} > \frac{\sin(h)}{h} > \cos(h).$$

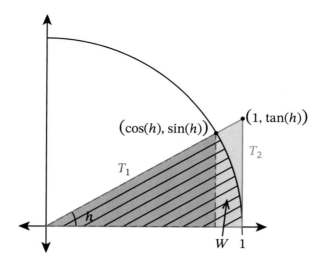

Figure 5.25. Two triangles and a slice of pie

As h gets closer and closer to 0, $\cos(h)$ and $\frac{1}{\cos(h)}$ both get closer and closer to 1. As $\frac{\sin(h)}{h}$ is trapped (or **squeezed** or **sandwiched**) between two numbers, both of which are close to 1, $\frac{\sin(h)}{h}$ must itself be close to 1.

So far we've only considered positive angles. To get our limit we have to also consider the possibility that $h < 0$. So suppose that $h = -\theta$, where θ is a small positive angle. Then

$$(5.91) \qquad \frac{\sin(h)}{h} = \frac{\sin(-\theta)}{-\theta} = \frac{-\sin(\theta)}{-\theta} = \frac{\sin(\theta)}{\theta} \approx 1.$$

Since $\frac{\sin(h)}{h}$ is close to 1 for all values of h that are close to 0, we have established that

$$(5.92) \qquad \lim_{h \to 0} \frac{\sin(h)}{h} = 1.$$

This also gives the first half of our small angle approximation. If x is small, $\frac{\sin(x)}{x}$ is close to 1, so $\sin(x) \approx x$.

To understand limits involving $\cos(h) - 1$, we first notice that

$$(5.93) \qquad \cos(h) - 1 = \frac{(\cos(h) - 1)(\cos(h) + 1)}{\cos(h) + 1} = \frac{\cos^2(h) - 1}{\cos(h) + 1} = \frac{-\sin^2(h)}{\cos(h) + 1}.$$

We then have

$$(5.94) \qquad \begin{aligned} \frac{\cos(h) - 1}{h} &= \frac{-\sin^2(h)}{h(\cos(h) + 1)} \\ &= -\left(\frac{\sin(h)}{h}\right)^2 \frac{h}{\cos(h) + 1}. \end{aligned}$$

As $h \to 0$, $\frac{\sin(h)}{h}$ goes to 1 (as we just proved), while $\frac{h}{\cos(h)+1}$ goes to 0, so the product goes to zero.

In fact, we can say more. By the exact same argument,

$$\frac{\cos(h) - 1}{h^2} = -\left(\frac{\sin(h)}{h}\right)^2 \frac{1}{\cos(h) + 1}.$$

As $h \to 0$, the right-hand side approaches $-(1)^2 \frac{1}{2} = -\frac{1}{2}$. Because $(\cos(h) - 1)/h^2 \approx -1/2$, we must have $\cos(h) - 1 \approx -h^2/2$, so $\cos(h) \approx 1 - \frac{h^2}{2}$. Replacing h with x, this is the second half of our small angle approximation.

5.9. Exercises

Derivative Rules

5.1. Find the derivative of each function.
  (a) $g(x) = 3x^5 - 2x + 5$
  (b) $f(x) = 3\sin(x) - 5e^x$
  (c) $h(x) = \tan(x) - \pi x^3 + ex$
  (d) $g(t) = 8\sqrt[3]{t^2} + \cos(t) - 6t + \pi^2$
  (e) $f(z) = 5^z + \sin(z) - \cos(z) + \sqrt{3}$
  (f) $h(r) = 5r^\pi - r^{4/3} + 5/2$

5.2. Use the product or quotient rule to find the derivative of each function.
  (a) $f(x) = x^4 \sin(x)$
  (b) $g(x) = 3x^3 e^x$
  (c) $h(x) = 3^x \tan(x)$
  (d) $f(x) = x^2 \cos(x) \sin(x)$
  (e) $g(s) = \sqrt[4]{s^7} \tan(s)$
  (f) $h(t) = \dfrac{t^4}{\tan(t)}$
  (g) $f(x) = \dfrac{\sin(x)}{(2x^4 - 1)}$
  (h) $g(t) = \dfrac{t^4 e^t}{5(t^4 - 2t)}$
  (i) $j(r) = \dfrac{3^r}{2\cos(r)\tan(r)}$
  (j) $f(x) = \dfrac{\tan(x)}{e^x(3x^{5/2} - 2x)}$

5.3. (a) Extend the product rule to express $(f(x)g(x)h(x))'$ in terms of $f(x)$, $g(x)$, $h(x)$, and their derivatives. (*Hint*: Think of $f(x)g(x)h(x)$ as $(f(x)g(x)) \times h(x)$ and apply the ordinary product rule twice.)
  (b) Explain what each term in your formula represents.
  (c) Suppose that $\ell(t)$, $w(t)$, and $h(t)$ are the length, width, and height of a box whose dimensions keep changing. At what rate would the volume of the box be changing if $\ell(t)$ changed but $w(t)$ and $h(t)$ were constant? If $w(t)$ changed but $\ell(t)$ and $h(t)$ were constant? If $h(t)$ changed but $\ell(t)$ and $w(t)$ were constant? If all three dimensions of the box changed?

5.4. Rewrite each of the following expressions in the form $f(u)$, where u is a function of x. (*Example:* $\sin^2(x) = u^2$, where $u = \sin(x)$.)

    (a) e^{x^2+1}

    (b) $\cos^3(x)$

    (c) $\sin^2(x) + 3\sin(x) + 7$

    (d) $\dfrac{e^x}{e^x + 1}$

5.5. Use the chain rule to find the derivative of each function with respect to x.

    (a) $9(x^{5/2} - 3x)^{30}$

    (b) $\cos^4(3x^2 - 5)$

    (c) $\sqrt[3]{\tan(x) - e^x}$

    (d) $9e^{x^4 + \pi x}$

    (e) $\sin(5x^2 + 4)$

    (f) $h(x) = (\sin^2(x) - 2.3x)^{7/4}$

5.6. Find the derivative of each function with respect to x or t. Indicate which rule(s) you are using.

    (a) $4x^4 \tan(x^3)$

    (b) $\dfrac{3^{x^2}}{\cos(4x)}$

    (c) $\sqrt{e^{3x} + \pi x^2}$

    (d) $\dfrac{\sin^2(x)}{(5x^3 - 3x + 5)^3}$

    (e) $\cos^4(2\sqrt{t} - 5)$

    (f) $(3t^{5/3} - 6t^{2/3} + 8t^{1/3})^5$

    (g) $e^{x^2}(3x^5 - 2x)^8 \tan^2(x)$

    (h) $\dfrac{3^{t^2}}{\cos(4t)}$

5.7. Let $h(x) = (2g(x))^3$, where $g(x)$ is a function with $g(5) = -6$ and $g'(5) = -10$. Find $h'(5)$.

5.8. Let $f(x) = g(2xh(x))$, where $g'(42) = -2$, $h'(3) = -1$, and $h(3) = 7$. Find $f'(3)$.

5.9. Let $h(t) = \tan(\cos(t))$. Find $h'(\pi/2)$ and $h'(3\pi/2)$.

5.10. Let $f(x)$ and $h(x)$ be functions satisfying

$$f(3) = 6, \qquad h(3) = 5, \qquad h(5) = 9,$$
$$f'(3) = 9, \qquad f'(5) = 10, \qquad h'(3) = h'(6) = 2.$$

Compute the derivative of each function with respect to x at $x = 3$.

    (a) $f(x) - h(x)$

    (b) $(f(x))^3$

    (c) $(h(x) - f(x))^2$

    (d) $f(h(x))$

    (e) $h(f(x))$

    (f) $f(x)h(x)$

    (g) $f(\tfrac{8}{9}x^2 - h(x))$

Derivatives and Microscopes

5.11.   Find the slope of the line tangent to the curve $y = 2x^2 - 4x$ at $x = 1$.

5.12.   Determine if $h(x) = \cos^2(x)$ is increasing or decreasing at $x = 3\pi/4$. At what rate is it changing?

5.13.   Consider the function $f(x) = \sin(2x) + 4$.
   (a) Find the equation of the line tangent to $y = f(x)$ at $x = 0$.
   (b) Write the microscope equation (linear approximation) for $f(x)$ near $x = 0$.
   (c) Use this linear approximation to estimate $g(0.05)$ and $g(0.1)$. Which of these approximations do you think is more accurate, and why?

5.14.   Let $f(x) = \sqrt[3]{x}$.
   (a) Write the microscope equation (linear approximation) for $f(x)$ near $x = 8$.
   (b) Use the microscope equation to estimate $\sqrt[3]{8.1}$ and $\sqrt[3]{7.8}$.
   (c) Evaluate $\sqrt[3]{8.1}$ and $\sqrt[3]{7.8}$ with a calculator. Which cube root was approximated more accurately, and by (approximately) what factor?

5.15.   Use the microscope equation to estimate $\sqrt{403}$ and $\sqrt[3]{1001}$.

5.16.   **Maximum Value.** Let $f(x) = \dfrac{1}{2x^2 - 3x + 6}$.
   (a) Using MATLAB, graph $f(x)$ on the interval $[-3, 3]$. By zooming in on the graph, find the point where $f(x)$ is maximized to two decimal place accuracy.
   (b) Find the derivative of $f(x)$.
   (c) Using part (b), find the exact value of x where $f(x)$ reaches its maximum value.
   (d) At what point is the graph of $f(x)$ rising most steeply? Describe how you determined the location of this point. (*Hint*: Graphing $f'(x)$ might give valuable insight.)

Exercises 5.17 and 5.18 are about **per capita** quantities, meaning how much there is **per person**. For instance, if a country with a population of 50 million has a gross domestic product (GDP) of $1 trillion, then its per capita GDP is $1 trillion/50 million = $20,000.

5.17.   The population of a U.S. state is currently 10,000,000 people and is shrinking by 10,000 people per year. In this state, the per capita yearly spending on energy is $2000 per person and is growing at a rate of $60 per year. Currently, how much is the whole state paying for energy? Is that total yearly energy expenditure growing or shrinking? How fast?

5.18.   In the same state as in Exercise 5.17, the total yearly personal income is $500 billion and is falling by $250 million per year. What is the current per capita personal income? Is this per capita income rising or falling? By how much?

Position and Velocity

5.19.   A ball is dropped at time $t = 0$ from the top of a tall building. As it falls, its height (in feet) from the ground is given by the formula $h(t) = 400 - 16t^2$, where t is measured in seconds.

    (a) Find a formula for the velocity $v(t) = h'(t)$ of the ball after t seconds.

    (b) Graph, on the same set of axes, the velocity and the height as functions of time. Your legend should include the different units for h and v.

    (c) At what time does the ball hit the ground? What is its velocity then?

5.20.   A mass hanging from a spring is oscillating up and down. Its height in inches, measured from a particular reference point, is $z(t) = 5\cos(t) + 20$, where t is measured in seconds.

    (a) Find a formula for the velocity $v(t)$ of the ball.

    (b) What is the mass doing when $v(t) < 0$? What is it doing when $v(t) > 0$?

    (c) In what two situations is $v(t) = 0$?

    (d) What is the maximum height of the mass?

    (e) When and where does the mass reach its maximum velocity?

Circumference, Area, and Volume

5.21.   The circumference of a circle of radius r is $C(r) = 2\pi r$.

    (a) If r is measured in meters, what are the units of C and C'?

    (b) Compute the derivative $C'(r)$. Does it have the correct units, according to part (a)?

    (c) How much does the circumference of a circle change if its radius increases from 100 meters to 101 meters?

    (d) Imagine that the Earth is a perfect sphere, and that a giant belt about 24,000 miles long is placed along the equator. How much fabric would have to be added to the belt to raise it one inch (everywhere) above the ground?

5.22.   The area of a circle of radius r is $A(r) = \pi r^2$.

    (a) If r is measured in meters, what are the units of A and A'?

    (b) Compute the derivative $A'(r)$. Does it have the correct units, according to part (a)?

    (c) Imagine a circular public park of radius 100 meters. The city decides to put a sidewalk, 1 meter wide, around the outer edge of the park. Using a linear approximation for area near $r = 100$ meters, estimate the area of the sidewalk.

    (d) The area of the sidewalk is also approximately the same as the width of the sidewalk times the length of the sidewalk. (This is only approximate as the the outside and inside of the sidewalk have slightly different lengths, as we saw in Exercise 5.21.) What is the approximate length of the sidewalk?

    (e) Repeat parts (c) and (d) for a circle of radius r and a sidewalk of width Δr. What does this tell us about how the area and circumference of a circle are related?

5.23.   The volume of a sphere of radius r is $V(r) = \frac{4}{3}\pi r^3$.

    (a) If r is measured in meters, what are the units of V and V'?

    (b) Imagine a sphere of radius 3 meters that we want to paint with a layer of paint that is 1 mm thick. The painted sphere will then have a radius of 3.001 meters. Using the microscope equation for V near $r = 3$ meters, estimate the volume of paint needed to do the job.

(c) The amount of paint needed is approximately the surface area of the sphere times the thickness of the paint. What is the approximate surface area of the sphere?

(d) Repeat parts (b) and (c) for a sphere of arbitrary radius r and paint of thickness Δr to derive a formula for the surface area of a sphere of radius r.

5.24.   If the radius of a spherical balloon is r inches, its volume is $\frac{4}{3}\pi r^3$ cubic inches.

(a) The radius of the balloon is changing. What is the change in the balloon's volume per change in its radius when the radius is 3 inches?

(b) When the radius is 3 inches, approximately how much does the radius have to increase to increase the volume by 25 cubic inches?

(c) Suppose that someone is inflating the balloon at the rate of 19 cubic inches of air per second. When the radius is 9 inches, at what rate is the radius increasing in inches per second?

Digestion. In 1986, Deborah Penry and Peter Jumars used ideas from chemical reactor theory to explore optimal digestion strategies[7]. In some animals, digestion can be modeled as a chemical reactor of volume V, where food enters the gut and is broken down into nutrient products, which are then absorbed across the gut lining. We assume the mixture in the gut is perfectly stirred and exits the gut at the same rate as it enters. Let S_f be the **concentration** (mass/volume) of a substrate (such as starch) in a food being consumed. It enters the gut, where the concentration of the substrate is $S(t)$, and it reacts with an enzyme (such as amylase) to produce a nutrient product (such as glucose) at concentration $P(t)$. To create a simple model, we make the following assumptions:

- In the gut, the rate of substrate breakdown into the nutrient product is given by kVS, where k is a rate constant and V is the (constant) volume of the gut. Note that, as the substrate is converted to product, $S(t)$ decreases and $P(t)$ increases.

- The nutrient product is absorbed across the gut wall at a rate of aVP, where a is the absorption constant.

- At all times, food enters the gut at flow rate q.

- The contents of the gut are thoroughly mixed and exit at the same flow rate q that food enters the gut.

5.25.   (a) The dependent variables S and P are measured in mass per volume (as they are concentrations), while the independent variable t is time. In what units are the parameters k, a, q, V, and S_f measured?

(b) Explain why the rate equations for the process described are

$$V\frac{dS}{dt} = qS_f - qS - kVS,$$
$$V\frac{dP}{dt} = kVS - aVP - qP.$$

(*Hint*: Think about your answers from part (a).)

(c) Suppose the organism continuously eats food containing the same initial concentration of the substrate, and the concentrations of S and P in the

[7]Penry & Jumars (1986), *Chemical Reactor Analysis and Optimal Digestion*, Bioscience Vol. 36 No. 5. Exercises 5.25 and 5.26 are inspired by material in "A First Course in Differential Equations", by J. David Logan, Springer, 2006.

gut become constant. When this occurs, what are the values of dS/dt and dP/dt? (When this occurs, we say the system is in a **steady state**).

(d) Show that the steady-state concentrations are given by

$$S_e = \frac{qS_f}{q + kV}$$

and

$$P_e = kVS_f \frac{q}{(q + kV)(aV + q)}.$$

5.26.   In Exercise 5.25, we found that the equilibrium solutions for the digestion model were given by

$$S_e = \frac{qS_f}{q + kV}$$

and

$$P_e = kVS_f \frac{q}{(q + kV)(aV + q)}.$$

(a) Many animals are able to control the flow rate q. For these animals P_e can be thought of as a function of q. In this case, show that the maximum equilibrium concentration P_e occurs when $q = \sqrt{akV}$.

(b) In creating the digestion model, we used aVP to model the **absorption rate** of the nutrient P across the gut wall. Some ecologists believe that animals control the flow rate q in order to maximize the absorption rate of nutrients. Use your result from part (a) to show that the maximum absorption rate of P is

$$\frac{akS_f V}{\left(\sqrt{a} + \sqrt{k}\right)^2}.$$

5.27.   **Scaling Time, Revisited.** In Chapter 2, Exercises 2.18–2.20, we considered what happens when we rescale the time variable in a rate equation. We picked a length of time T and defined the dimensionless parameter $\tau = t/T$. Use the chain rule to show that, for any quantity P,

$$\frac{dP}{d\tau} = T\frac{dP}{dt}.$$

Territory Defense. Many animals defend a territorial range from other members of their own (and other) species in order to make sure they have adequate resources to ensure their own survival. They gain energy by consuming the resources within their defended territory. We model territory defense by assuming the defended territory is circular with radius r miles and add two more assumptions:

- The energy $G(r)$ gained per day from consuming food from the territory is directly proportional to the *radius* of the territory being defended.

- Defending a territory comes with an energy cost to the animal. The energy $E(r)$ spent foraging for food and defending the territory per day is directly proportional to the *area* of the region being defended.

5.28.   (a) Discuss whether modeling the territorial range as circular, and making the above assumptions, is reasonable.

(b) Suppose that $E(1) = 2500$ calories and $G(1) = 3000$ calories for a certain animal. Find the territorial size this animal should defend in order to maximize the energy gain $G - E$.[8]

5.29. The graph below gives the total prey encountered per day, $e = P(x_m)$, for a certain predator, where x_m is metabolic energy (perhaps measured in calories/day) spent foraging for prey.

(a) Interpret the statement $P(500) = 10$.

(b) Sketch a graph of $P'(x_m)$. This is called the **marginal encounter rate**. It tells us how many extra prey the predator can catch per extra unit of energy it puts into predation.

(c) Describe in words how the marginal encounter rate changes as the amount of energy spent on foraging increases.

(d) Sketch a graph of $P''(x_m)$. Explain the foraging significance of the point of inflection of the original function $P(x_m)$. (*Hint*: look up "diminishing returns".)

Derivatives and Graphs

5.30. Let $f(x) = x^3 + 3x$.
(a) Make a sign chart for $f(x)$.
(b) Compute $f'(x)$ and make a sign chart for $f'(x)$.
(c) Compute $f''(x)$ and make a sign chart for $f''(x)$.
(d) Sketch the graph of $f(x)$ between $x = -2$ and $x = 2$.

5.31. Let $f(x) = x^4 - 4x^2 + 3$.
(a) Make a sign chart for $f(x)$.
(b) Compute $f'(x)$ and make a sign chart for $f'(x)$.
(c) Compute $f''(x)$ and make a sign chart for $f''(x)$.
(d) Sketch the graph of $f(x)$ between $x = -2.5$ and $x = 2.5$.

5.32. Let $f(x) = xe^{-x}$.
(a) Make a sign chart for $f(x)$.
(b) Compute $f'(x)$ and make a sign chart for $f'(x)$.
(c) Compute $f''(x)$ and make a sign chart for $f''(x)$.
(d) Sketch the graph of $f(x)$ between $x = -1$ and $x = 4$.

---

[8]This idea was originally explored by M. R. Cullen in *Mathematics for the Biosciences* (1983).

5.33.   Let $f(x) = e^{-x^2}$. This function is always positive, so we don't have to bother with a sign chart for $f(x)$.
   (a) Compute $f'(x)$ and make a sign chart for $f'(x)$.
   (b) Compute $f''(x)$ and make a sign chart for $f''(x)$.
   (c) Sketch the graph of $f(x)$ between $x = -2$ and $x = 2$.

5.34.   Answer the following questions for each graph.
   • At the indicated point x^*, what is the sign ($+$, $-$, or 0) of f'?
   • Is f' (not f) increasing or decreasing at x^*?
   • What is the sign of f'' (the derivative of f') at x^*?

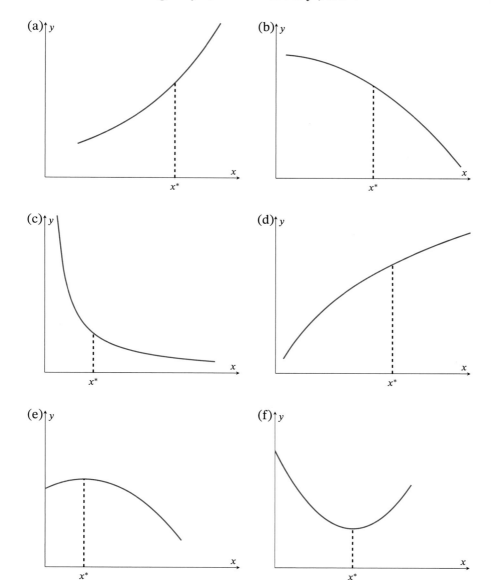

5.35.   Sketch the graph of a function $f(x)$ that satisfies each set of conditions. (There are many such functions to choose from.) On each graph, mark any critical points or extrema.

(a) $f'(x) < 0$ for $x < 0$; $f'(0) = 0$; $f'(0) > 0$ for $0 < x < 3$;
    $f'(3) = 0$; $f'(x) < 0$ for $x > 3$.

(b) $f'(x) < 0$ for $x < 1$; $f'(1) = 0$; $f'(x) > 0$ for $x > 1$.

(c) $f'(-2) = 0$; $f''(-2) < 0$; $f'(2) = 0$; $f''(2) > 0$.

(d) $f(5/2) = 1$; $f'(x) > 0$ everywhere; $f''(x) > 0$ when $x < 5/2$;
    $f''(x) < 0$ when $x > 5/2$.

5.36.   Copy each graph and mark any critical points or extrema. Indicate which extrema are local and which are global. (Assume that at their ends the curves continue in the direction they are headed.)

(a)

(b)

(c)

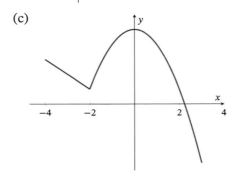

5.37. Consider the following graph of a function $f(x)$.

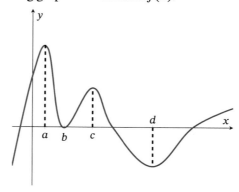

(a) Is the second derivative greater at point b or point d? Explain.

(b) On a copy of this curve, indicate where the curve is concave up and where it is concave down.

(c) Sketch a graph of the derivative $f'(x)$. Label the points corresponding to a through d on your sketch.

(d) Sketch a graph of the second derivative $f''(x)$. Label the points corresponding to a through d on your sketch.

Maxima and Minima

5.38. At what positive value of x does $f(x) = \frac{1}{4}x + \frac{1}{x^2}$ attain its minimum value? Explain how you found this value. (You should *not* use MATLAB or a graphing calculator to do this problem.)

5.39. What is the minimum value of the function $f(x) = \frac{1}{4}x + \frac{1}{x^2}$ on the interval $[3, 5]$? Where does this minimum occur? Explain how you found this value of x. (You should *not* use MATLAB or a graphing calculator to do this problem.)

Newton's Method

5.40. Newton used the example $x^3 - 2x - 5 = 0$ to introduce his method for finding roots.

(a) Explain why there must be a root somewhere between $x = 2$ and $x = 3$.

(b) Starting at $x_0 = 2$, find this root to an accuracy of at least four decimal places.

5.41. Use Newton's method to find a solution of $x^3 + 2x^2 = 6$ near the point $x = 1$, accurate to at least four decimal places.

5.42. Consider the equation $x^4 - 8x^2 + 5 = 0$.

(a) Starting at $x_0 = \frac{1}{2}$, find a root to an accuracy of at least four decimal places.

(b) Starting at $x_0 = -\frac{1}{2}$, find a different root to an accuracy of at least four decimal places.

(c) Find the two other roots, again to at least four decimal places, by starting at $x_0 = 3$ and $x_0 = -3$.

(d) What happens if you start at $x_0 = 0$, at $x_0 = 2$, or at $x_0 = -2$? Why are these really bad initial guesses?

Food for Thought: While there is no general formula for the roots of a polynomial of fifth order or higher, and the formulas for third and fourth-order polynomials are too complicated to be useful, we can sometimes solve higher-order polynomial equations by other means. For instance, the equation $x^4 - 8x^2 + 5$ can be solved exactly by defining $y = x^2$, which turns the equation into $y^2 - 8y + 5 = 0$. The quadratic formula then tells us that $y = 4 \pm \sqrt{11}$, so $x = \pm\sqrt{4 \pm \sqrt{11}}$. You should check that these are in fact the four roots that you approximated in Exercise 5.42.

Chapter 6

Models of Growth and Oscillation

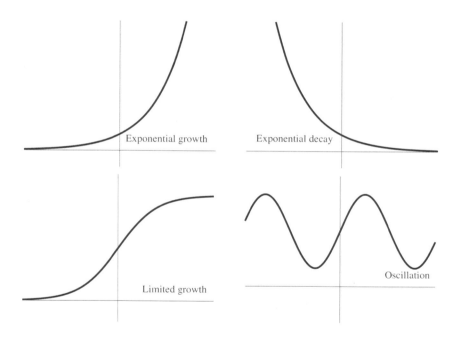

Figure 6.1. Four ways that a quantity can evolve

Figure 6.1 shows four kinds of graphs that you may see when studying real-world phenomena. The top left graph might represent the number of bacteria on a Petri dish, the value of a bank account, the number of infected people at the beginning of an epidemic, or the population of an invasive species. The top right graph might represent a quantity of radioactive material, the amount of pollution in a lake that is gradually flushing itself clean, or the population of a native species that is being wiped out by an invasive species. The bottom left graph might represent the later stages of the top left picture. The population of a bacteria colony, or of an invasive species, can't keep growing forever. Eventually it levels out. Finally, the bottom right graph could represent any quantity that goes in a repeating cycle, from the motion of a wheel to a pendulum to the seasons to the business cycle. These four patterns appear over and over again in the real world. In this chapter we will study how and why that happens.

The world operates according to laws, but these laws typically don't tell us how *much* of something there will be at time t. Instead, these laws tell us how fast that quantity (or those quantities) are *changing*. Population biology doesn't determine how many rabbits are in a colony, but it does say how the population changes with time. The law of gravity doesn't determine the velocity of a falling object, but it does say how fast that velocity is changing. Your bank's policies don't determine how much money is in your bank account, but they do determine how fast your balance is increasing.

In other words, most of the laws that govern the natural world are expressed mathematically as **differential equations**. These are equations that relate the derivative (usually with respect to time) of one or more quantities to the current levels of those quantities and to the time.[1] For instance, if we invest a certain amount of money at 6% annual interest (compounded continuously), then our balance $y(t)$ follows the differential equation

(6.1)
$$\frac{dy}{dt} = 0.06y.$$

We have also already seen that the numbers of Susceptible, Infected and Recovered individuals in an epidemic follow the SIR **system** of differential equations

(6.2)
$$\frac{dS}{dt} = -aSI,$$
$$\frac{dI}{dt} = aSI - bI,$$
$$\frac{dR}{dt} = bI.$$

It turns out that very many phenomena are described, at least approximately, by a few very simple differential equations. The solutions to these differential equations then give the pictures in Figure 6.1.

In this chapter,

- We will see how to take the laws of the world (described in words) and express them as differential equations. This is the essence of modeling. Most of the time, as with modeling the spread of disease, the resulting differential equations will be too complicated for us to find a formula for the solution.

---

[1]Before we learned what derivatives were, we called such things *rate* equations, but almost everybody calls them *differential* equations, and we're going to follow suit.

- We will learn about the simplest differential equations, so simple that we *can* solve them exactly.

(6.3)
$$\frac{dy}{dt} = ry; \qquad \text{(exponential growth)}$$
$$\frac{dy}{dt} = -ry; \qquad \text{(exponential decay)}$$
$$\frac{dy}{dt} = rx; \qquad \frac{dx}{dt} = -ry; \qquad \text{(circular motion)}.$$

- We will learn about (or review) the kinds of functions that show up as solutions to these equations. These include exponentials, logs and trigonometric (trig) functions. Exponentials and logs are explained in Section 6.2. Trig functions are reviewed in the supplemental material at the end of the chapter, titled "A Crash Course in Trigonometry".

- We will learn about more realistic models that take more features into account. In particular, in Section 6.5 we will study the **logistic equation**

(6.4)
$$\frac{dy}{dt} = ry\left(1 - \frac{y}{L}\right)$$

that describes the growth of a single population with limited resources, as well as the **Lotka-Volterra** equations that describe the interaction of predators with prey. Here the emphasis isn't on finding a formula for the solution, but on understanding what is happening qualitatively.

- We will learn how to model a number of other phenomena, such as symbiosis and chemical reactions.

6.1. Modeling with Differential Equations

There are several ingredients to a model involving differential equations.

(1) The **differential equation** itself describes the rules of the game.

(2) The **initial conditions** describe our starting point. A differential equation together with an initial condition is called an **initial value problem**, or IVP.

(3) A **solution** to the IVP is a function whose derivative follows the rules established by the differential equation and whose starting value matches the initial condition.

(4) A **numerical solution** to the IVP. Most of the time we can't find a formula for the solution to the IVP, but that's OK. As long as we can approximate it on a computer, we can make useful predictions.

(5) **General solutions** to a differential equation are families of functions that solve a differential equation, involving parameters that can be adjusted to match whatever initial conditions we want. If we can find a general solution (and that's a big if!), then we can understand what happens without relying on a computer.

Let's go over these ingredients, one at a time.

The Rules of the Game.

Fish populations. If we are tracking the population of trout in a lake, what are the rules of the game? Under normal conditions, the number of fish that are born each year is proportional to the number of fish that are already there. Likewise, the number of fish who die is proportional to the current population, so *the rate at which the population is changing is proportional to the current population.* This proportionality is expressed as a **rate** r, which might vary from lake to lake. Like the coefficients a and b in the SIR model, r is a **parameter** of the system. We can summarize the rule in two ways:

The population grows at a rate r times the current population.

(6.5)
$$\frac{dy}{dt} = ry.$$

The differential equation and the sentence above it mean *exactly* the same thing! The only difference is that one is written in words and the other is written with a formula. dy/dt is a very efficient way of saying "the rate at which the population is growing" and ry is an efficient way of saying "the rate r times the current population".

Now suppose that fishermen catch fish at a rate of F/year. Then the rules of the game would be different, and our differential equation would be different. The rules would then be

Births minus natural deaths occur at a rate proportional to the population (ry) and fish are harvested at a constant rate (F).

(6.6)
$$\frac{dy}{dt} = ry - F.$$

Each term in the differential equation corresponds to a feature of the rule. Extra features mean extra terms.

Some lakes have rules that are more complicated, in which case our differential equation will be more complicated. For instance, suppose that fishing is only allowed when the population of the lake is over 50,000. Our new rule, in both words and equations, is

If $y > 50,000$, then fish are created at rate ry and removed at rate F.
If $y \leq 50,000$, then fish are only created at rate ry.

(6.7)
$$\frac{dy}{dt} = \begin{cases} ry - F & \text{if } y > 50,000, \\ ry & \text{if } y \leq 50,000. \end{cases}$$

Beware of falling rocks. Next let's consider the rules of a completely different game: a rock that is falling to the ground. We don't know the rock's velocity $v(t)$, but (ignoring air resistance) the law of gravity tells us how fast that velocity is changing:

(6.8)
$$\frac{dv}{dt} = -32\frac{\text{feet}}{\text{second}^2}.$$

Because it ignores air resistance, this differential equation does not describe reality exactly. It is only an approximation, but as we already know, *close is good enough*. As long as the rock is reasonably heavy and isn't moving too fast, equation (6.8) is accurate. If we are talking about a falling feather instead of a rock, or if the rock is a meteorite moving at thousands of feet per second, then we have to add an additional term to the equation to account for air resistance.

You may have noticed that $v(t)$ is itself a rate of change. It is the derivative of the height $z(t)$ of our rock above the ground. The law of gravity tells us about the *second* derivative of z. That is,

$$(6.9) \qquad \frac{d^2 z}{dt^2} = -32 \frac{\text{feet}}{\text{second}^2}.$$

Equation (6.9) is called a **second-order** differential equation, because it describes a second derivative, while (6.8) is called a **first-order** differential equation. Second-order differential equations come up a lot in physics, where Newton's 2nd Law of Motion, $F = ma$, relates the acceleration a of an object (that is, the second derivative of the position) to the force F applied to that object. However, every second-order system of equations can be converted to a first-order system of equations by defining additional variables such as the velocity. In the case of a falling rock, the second-order differential equation (6.9) is equivalent to the system of two first-order equations

$$(6.10) \qquad \begin{aligned} \frac{dz}{dt} &= v, \\ \frac{dv}{dt} &= -32 \text{ ft/s}^2. \end{aligned}$$

In this chapter we will mostly concentrate on first-order equations of the form

$$(6.11) \qquad \frac{dy}{dt} = f(y, t),$$

for a single function $y(t)$, and on sets of differential equations of the form

$$(6.12) \qquad \begin{aligned} \frac{dy_1}{dt} &= f_1(y_1, \ldots, y_n, t), \\ \frac{dy_2}{dt} &= f_2(y_1, \ldots, y_n, t), \\ &\vdots \\ \frac{dy_n}{dt} &= f_n(y_1, \ldots, y_n, t), \end{aligned}$$

involving a larger set $\{y_1(t), \ldots, y_n(t)\}$ of functions. In our examples, $f(y, t)$ and $f_j(y_1, \ldots, y_n, t)$ will be fixed expressions such as $0.05y - 100$ or $y_1^2 - y_2^2 + 3t$.

Initial Conditions and Initial Value Problems. By themselves, the rules of the game don't determine the outcome. Imagine three nearly identical lakes. In all three lakes, the fish naturally grow at a rate $r = 0.25$/year, that is 25% per year. Fishing is allowed at all three lakes, and fishermen catch 2500 fish/year at each one. However, the initial fish population is 5000 in the first lake, 10,000 in the second lake, and 20,000 in the third lake. The populations of all three lakes are described by the same differential equation:

$$(6.13) \qquad \frac{dy}{dt} = 0.25y - 2500.$$

The population in the first lake drops, as the natural increase of 1250 fish/year isn't enough to cover the 2500 fish/year that are removed. The population stays steady in the second lake, as the natural increase of 2500 exactly matches the rate of fishing. Meanwhile, the population of the third lake steadily grows. See Figure 6.2.

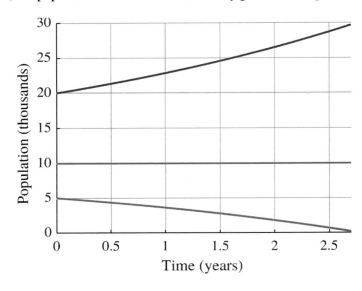

Figure 6.2. A tale of three lakes. Solutions to the differential equation (6.13)

In other words, the result of our model depends not just on the differential equation itself, but also on the starting point

$$(6.14) \qquad y(0) = 5000 \quad \text{or} \quad y(0) = 10{,}000 \quad \text{or} \quad y(0) = 20{,}000.$$

We call the starting point an **initial condition**. For our fish problem, that means specifying the fish population y at some initial time t_0. (We often start things at $t_0 = 0$, but not always.)

Similarly, our falling rock problem involves two quantities, z and v, so our initial conditions have to specify both $v(t_0)$ and $z(t_0)$. For the SIR equations, we need to specify $S(t_0)$, $I(t_0)$ and $R(t_0)$. In general, we need to specify one piece of data for each variable in the problem.

Same Equation, Different Setting. A bank that offers interest but charges fees follows essentially the same differential equation as a popular fishing lake. If the bank gives 5% interest and charges $100/year in fees, then an investor's bank balance follows the differential equation

$$(6.15) \qquad \frac{dy}{dt} = 0.05y - 100.$$

That is, the investor gets money from the bank at a rate $0.05y$ per year, but gives money back to the bank at a rate of 100 dollars/year. A rich investor who deposits $10,000 would make money, as he would get more in interest than he pays in fees. A poor investor who deposits $1,000 would lose money, as his yearly interest of $50 would only cover half of his fees.

Do you think that's fair? It would be nice if life were fair, with everybody playing by the same rules and starting out with the same initial conditions, but the real world isn't like that. Some people are born rich, others poor. Some are born with special abilities or with disabilities. Some have more educational opportunities than others. The list goes on.

Almost everybody believes that things should be as fair as possible, but people disagree about what "fair" means. Some take the differences in initial conditions as given and want to make the rules of the game exactly the same for everybody. Others think that the rules should include handicaps to help those whose initial conditions are worse. Still others say that we should redistribute resources so that everybody has roughly the same initial conditions.

So what do *you* think is fair? Math can't provide the answer, but at least it can frame the question. The better you understand how outcomes depend both on rules (differential equations) and on initial conditions, the better you'll be at pursuing whatever kind of fairness you believe in.

A differential equation, together with a set of initial conditions, is called an **initial value problem**, or IVP. Here are some examples of IVPs:

- The IVP

$$(6.16) \qquad \frac{dy}{dt} = -0.1(y - 70); \qquad y(0) = 170$$

might describe the temperature $y(t)$ of a cup of coffee (in degrees Fahrenheit) t minutes after it was brewed. It loses heat at a rate proportional to how much hotter the coffee is than the room around it. The initial condition is the temperature of the coffee when it was brewed.

- The IVP

$$(6.17) \qquad \frac{dy}{dt} = 0.1(70 - y); \qquad y(0) = 20$$

might describe the temperature of a container of ice cream t minutes after it was taken out of the freezer. It heats up at a rate proportional to how much hotter the room is than the ice cream, and the initial condition is its temperature when it came out of the freezer. As $0.1(70 - y) = -0.1(y - 70)$, this is actually the same differential equation as the previous example, but the initial condition is different. The coffee started off hot, but the ice cream started off cold.

- The IVP

(6.18) $$\frac{dv}{dt} = -32 + .001v^2; \qquad v(0) = 0$$

describes the vertical velocity of a skydiver, in feet/second, t seconds after jumping out of an airplane. The differential equation describes the forces involved: gravity pulling downwards and air resistance pushing upwards. The initial condition reflects the fact that the airplane is neither rising nor falling when the skydiver jumps.

Solutions. A function that satisfies the differential equation, and whose starting value matches the initial condition, is said to be a **solution** to the initial value problem. The IVP describes the rules of the game and the starting point, but the solution is what actually happens! *Building* a model involves figuring out what the IVP should be. *Running* the model means finding the solution.

As with all functions, solutions can be expressed via graphs, formulas, tables, or algorithms. For instance, the three graphs in Figure 6.2 show the solutions to three IVP's. The bottom graph is the solution to (6.13) with initial condition $y(0) = 5000$, the middle graph is the solution to (6.13) with initial condition $y(0) = 10,000$, and graph at the top is the solution to (6.13) with initial condition $y(0) = 20,000$.

Sometimes we describe our solutions with formulas. The functions $z(t) = 37 + 20t - 16t^2$ and $v(t) = 20 - 32t$ are solutions to (6.10) with initial conditions $z(1) = 41$ (feet) and $v(1) = -12$ (feet/second), as you can (and should!) check by taking the derivatives of $37 + 20t - 16t^2$ and $20 - 32t$ (to check the equations), and by plugging $t = 1$ into these expressions (to check the initial conditions).

It is theoretically possible to cook up IVP's that don't have solutions, or that have multiple solutions, but these require contrived (some would say *pathological*) choices of the expression $f(y, t)$. As long as $f(y, t)$ is remotely reasonable (and in particular, whenever $f(y, t)$ is a differentiable function of y and t), we have the

---

Existence and Uniqueness Principle: *Under normal circumstances, there is one and only one solution to the initial value problem*

(6.19) $$\frac{dy}{dt} = f(y, t); \qquad y(t_0) = y_0.$$

Numerical Solutions. Sometimes, when an IVP is particularly simple, we can find a formula for its solution. Some of these easy-to-solve problems are very important, and we will go over a few of them in the coming sections. The bad news, however, is that *most* real-world IVPs just can't be solved by finding a magic formula.

But there's good news, too. Even if we can't find a formula for the solution, we can solve the IVP on a computer anyway, using the ideas that we already used in Chapter 3 to tackle compound interest and the SIR equations. Here we repeat the algorithm, called **Euler's method**, to get a solution to any IVP of the form (6.19), with t running from an initial time t_0 to a final time t_f:

(1) Decide on a number N of steps we are going to take. This breaks the interval $[t_0, t_f]$ into subintervals of size $\Delta t = (t_f - t_0)/N$. For $j = 1, 2, \ldots, N$, let $t_j = t_0 + j\Delta t$. Set $y(t_0) = y_0$.

(2) For $j = 0, 1, \ldots, N-1$, use the known value of $y(t_j)$ to compute $y'(t_j) = f(y(t_j), t_j)$. Then use the linear approximation $y(t_{j+1}) \approx y(t_j) + y'(t_j)\Delta t$ to approximate $y(t_{j+1})$. These two steps can be combined into a single equation:

(6.20) $$y(t_{j+1}) \approx y(t_j) + f(y(t_j), t_j)\Delta t.$$

(3) Plot the array $[y(t_0), \ldots, y(t_f)]$ against the array $[t_0, \ldots, t_f]$ to get a graph of $y(t)$. If we prefer, we can print out a table of values.

(4) If we need more accuracy, we can repeat the process with a larger value of N and a smaller value of Δt. As N gets bigger and bigger, our approximate solution will get closer and closer to the true solution to (6.19).

Euler's method is a reliable crystal ball. With its help, and preferably with the help of a computer, we can solve even the ugliest-looking IVPs and get clean graphical results.[2] In many cases, solving an IVP numerically is the best way, or even the only practical way, to predict the future.

However, solving an IVP numerically has its drawbacks. If we want to understand how the final value $y(t_f)$ depends on the initial value $y(t_0) = y_0$, we have to run the program for a particular value of y_0, then run it again for a slightly different value, and so on, until we have enough data to see the pattern. If we want to understand how the solutions depend on the parameters of the differential equation (e.g., the growth rate of fish or the amount of fishing allowed), that's even more work. Computing $y(t_j)$ for (say) 100 different initial conditions and 100 different parameters requires 100×100 = 10,000 runs. That's a lot!

General Solutions. Sometimes there's a better way. Not always, but sometimes.

If a and b are any two numbers, then the functions

(6.21) $$z(t) = a + bt - 16t^2; \qquad v(t) = b - 32t$$

are solutions to the differential equations (6.10), namely $dz/dt = v$ and $dv/dt = -32$. This is because the derivative of $z(t) = a + bt - 16t^2$ is $b - 32t$, which equals $v(t)$, and the derivative of $v(t) = b - 32t$ is -32, as required. If somebody gives us initial conditions, we can use them to determine what the parameters a and b should be. For instance, if $v(1) = -12$ and $z(1) = 41$, then

(6.22)
$$-12 = v(1) = b - 32 \text{ so } b = 20,$$
$$41 = z(1) = a + b - 16 = a + 4 \text{ so } a = 37,$$

and our solution is $v(t) = 20 - 32t$, $z(t) = 37 + 20t - 16t^2$. (6.21) is called a **general solution** to the differential equations (6.10), while $v(t) = 20-32t$, $z(t) = 37+20t-16t^2$ is a **particular solution** to the IVP with $v(1) = -12$, $z(1) = 41$.

---

[2]There do exist much more efficient algorithms for solving IVPs numerically. However, those methods are more complicated, so we're not going to study them here.

You're probably wondering where the general solution (6.21) comes from. We'll learn more about this solution in Chapter 7, but at this stage of the game it isn't important! All we need to know is that *it works*, which is something that you can check for yourself by computing a few derivatives. As the solutions to IVP problems are unique, any solution that works must be the one and only solution. "Guess and check" is a perfectly valid strategy for solving IVPs!

As another example, the general solution to the equation $dy/dt = ry$ is $y(t) = Ae^{rt}$. It solves the equation because

$$(6.23) \qquad \frac{dy}{dt} = \frac{d(Ae^{rt})}{dt} = Are^{rt} = ry(t),$$

and because we can use the parameter A to fit any initial condition. If $y(0) = y_0$, then

$$(6.24) \qquad y_0 = Ae^{r0} = Ae^0 = A, \quad \text{so } A = y_0 \quad \text{and } y(t) = y_0 e^{rt}.$$

If we want to understand the effect of changing y_0, we just adjust A. What's more, our solution has the growth rate built into it as the parameter r. If we want to understand the effect of changing the growth rate from 25% to 26%, we just change $y_0 e^{0.25t}$ to $y_0 e^{0.26t}$.

All of the simple differential equations that we consider in this chapter have general solutions that you should learn to recognize. Understanding the functions (like e^{rt}) that appear in these solutions goes a long ways towards understanding the real-world systems that these differential equations model.

However, we must be careful not to overstate the importance of general solutions. They are often very hard to obtain. Whole textbooks have been written on techniques for finding them, and even they only touch on a small subset of the relevant differential equations out there. Most complicated differential equations don't even have general solutions that can be easily written down! To understand solutions to differential equations, we should be ready to

(1) Use general solutions when we can find them, both to solve specific IVPs and to get insight into how the solutions depend on initial conditions and on parameters, and

(2) Use other techniques, such as Euler's method, when we can't find a general solution, or can't even find a formula for the solution to a specific IVP.

6.2. Exponential Functions and Logarithms

Exponential functions, logarithms, and trigonometric functions are needed to understand all of the behaviors seen in Figure 6.1, but their properties take some getting used to. Most calculus students have already seen their definitions in previous courses, but fairly few are comfortable with equations such as

$$(6.25) \qquad \ln(5e^{x^2}) - e^{2\ln(x)} = \ln(3)\log_3(5).$$

If you are already comfortable with equation (6.25), feel free to skip ahead to Section 6.3 and to use this section as a reference. If not, read on for an overview of exponentials and logs. Trig functions are covered at the end of the chapter in the Supplemental Materials.

What Does a^x **Mean?** You were probably introduced to exponentials back in middle school, where you learned that 2^3 means $2 \times 2 \times 2$, and that 2^5 means $2 \times 2 \times 2 \times 2 \times 2$. In general, a^x means "a multiplied by itself x times". Based on that, we have

$$(6.26) \qquad 2^3 \times 2^5 = (2 \times 2 \times 2) \times (2 \times 2 \times 2 \times 2 \times 2).$$

The product of 2^3 and 2^5 has $3 + 5 = 8$ factors of 2, three from 2^3 and five from 2^5, so $2^3 \times 2^5 = 2^{3+5}$. Similarly, if a is any real number and b and c are positive integers, then

$$(6.27) \qquad a^b \times a^c = a^{b+c}.$$

We can also look at $(2^3)^5 = (2^3) \times (2^3) \times (2^3) \times (2^3) \times (2^3)$. This has 3 factors of 2, repeated 5 times, for a total of 3×5 factors of 2. In general,

$$(6.28) \qquad (a^b)^c = a^{bc}.$$

Our naive definition that a^x means "a multiplied by itself x times" works fine for 2^3 or 2^{17}, but what do 2^0, 2^{-3}, and $2^{1.363}$ mean? You can't multiply 2 by itself 0 times, or -3 times, or 1.363 times! Our challenge is to expand our definition while preserving the rules (6.27) and (6.28).

We begin with a^0. Because $0 + b = b$, we must have

$$(6.29) \qquad a^b = a^{0+b} = a^0 a^b,$$

by rule (6.27). The only way for that to work is if $a^0 = 1$. So we define

$$(6.30) \qquad a^0 = 1 \text{ for all } a \neq 0.$$

(The expression 0^0, just like the ratio $0/0$, is not defined.)

Next we look at negative powers. As $b + (-b) = 0$, we must have

$$(6.31) \qquad a^b a^{-b} = a^{b+(-b)} = a^0 = 1.$$

For this to work, a^{-b} must equal $1/a^b$. Put another way, "multiply a by itself -3 times" essentially means "start at 1 and divide by a 3 times". This makes sense for all $a \neq 0$.

Next we look at fractional powers. If p and q are integers, with $q \neq 0$, we have

$$(6.32) \qquad \left(a^{p/q}\right)^q = a^p,$$

so $a^{p/q}$ is the qth root of a^p, so $2^{1.363}$ means the 1000th root of $2^{1,363}$. It's also $(a^{1/q})^p$, which is the pth power of the qth root of a, so $2^{1.363}$ is the 1363rd power of the 1000th root of 2.

Finally, to make sense of an irrational power like 2^π, we have to use approximations. *Close is good enough!* Because $3.14159 < \pi < 1.314160$, 2^π is somewhere between $2^{3.14159}$ and $2^{3.14160}$. If $r_1, r_2, \ldots$ is a sequence of rational numbers getting closer and closer to π, then $2^{r_1}, 2^{r_2}, \ldots$, is a sequence of real numbers getting closer and closer to 2^π.

In this way, we've defined what a^x means for every positive real number a and for every real number x. Specifically:

- If x is a positive integer, a^x is the product of a with itself x times.

- $a^0 = 1$.
- If x is a positive integer, then $a^{-x} = 1/a^x$.
- If $x = p/q$ is a rational number, then a^x is the qth root of a^p, and is also the pth power of the qth root of a.
- If x is an irrational number, then we can approximate x arbitrarily well by a rational number r. a^x is the limit of a^r as r approaches x.

These last two definitions generally require $a > 0$, since if $a < 0$, then the qth root of a may not exist. Exponentials follow the

---

Three Laws of Exponents. If $a > 0$ and b and c are any real numbers, then

$$a^b \times a^c = a^{b+c},$$

(6.33)
$$a^b/a^c = a^{b-c},$$

$$\left(a^b\right)^c = a^{bc}.$$

The Graph of a^x. When $a > 1$, the function a^x grows very quickly with x. For instance, 2^{20} is a little over a million, but 2^{19} is only about half a million. Half of the growth of the function 2^x between $x = 0$ and $x = 20$ occurs between $x = 19$ and $x = 20$. And then 2^{21} is over *two* million. In the other direction, 2^x shrinks rapidly as x goes negative. 2^{-20} is a little less than a millionth. 2^{-21} is less than a two-millionth.

Figure 6.3 shows the graphs of three exponential functions, namely $f(x) = 2^x$, $g(x) = (1/2)^x$ and $h(x) = 4^x$. Notice that all three graphs go through $(0,1)$, as $a^0 = 1$ no matter what a is (as long as $a \neq 0$). All three graphs lie above the x axis, as a^x is always positive. All three graphs have similar shapes. In fact, the graphs of $f(x)$ and

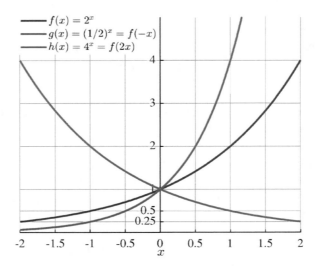

Figure 6.3. The graph of a^x for $a = 2, \frac{1}{2}$ and 4

$g(x)$ are mirror images of each other, as

$$(6.34) \qquad g(x) = (1/2)^x = (2^{-1})^x = 2^{-x} = f(-x).$$

The graph of $f(x)$ rises quickly to the right and dies off quickly to the left. The graph of $g(x)$ rises quickly to the left and dies off quickly to the right. We say that $f(x)$ shows **exponential growth**[3] while $g(x)$ shows **exponential decay**, but they're two sides of the same coin. Decay is just growth run backwards.

The function $h(x)$ behaves like $f(x)$, except that its growth (to the right) and decay (to the left) is twice as fast.

$$(6.35) \qquad h(x) = 4^x = (2^2)^x = 2^{2x} = f(2x).$$

In fact, we'll soon see that *all* of the functions a^x (with different values of $a \neq 1$) are essentially the same, up to reflection and horizontal scale.

Logarithms. The function $\log_a(x)$, usually pronounced "logarithm (or log) base a of x", is the inverse of the function a^x. That is,

$$(6.36) \qquad \begin{aligned} y &= \log_a(x) \text{ means } x = a^y, \\ x &= \log_a(y) \text{ means } y = a^x. \end{aligned}$$

In particular, $a^{\log_a(x)} = x$ (that's the first line) and $\log_a(a^x) = x$ (that's the second).

If that doesn't make sense to you right away, think about the inverse functions "square" and "square root". If x and y are nonnegative, $y = \sqrt{x}$ means $x = y^2$ and $x = \sqrt{y}$ means $y = x^2$. Both $(\sqrt{x})^2$ and $\sqrt{x^2}$ are equal to x. If you apply a function and then apply its inverse, or if you first apply the inverse function and then apply the original function, you get back to where you started.

The graph of $\log_a(x)$, shown in Figure 6.4, is just like the graph of a^x, only with the roles of x and y reversed. Instead of going through $(0, 1)$, it goes through $(1, 0)$. Instead of lying above the x axis, it lies to the right of the y axis. This means that $\log_a(x)$ *is only defined when $x > 0$*. Instead of shooting up extremely quickly, with $2^{21} \approx 2,000,000$ being twice as big as $2^{20} \approx 1,000,000$, it grows extremely slowly, with $\log_2(2,000,0000) \approx 21$ being only one greater than $\log_2(1,000,000) \approx 20$. Instead of a^x being greater than 1 for $x > 0$ and between 0 and 1 for $x < 0$, $\log_a(x)$ is positive for $x > 1$ and negative for $0 < x < 1$.

Let's compute some examples:

$$(6.37) \qquad \begin{aligned} 64 &= 2^6 \text{ so } \log_2(64) = 6 \\ 1/16 &= 2^{-4} \text{ so } \log_2(1/16) = -4 \\ 64 &= 4^3 \text{ so } \log_4(64) = 3 \\ 1/16 &= 4^{-2} \text{ so } \log_4(1/16) = -2. \end{aligned}$$

Note that for $x = 64$ and for $x = 1/16$, $\log_4(x)$ is half as big as $\log_2(x)$. This is because if $y = \log_4(x)$, then $x = 4^y = 2^{2y}$, so $\log_2(x) = 2y$. That is, the functions $\log_2(x)$ and

---

[3]When things grow quickly, people often describe the growth as "exponential", but true exponential growth is much faster than what people usually mean.

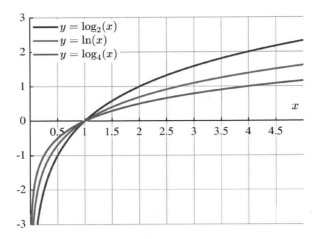

Figure 6.4. The graph of $\log_a(x)$ for $a = 2, e$ and 4

$\log_4(x)$ are the same, up to a constant multiple:

$$(6.38) \qquad\qquad \log_2(x) = 2\log_4(x).$$

We will soon see that all of the functions $\log_a(x)$ are the same, up to a constant factor, so we only really need to learn about one log function.

A particularly important logarithm is the log base e. This is called the **natural log** and is usually written $\ln(x)$ rather than $\log_e(x)$. Some people pronounce $\ln(x)$ as "lin x", while others just call it "log x", with the "natural" part understood.

Closely related to the laws of exponents are the

> **Laws of Logarithms:** If x, y, and a are positive numbers with $a \neq 1$, and if r is any real number, then
>
> $$\log_a(xy) = \log_a(x) + \log_a(y),$$
>
> $$\log_a(x/y) = \log_a(x) - \log_a(y),$$
>
> (6.39) $\qquad\log_a(x^r) = r\log_a(x),$
>
> $$\log_a(x) = \log_a(b)\log_b(x), \qquad \text{and}$$
>
> $$\log_a(x) = \frac{\ln(x)}{\ln(a)}.$$

The first law of logs comes from the first law of exponents (6.33). Let $b = \log_a(x)$, so $x = a^b$, and let $c = \log_a(y)$, so $y = a^c$. Then

$$xy = a^b a^c = a^{b+c}, \qquad \text{so}$$

(6.40)

$$\log_a(xy) = b + c = \log_a(x) + \log_a(y).$$

Similarly, the second law of logs comes from the second law of exponents.

$$\frac{x}{y} = \frac{a^b}{a^c} = a^{b-c}, \qquad \text{so}$$

(6.41)

$$\log_a(x/y) = b - c = \log_a(x) - \log_a(y).$$

The third law of logs comes from the third law of exponents. If $x = a^b$, then $x^r = a^{rb}$, so $\log_a(x^r) = rb = r\log_a(x)$.

The fourth and fifth laws of logs, sometimes called the "change of base" rules, are a little trickier. By definition, $x = b^{\log_b(x)}$ and $b = a^{\log_a(b)}$. But then

$$x = b^{\log_b(x)}$$

(6.42)

$$= \left(a^{\log_a(b)}\right)^{\log_b(x)}$$

$$= a^{\log_a(b)\log_b(x)},$$

so $\log_a(x) = \log_a(b)\log_b(x)$. Plugging in $x = a$ also gives the useful fact that $\log_a(b)$ and $\log_b(a)$ are reciprocals. In particular, $\log_a(e) = 1/\ln(a)$. Finally,

(6.43)

$$\log_a(x) = \log_a(e)\log_e(x) = \ln(x)/\ln(a).$$

Look back at equation (6.25) at the beginning of this section. With our rules in hand, we can make sense of it.

$$\ln(5e^{x^2}) = \ln(5) + \ln(e^{x^2})$$

$$= \ln(5) + x^2$$

$$e^{2\ln(x)} = e^{\ln(x^2)}$$

$$= x^2,$$

(6.44)

$$\ln\left(5e^{x^2}\right) - e^{2\ln(x)} = \ln(5) + x^2 - x^2$$

$$= \ln(5)$$

$$= \ln(3)\frac{\ln(5)}{\ln(3)}$$

$$= \ln(3)\log_3(5).$$

Derivatives of Exponentials and Logarithms. Next we turn to the derivative of a^x. We already saw that

$$(a^x)' = \lim_{h \to 0} \frac{a^{x+h} - a^x}{h}$$

(6.45)

$$= \lim_{h \to 0} \frac{a^h - 1}{h} a^x$$

$$= k_a a^x.$$

That is,

> *The rate at which an exponential function grows is proportional to how much is already there.*

Another way of putting this is that a^x is the (unique) solution to the IVP

(6.46)
$$\frac{dy}{dx} = k_a y; \qquad y(0) = 1.$$

At this point, we don't understand much about the constant k_a, except that $k_2 \approx 0.69$, $k_3 \approx 1.1$, and that there is a number called e for which $k_e = 1$.

However, the function $e^{k_a x}$ also solves the IVP (6.46), as

(6.47)
$$(e^{k_a x})' = e^{k_a x}(k_a x)' = k_a e^{k_a x} \text{ and } e^{k_a 0} = e^0 = 1.$$

This means that a^x and $e^{k_a x}$ must be the same function! In particular,

(6.48)
$$a = a^1 = e^{k_a(1)} = e^{k_a} \quad \text{so} \quad k_a = \log_e(a) = \ln(a).$$

There are two lessons here. First, our mystery constants k_a are really natural logs. Second, all exponential functions are the same, up to a horizontal scaling:

(6.49)
$$a^x = (e^{\ln(a)})^x = e^{\ln(a)x}.$$

If we understand the functions e^{kx}, where k is an arbitrary constant, then we understand *all* exponentials.

Finally, we consider the derivatives of our log functions. As $x = e^{\ln(x)}$, the chain rule tells us that

$$(x)' = \left(e^{\ln(x)}\right)'$$

$$1 = e^{\ln(x)}(\ln(x))'$$

(6.50)

$$1 = x(\ln(x))'$$

$$(\ln(x))' = \frac{1}{x}.$$

As $\log_a(x)$ is just $\frac{\ln(x)}{\ln(a)}$, the derivative of $\log_a(x)$ is just $\frac{1}{x \ln(a)}$. Putting everything together, we have the

Derivatives of exponentials and logs:

(6.51)

$$\frac{de^x}{dx} = e^x$$

$$\frac{da^x}{dx} = \ln(a)a^x$$

$$\frac{d\ln(x)}{dx} = \frac{1}{x}$$

$$\frac{d\log_a(x)}{dx} = \frac{1}{x \ln(a)}.$$

6.3. Simple Models of Growth and Decay

There are many situations where the rate at which a real-word quantity changes is a linear function of that quantity. These include the following.

(1) *Banks.* If we have $y(t)$ dollars in a bank that offers an interest rate of r, then $dy/dt = ry$. If the bank also charges a fee of \$F/year, then $dy/dt = ry - F$, which is still a linear function of y. If we invest an additional \$I/year in the account (e.g., saving for retirement), then $dy/dt = ry + I - F$, which is still a linear function.

(2) *Populations.* The same ideas apply to the number of humans in a city, the number of rabbits on an island, the number of fish in a lake or the number of bacteria in a Petri dish. Both the number of births and the number of deaths per unit time are roughly proportional to how many individuals are already there, so $dy/dt = ry$. If additional individuals are joining or leaving the population at a constant rate (e.g., people migrating to the city or rabbits being hunted), then $dy/dt = ry + M$, where M (for "migration") is the influx minus outflux.

(3) *Early stage epidemics.* At the beginning of an epidemic, almost everybody is a Susceptible, so $I'(t) = aSI - bI = (aS - b)I$ is well-approximated by $(aT - b)I$, where T is the total population.

(4) *Late stage epidemics.* If an epidemic is severe, it will eventually run out of Susceptibles, after which the number of Infecteds will decrease by the rule $I'(t) = -bI$.

(5) *Radioactive decay.* The rate at which radioactive nuclei disintegrate is proportional to how many are still left. If $y(t)$ is the mass of radioactive material remaining at time t, then $dy/dt = -ry$.

(6) *Heating and cooling.* If the temperature of a cup of coffee is $y(t)$, then the cup will lose heat at a rate proportional to the difference between the coffee's temperature and room temperature (say, 70 degrees Fahrenheit). The temperature will drop at a rate proportional to the temperature difference, so $dy/dt = -r(y - 70)$. If we have something colder than room temperature, like a cold drink or a dish of ice cream, it will heat up at rate $r(70 - y) = -r(y - 70)$. The same differential equation describes both heating and cooling.

Once we figure out how to solve the initial value problems

$$(6.52) \qquad dy/dt = ry, \qquad y(0) = y_0$$

and

$$(6.53) \qquad dy/dt = ry + p, \qquad y(0) = y_0,$$

where r and p are constants, we will be able to understand *all* of these real-world systems.

Exponential Growth and Decay. The solution to (6.52) is easy. We have already seen that the derivative of $y(t) = Ae^{rt}$ is $r(Ae^{rt})$, and that $y(0) = Ae^0 = A$, so $A = y_0$. In other words, the solution to (6.52) is

$$(6.54) \qquad y(t) = y_0 e^{rt}.$$

The IVP (6.52) and the solution (6.54) are *incredibly* common. Not only should you memorize them, but getting from one to the other should become a spinal reflex for you.

If our initial condition occurs at a time $t_0 \neq 0$, we just adjust our formula to count the time since we started. If $dy/dt = ry$ and $y(t_0) = y_0$, then

$$(6.55) \qquad\qquad y(t) = y_0 e^{r(t-t_0)}.$$

For instance, suppose that a city's population is growing at 2.5%/year, and that its population in the year 2020 was 1,000,000 people. How many people will be in the city in the year 2030? When will the population hit 2,000,000?

To figure out $y(2030)$, we just plug $y_0 = 1,000,000$, $r = 0.025$ and $t = 2030$ into equation (6.55).

$$(6.56) \qquad y(2030) = 1,000,000 e^{0.025(2030-2020)} = 1,000,000 e^{0.25} \approx 1,284,000.$$

To figure out when the population will hit 2 million, we solve

$$
\begin{aligned}
y(t) = &\quad 2,000,000, \\
1,000,000 e^{r(t-t_0)} = &\quad 2,000,000, \\
e^{r(t-t_0)} = &\quad 2, \\
r(t - t_0) = &\quad \ln(2), \\
t = &\quad t_0 + \frac{\ln(2)}{r}, \\
\approx &\quad 2020 + \frac{0.693}{0.025}, \\
= &\quad 2047.7.
\end{aligned}
$$

(6.57)

When a quantity is growing exponentially, we call the time it takes for the quantity to double the **doubling time**. When a quantity is shrinking exponentially, we call the time it takes to drop to half its initial value the **half-life**. The solutions to $e^{rt} = 2$ and $e^{-rt} = 1/2$ are both $t = \ln(2)/r$. Because $\ln(2)$ is approximately 0.70, this leads to a very simple rule of thumb.

> **The law of 70.** The doubling time (or half-life) is approximately 70 divided by the growth (or decay) rate in percent.

If a city grows at 2%/year, it will double in about 35 years. If it grows at 7%, it will double in 10 years. If a quantity decays at 10%/year, it will be down to half of its initial value in 7 years, a quarter of its initial value in 14 years, and an eighth of its initial value in 21 years.

The same methods work for all problems involving exponential growth. To figure out how much we'll have at time t, just plug t into (6.55). To figure out when we'll reach a certain level, set up equation (6.55), divide by y_0, take the natural log of both sides, and do a little algebra.

We can also use logs to compute growth (or decay) rates from data. For instance, suppose that the population of a city is growing exponentially. In 2020 there are 1.3 million people. In 2030 there are 1.65 million people. What is the growth rate r?

We set this up almost exactly like the previous problem, except that we're solving for r instead of t.

$$y(2030) = 1,650,000,$$

$$y(2020)e^{r(2030-2020)} = 1,650,000,$$

(6.58)
$$1,300,000e^{10r} = 1,650,000,$$

$$e^{10r} = 1.65/1.3 \qquad = 1.269,$$

$$10r = \ln(1.269) \qquad = 0.2384,$$

$$r = 0.02384.$$

Our city is growing at between 2.3% and 2.4% per year.

To recap:

- The IVP (6.52) comes up in many applications. The differential equation describes the growth process, and the initial condition sets a starting point.

- The *solution* to this IVP is the exponential function $y(t) = y_0 e^{rt}$. If the initial condition is $y(t_0) = y_0$ instead of $y(0) = y_0$, then the solution is $y(t) = y_0 e^{r(t-t_0)}$. These solutions are worth memorizing.

- If we know y_0, r and t, then we use exponentials to find $y(t)$. If we know y_0, $y(t)$ and t, then we use logs to find r. If we know y_0, $y(t)$ and r, we use logs to find t.

Growth or Decay with an Offset. We finish this section with the slightly more complicated differential equation (6.53). The expression for dy/dt, namely $ry + p$, can be rewritten as $r(y + \frac{p}{r})$. If we define a new variable $z = y + \frac{p}{r}$, then z and y only differ by a constant, so their derivatives are the same.

(6.59)
$$\frac{dz}{dt} = \frac{dy}{dt} = r\left(y + \frac{p}{r}\right) = rz.$$

But that's something we already know how to solve! By now, the solution (6.54) should be a spinal reflex. Combining that with a little algebra, we have

(6.60)
$$z(t) = z(t_0)e^{r(t-t_0)},$$

(6.61)
$$y(t) + \frac{p}{r} = \left(y_0 + \frac{p}{r}\right)e^{r(t-t_0)},$$

(6.62)
$$y(t) = \left(y_0 + \frac{p}{r}\right)e^{r(t-t_0)} - \frac{p}{r}.$$

Although it comes up fairly often, the solution (6.62) is *not* worth memorizing. Instead, you should remember the method we used to get it. Define a new variable z that turns our IVP into the spinal reflex problem (6.52). Then use a little algebra.

Returning to our coffee example, if the temperature $y(t)$ of a cup of coffee satisfies $dy/dt = -0.1(y - 70)$ and $y(0) = 170$, where we're measuring time in minutes, then we define $z = y - 70$ to get

$$dz/dt = -0.1z \text{ and } z(0) = 100,$$

(6.63) $$z(t) = 100e^{-0.1t},$$

$$y(t) = 70 + 100e^{-0.1t}.$$

The coffee's temperature $y(t)$ doesn't decay exponentially, but the *difference* $z(t)$ between the coffee's temperature and room temperature does. Every $70/10 = 7$ minutes, the coffee gets halfway towards room temperature.

A final note. Don't get hung up on the sign of r or the sign of p. If the quantity of radioactive material follows the rule $dy/dt = -0.02y$, then we have described the situation as $dy/dt = rt$ with $r = -0.02$, whose solution is $y(t) = y_0e^{r(t-t_0)} = y_0e^{-0.02(t-t_0)}$. Others describe it as $dy/dt = -rt$ with $r = +0.02$, whose solution is $y(t) = y_0e^{-r(t-t_0)} = y_0e^{-0.02(t-t_0)}$. In the end, the solutions are the same! It's purely a matter of taste whether we talk about the material growing at a negative rate or shrinking at a positive rate.

6.4. Two Models of Oscillation

The models in the last section only went in one direction. Either the quantity in question grew forever, or it shrank forever. However, many things in the real world go in cycles. These include everything from sunrise and sunset to seasons to the proverbial business cycle. In this section we explore two very simple models that give rise to sine and cosine functions, and in Section 6.5 we consider some more complicated models of growth and oscillation.

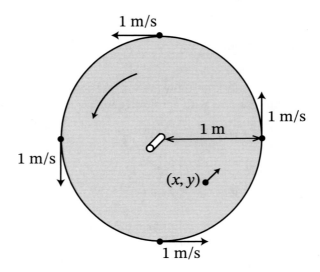

Figure 6.5. A rotating wheel

Wheels. Imagine that a wheel of radius 1 meter, centered at the origin, is rotating counterclockwise, as in Figure 6.5, and that the rim of the wheel is moving at a speed of 1 meter/second. The right side of the wheel is moving up, the top of the wheel is moving to the left, the left side of the wheel is moving down, and the bottom of the wheel is moving to the right. The rate at which a point on the wheel is moving up is proportional to its x-coordinate, and the rate at which it is moving to the left is proportional to its y coordinate. Because a point at position $(1,0)$ is moving up at speed 1 and a point at position $(0,1)$ is moving left at speed 1, those proportionality constants must both be 1. In other words, the coordinates $x(t)$ and $y(t)$ of a point on the wheel satisfy the differential equations

$$(6.64) \qquad \frac{dx}{dt} = -y, \qquad \frac{dy}{dt} = x.$$

The derivative of $\cos(t)$ is $-\sin(t)$, and the derivative of $\sin(t)$ is $\cos(t)$, so one solution to these equations is

$$(6.65) \qquad x(t) = \cos(t), \qquad y(t) = \sin(t).$$

This describes the trajectory of a spot that starts at $(1,0)$. After time t it has rotated t radians counterclockwise, so its new coordinates are $(\cos(t), \sin(t))$.

Because the derivative of $-\sin(t)$ is $-\cos(t)$ and the derivative of $\cos(t)$ is $-\sin(t)$, another solution is

$$(6.66) \qquad x(t) = -\sin(t), \qquad y(t) = \cos(t).$$

This describes the trajectory of a spot that starts at $(0,1)$. (Remember to "guess and check". We don't need to know where the solutions (6.65) and (6.66) come from to see that they work.) Combining this with our other solution, we get the *general* solution to our equations:

$$(6.67) \qquad x(t) = a\cos(t) - b\sin(t), \qquad y(t) = a\sin(t) + b\cos(t).$$

You should check that these functions really do satisfy the differential equations (6.64). Because $\cos(0) = 1$ and $\sin(0) = 0$, we have $x(0) = a$ and $y(0) = b$, so this describes the trajectory of a spot on the wheel that starts at position (a, b).

Now suppose that our starting point is on the rim, at an angle A above the x axis, as in Figure 6.6. That is, $x(0) = \cos(A)$ and $y(0) = \sin(A)$. After time t, our wheel has rotated $B = t$ radians counterclockwise, so our new angle should be $A + B = A + t$, and our new coordinates should be $(\cos(A+B), \sin(A+B))$. Comparing this to our general solution with $a = \cos(A)$ and $b = \sin(A)$ gives

$$(6.68) \qquad \begin{aligned} \cos(A + B) &= \cos(A)\cos(B) - \sin(A)\sin(B), \\ \sin(A + B) &= \cos(A)\sin(B) + \sin(A)\cos(B). \end{aligned}$$

These are the **addition of angle** formulas for sine and cosine.[4]

---

[4]This approach to addition of angles is something of a cheat, because we previously *used* the addition of angle formulas to compute the derivatives of $\sin(x)$ and $\cos(x)$. To avoid circular reasoning (pun intended), we will derive the formulas (6.68) from scratch in the "Crash course in trigonometry" in Section 6.8.

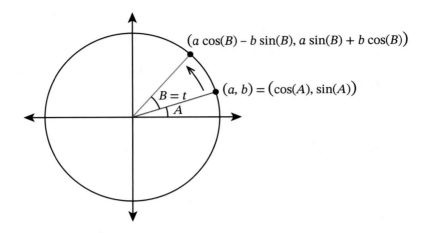

Figure 6.6. Understanding addition of angles through differential equations

We have just described a wheel that is rotating at 1 radian/second. More generally, a wheel that is rotating at angular speed ω (measured in radians/second) is governed by the differential equations

$$(6.69) \qquad \frac{dx}{dt} = -\omega y, \qquad \frac{dy}{dt} = \omega x,$$

whose general solution is

$$(6.70) \qquad x(t) = a\cos(\omega t) - b\sin(\omega t), \qquad y(t) = a\sin(\omega t) + b\cos(\omega t).$$

The example of the rotating wheel shouldn't come as a surprise. The sine and cosine functions were defined in terms of positions on a circle, so we should expect to see them whenever we see circles. The next example shows how we get sines and cosines from oscillations that don't seem to have anything to do with circles.

Springs. Figure 6.7 shows a block of mass m connected to a spring. Hooke's law says that if the spring is stretched a distance x, then it will pull the block back towards equilibrium with a force proportional to x. If the spring is compressed, it will push the block. In general, the force F on the spring is

$$(6.71) \qquad F = -kx,$$

where x is the displacement from equilibrium. The proportionality constant k is called the **spring constant**. Meanwhile, Newton's second law $F = ma$ says that the acceleration of the block is proportional to the force applied to it. Combining these, we get that

$$(6.72) \qquad \frac{d^2 x}{dt^2} = a = \frac{F}{m} = -\frac{k}{m}x = -\omega^2 x,$$

where $\omega = \sqrt{k/m}$. That is, the oscillations of the block and spring are governed by the second-order differential equation (6.72). This is similar to (6.52), except that it involves a second derivative instead of a first derivative.

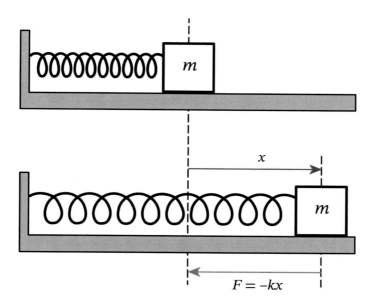

Figure 6.7. A mass on a spring

The functions e^{kt} have the property that their first derivatives are proportional to themselves, and their second derivatives are positive multiples of themselves: $\left(e^{kt}\right)'' = k^2 e^{kt}$. In contrast, the functions $\sin(\omega t)$ and $\cos(\omega t)$ have the property that their second derivatives are *negative* multiples of themselves:

(6.73)
$$\frac{d^2}{dt^2}\sin(\omega t) = \frac{d}{dt}\omega\cos(\omega t) \qquad = -\omega^2\sin(\omega t)$$

$$\frac{d^2}{dt^2}\cos(\omega t) = \frac{d}{dt}(-\omega\sin(\omega t)) \quad = -\omega^2\cos(\omega t).$$

In fact, the general solution to (6.72) is

(6.74)
$$x(t) = a\cos(\omega t) + b\sin(\omega t).$$

At time $t = 0$ we have $x = a$ and $\frac{dx}{dt} = \omega b$, so we can relate a and b to the initial position x_0 and the initial velocity v_0 and write

(6.75)
$$x(t) = x_0\cos(\omega t) + \frac{v_0}{\omega}\sin(\omega t).$$

The position of the block as a function of time is exactly the same as the x-coordinate of a spot on a wheel that is rotating at ω radians/second. What's more, it takes $2\pi/\omega$ seconds to complete a cycle, regardless of the size of the oscillation. This allows us to use oscillations to count time. Until electronics took over in the the mid-20th century, most clocks and wrist watches were based on this principle.

Hooke's Law is specific to springs, but most physical oscillations follow similar rules. For instance, the force pulling a pendulum back towards the center is (approximately) proportional to the angle of displacement, so the period of a pendulum doesn't depend on the size of the oscillation. Even in the electronics age, clocks still count oscillations, but the thing oscillating is more complicated than a mass on a spring. In fact, small fluctuations around almost *any* stable equilibrium are governed by a rule similar to Hooke's Law. Each system exhibits a characteristic frequency ω, and fluctuations are described by (6.75). This pattern is called **simple harmonic motion**.

6.5. More Sophisticated Population Models

In this section we will consider four problems, three realistic and one fantasy, where the behavior of the system is more complicated than simple exponential growth, exponential decay, or sines and cosines. The first two lead to **limited growth**, as in the third graph in Figure 6.1. The third leads to oscillations, as in the last graph in Figure 6.1. The last one addresses symbiosis.

The Australian Rabbit Problem. Rabbits are known for three things. They are very cute. They deliver Easter baskets. And they breed like rabbits.

In 1859, Thomas Austin released 24 rabbits in Australia, in the hope that they would breed and make for good hunting. As he put it,

> The introduction of a few rabbits could do little harm and might provide a touch of home, in addition to a spot of hunting.

He was wrong. The rabbit population increased at a rate of about 10%/month, and after 10 years (120 months) there were roughly

$$(6.76) \qquad\qquad 24e^{0.1(120)} = 24e^{12} = 3.9 \text{ million}$$

rabbits in that part of Australia.[5] So how many rabbits were there after 20 years? After 30 years?

If the rabbit population continued to grow exponentially, there would have been $24e^{24} = 636$ billion rabbits in Australia in 1879, or about 214,000 rabbits per square mile. By 1889 there would have been about 100 quadrillion rabbits, or about 1250 rabbits per square *foot*.

Those numbers are absurd. Exponential growth is so fast that it simply *can't go on forever*. At some point, the presence of so many rabbits changes the rules of the game and the simple differential equation $dy/dt = ry$ no longer applies. To model situations like the Australian rabbit problem, we need a more sophisticated model.

When there is a shortage of resources such as food and nesting sites, rabbits are forced to compete with one another. When two rabbits want the same piece of food, at least one of them will go away hungry (or perhaps injured or dead, if they fight over it). This diminishes the population in proportion to how often two rabbits meet, which is proportional to the *square* of the population. That is,

$$(6.77) \qquad\qquad \frac{dy}{dt} = ry - \alpha y^2,$$

---

[5]That figure is just a rough estimate. All that is known for sure is that millions of rabbits were being hunted each year, yet their population still grew.

where we have introduced a second parameter α that gives the strength of the competition. Equation (6.77) is called the **logistic equation**, and it is usually written in a slightly different form,

(6.78)
$$\frac{dy}{dt} = ry - \alpha y^2$$
$$= ry\left(1 - \frac{\alpha y}{r}\right)$$
$$= ry\left(1 - \frac{y}{L}\right),$$

where $L = r/\alpha$. The parameter L is called the **carrying capacity** of the system. For Australian rabbits, it is about 600 million.

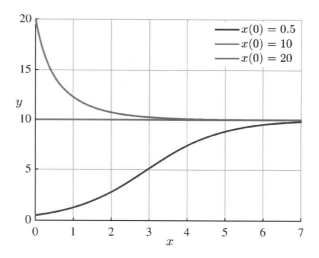

Figure 6.8. Three solutions to the logistic equation with $r = 1$ and $L = 10$.

Notice that dy/dt is positive if $0 < y < L$ and negative if $y > L$. In the logistic model, also called the **limited growth model**, the population grows until it approaches the carrying capacity, but it never goes above the carrying capacity. If for some reason we start with $y > L$, as in the top solution in Figure 6.8, then the population shrinks until it approaches the carrying capacity.

The formula for the general solution to (6.78) is known, but isn't very important. (It happens to be $y(t) = LAe^{rt}/(1 + Ae^{rt})$.) Much more important is understanding its qualitative behavior.

When $y \ll L$, then $(1 - \frac{y}{L}) \approx 1$ and $\frac{dy}{dt} \approx ry$. In other words, the early stages look just like exponential growth, proportional to e^{rt}. After y reaches $L/2$, the system slows down, and growth gets slower and slower. When y is close to L, we can approximate $ry(1-\frac{y}{L})$ as $rL(1-\frac{y}{L}) = r(L-y)$. We have already seen how to solve $dy/dt = r(L-y)$. As the derivative of $L-y$ is approximately $-r(L-y)$, the quantity $L-y$ decays exponentially, like a constant times e^{-rt}. In this way, the limited growth model combines the features of exponential growth (at the beginning) and exponential decay (at the end).

Humans versus Zombies. Suppose that a town of 1000 people is attacked by zombies. At time t, there are $Z(t)$ zombies and $H(t) = 1000 - Z(t)$ humans. On any given day, each human has a 1% chance of running into each of the zombies. If they meet, there is a 10% chance of the zombie turning the human into a zombie, and a 90% chance of the human getting away. If zombies never die, and if we start with 20 zombies and 980 humans, how many humans and zombies will there be after t days?

This is exactly the SIR model, only with no recovery, as zombies stay zombies forever. The humans are all Susceptible to becoming zombies, and the zombies are Infected. The transmission coefficient is $a = (0.01)(0.1) = 0.001$. We then have

(6.79)
$$\frac{dH}{dt} = -0.001HZ,$$
$$\frac{dZ}{dt} = 0.001HZ$$
$$= 0.001Z(1000 - Z)$$
$$= Z\left(1 - \frac{Z}{1000}\right).$$

The equation for Z in the SIR model with no recovery is *exactly* the logistic equation with $r = 1$ and $L = 1000$. In the early stages of the zombie apocalypse, the number of zombies grows exponentially as e^t. In the late stages, the number of humans shrinks exponentially as e^{-t}, as shown in Figure 6.9.

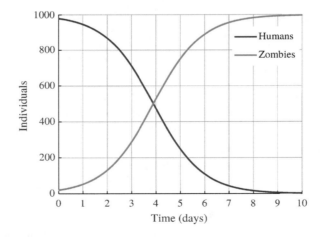

Figure 6.9. The zombie apocalypse. Starting on day 0 with 20 zombies and 980 humans, the human population is reduced to just two on day 10.

Predators and Prey. We are mostly interested in differential equations where the rules of the game don't change with time, so the right-hand side $f(y, t)$ is actually only a function of y. First-order differential equations of this sort, with a single variable, never lead to oscillations. To get oscillations, we either need a second-order equation, as with the mass on a spring (6.72), or first-order equations in two or more variables, as with the rotating wheel (6.64).

The dynamics of predators and prey provides a good example of the second kind of system. Suppose that a predator that eats rabbits, such as foxes, takes hold in Australia. At first the fox population would grow quickly, as the foxes would have lots of food. The large fox population would then cause the rabbit population to drop. The low rabbit population would then cause the foxes to starve, and the fox population would drop. The low fox population would allow the rabbits to recover. And the cycle would repeat.

Reality check: Using foxes to control rabbits in Australia is a *very* bad idea, as foxes eat more than just rabbits. Foxes are themselves considered an invasive species, and threaten a number of native species, such as the bilby. (Easter celebrations in Australia sometimes feature an Easter bilby instead of the hated rabbit.)

Just as the logistic model applies to all sorts of limited growth problems, there is a model, called the Lotka-Volterra model, that applies to all sorts of predator-prey problems. We can talk about foxes and rabbits, or owls and mice, or wolves and elk, or deer and the plants that they eat. Mathematically these are all essentially the same.

Let $x(t)$ be the population of a prey (e.g., rabbits) and $y(t)$ be the population of a predator (e.g., foxes). The simplest version of the Lotka-Volterra model says that

(6.80)
$$\frac{dx}{dt} = \alpha x - \beta xy,$$
$$\frac{dy}{dt} = \gamma xy - \delta y,$$

where $\alpha, \beta, \gamma,$ and δ are positive parameters. In the absence of predators, the prey population would grow exponentially as $e^{\alpha t}$. In the absence of prey, the predator population would decay as $e^{-\delta t}$. However, every time a predator meets a prey animal, something that happens with a frequency proportional to xy, it's good news for the predator (add γxy to dy/dt) and bad news for the prey (subtract βxy from dx/dt).

A typical trajectory for this system, with $(\alpha, \beta, \gamma, \delta) = \left(\frac{2}{3}, \frac{4}{3}, 1, 1\right)$ and initial conditions $x(0) = 0.5$ and $y(0) = 0.25$ is shown in Figure 6.10.

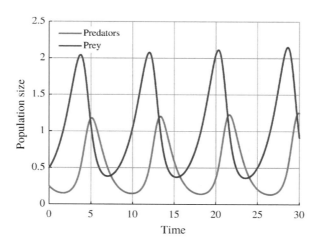

Figure 6.10. A solution to the Lotka-Volterra equations with $(\alpha, \beta, \gamma, \delta) = \left(\frac{2}{3}, \frac{4}{3}, 1, 1\right)$.

The functions aren't exactly sines and cosines, but they have the same qualitative behavior. The solutions are periodic, with a cycle of growth of the prey, followed by growth of the predator, followed by decline of the prey, followed by decline of the predator. As ecosystems always have many more prey than predators, the specific numbers for the populations should not be taken literally, and should perhaps be thought of as tens of predators and thousands of prey.

After learning about Australian rabbits, you might think that we should also take competition among prey, or among predators, into account. In fact, a more sophisticated version of the Lotka-Volterra model does add a small competition term $-\mu x^2$ to dx/dt and sometimes adds a competition term $-\nu y^2$ to dy/dt. With those terms in place, the populations of predators and prey gradually spiral inward toward a steady state, with x settling close to δ/γ and y settling close to α/β.

In the real world, populations of predators and prey do tend to stabilize over time. However, random events like forest fires or a hard winter eventually knock things out of equilibrium, after which the oscillations appear all over again.

Symbiosis: Mutualism and Competition. In a predator-prey situation, the presence of one species (prey) is good for the other (predator), while the presence of the predator is bad for the prey. There are also ecosystems where two species hurt each other (competition) or where they help each other (mutualism). The word **symbiosis**, from Greek roots meaning "living together", describes all of these situations.[6]

When modeling symbiotic ecosystems, the equation for dx/dt contains a term proportional to xy, as the rate at which the two species interact is proportional to the product of their populations. If the presence of y helps x, then the proportionality constant is positive. If y hurts x, then the proportionality constant is negative. Likewise, there is an xy term in the equation for dy/dt, positive if x helps y and negative if x hurts y. In **mutualism**, both constants are positive. In **parasitism** or **predator-prey**, one is positive and the other is negative. In **competition**, both are negative.

For instance, suppose that we are modeling the population of bees (B) and flowering plants (F) in a certain area. The bees need flowers to provide nectar and pollen, while the flowering plants need bees to pollinate. Neither the bees nor the flowers can thrive without the other. However, the bee population will increase if there are enough flowers, and the flower population will increase if there are enough bees. Both populations together can be modeled by the equations

(6.81)
$$\frac{dB}{dt} = -\alpha B + \beta BF,$$
$$\frac{dF}{dt} = \gamma BF - \delta F,$$

where α, β, γ, and δ are positive numbers. This is mutualism, because the terms containing β and γ are positive. Meanwhile, the terms containing α and δ are negative, because neither species can live for long without the other. These equations are just like the Lotka-Volterra equations (6.80), only with the signs of the α and β terms flipped.

---

[6]In older texts, the word *symbiosis* often meant species that helped one another (mutualism). Over time, its meaning has expanded to include competition and parasitism.

Sometimes we have two species that would do just fine by themselves, but are each hurt by the presence of the other. For instance, woodpeckers (W) and squirrels (S) might seek out nesting sites in the same trees. In that case, the woodpeckers and squirrels in **competition** for nesting sites. The number of woodpeckers and squirrels in the area might be described by the equations

(6.82)
$$\frac{dW}{dt} = \alpha W - \beta W S,$$
$$\frac{dS}{dt} = -\gamma W S + \delta S,$$

This is just like (6.80), only with the signs of γ and δ flipped.

As with the Lotka-Volterra equations, we have ignored the effects of competition within each species. A more accurate model might include a term proportional to $-B^2$ in dB/dt, a term proportional to $-F^2$ in dF/dt, and so on. We will examine competition, and its effect on natural selection, in Exercises 6.41 through 6.52.

6.6. Chemistry

Modeling chemical reactions uses exactly the same sorts of arguments that we used to build the SIR model.

The simplest chemical processes occur when a single molecule can take two (or more) forms. For instance, cyclohexane (C_6H_{12}, with the carbon atoms arranged in a ring) can either sit in a "boat" (B)[7] or "chair" (C) configuration. There is a certain probability per unit time of a boat turning into a chair, or vice-versa. That is, if we denote the concentration of the two kinds of cyclohexane as $[B]$ and $[C]$, there are constants k_1 and k_2 such that

(6.83)
$$\frac{d[B]}{dt} = -k_1[B] + k_2[C]; \qquad \frac{d[C]}{dt} = k_1[B] - k_2[C].$$

This is because boats turn into chairs at rate $k_1[B]$ and chairs turn into boats at rate $k_2[C]$. This is like the recovery phase of the SIR model, in which Infecteds turn into Recovereds at rate bI, only with both directions being possible.

At equilibrium, the rate of change is zero. *What goes up has to stop before it comes down!* This means that at equilibrium, $k_1[B] = k_2[C]$, so

(6.84)
$$\frac{[B]}{[C]} = \frac{k_2}{k_1}.$$

No matter what overall concentration of cyclohexane we are dealing with, the ratio of boats to chairs is always k_2/k_1.

The next simplest kind of reaction is where two molecules (A and B) join to form a third (C), as in the combination of CO_2 and water to form carbonic acid:

(6.85)
$$CO_2 + H_2O \Leftrightarrow H_2CO_3.$$

In such **bimolecular** $A + B \Leftrightarrow C$ reactions, the rate at which the reaction goes forward is proportional to $[A][B]$, as it depends on a chance meeting of an A and a B. This is just

---

[7]including the so-called "twist-boat", which has slightly lower energy than the regular boat

like the infection step of the SIR model, where transmission depends on a Susceptible meeting an Infected. However, the backward direction runs at a rate proportional to $[C]$, as each C molecule has a certain probability per unit time of dissociating into A and B. The upshot is that

$$(6.86) \qquad \frac{d[A]}{dt} = \frac{d[B]}{dt} = -k_1[A][B] + k_2[C]; \qquad \frac{d[C]}{dt} = k_1[A][B] - k_2[C].$$

At equilibrium, $k_1[A][B] = k_2[C]$, so

$$(6.87) \qquad \frac{[A][B]}{[C]} = \frac{k_2}{k_1}.$$

This is a special case of the **law of mass action**. In general, this law says that the product of the concentrations of all of the molecules on the left-hand side of the equation, counted as many times as they appear, divided by the similar product of the concentrations on the right-hand side, is a fixed number. For instance, when describing the oxidation of glucose ($C_6H_{12}O_6 + 6O_2 \Leftrightarrow 6H_2O + 6CO_2$), equilibrium occurs at a fixed value of $[C_6H_{12}O_6][O_2]^6/([H_2O]^6[CO_2]^6)$.

An especially important $A + B \Leftrightarrow C$ reaction involves the ions H^+ and OH^- combining to form water. The law of mass action then says that

$$(6.88) \qquad \frac{[H^+][OH^-]}{[H_2O]} = \frac{k_2}{k_1}, \quad \text{so} \quad [H^+][OH^-] = \frac{k_2}{k_1}[H_2O].$$

In most situations, we are talking about a solution that is almost entirely water, so the concentration $[H_2O]$ is fixed. If we are measuring concentrations in molar (moles/liter), then $[H_2O]$ is around 55 and, at room temperature, k_2/k_1 is a little under 2×10^{-16}, meaning that

$$(6.89) \qquad [H^+][OH^-] \approx 10^{-14}.$$

So if $[H^+] = 10^{-7}$, then $[OH^-] = 10^{-7}$. In an acid, $[H^+]$ is bigger than 10^{-7} and $[OH^-]$ is less than 10^{-7}, while in a base it's the other way around. We measure the acidity of a solution by taking the log (base 10) of $[H^+]$:

$$(6.90) \qquad pH = -\log_{10}([H^+]).$$

In distilled water at room temperature, $[H^+] = [OH^-]$, so the pH of 7. Most outdoor water sources, however, have a pH between 6 and 6.5, thanks to the presence of dissolved CO_2, which combines with water to form carbonic acid. Coffee is slightly acidic, with a pH of 5 or so. Orange juice and carbonated soft drinks are even more acidic, with regular Coke and Pepsi both having pH values less than 3. That's almost as acidic as lemons (pH 2.2) and stomach acid (pH around 1.5). In the other direction, blood has a pH of around 7.4 and seawater is around 8.1. Beyond that are toxic chemicals like ammonia and bleach and lye.

Most living things tolerate acids much more than bases, with strong bases being flat-out poisonous. That's reflected in our sense of taste. Acids taste sour, and many of us like that tangy flavor. But bases taste bitter and we immediately spit them out for our protection.

6.7. Chapter Summary

The Main Ideas.

- Many things in the real world are described by differential equations. Each term in a differential equation quantifies something that makes a quantity grow or shrink. Modeling a system with a differential equation amounts to writing these factors down.

- Most differential equations are too complicated to be solved with a formula. Instead, we solve them numerically with Euler's method and similar techniques.

- A few very simple differential equations occur repeatedly in the real world, and their solutions *are* given by simple formulas. These include
    - $dy/dt = ry$, which gives exponential growth (if $r > 0$) or exponential decay (if $r < 0$).
    - $dx/dt = -\omega y$, $dy/dt = \omega x$, which gives circular motion (sines and cosines of ωt).
    - $d^2x/dt^2 = -\omega^2 x$, simple harmonic motion, which also gives sines and cosines of ωt.

- These simple models can be used to approximate more complicated equations. For instance, the logistic model looks like exponential growth of x at the beginning, and exponential decay of $L - x$ at the end.

- The exponential function $a^x = e^{x \ln(a)}$
    - Means more than just "multiply a by itself x times", which only makes sense when x is a positive integer.
    - Grows extremely quickly in one direction and decays extremely quickly in the other.
    - Follows the three laws of exponents.

- A logarithm is the inverse of an exponential. That is, $y = \log_a(x)$ means $x = a^y$.

- The natural log function $\ln(x)$ is the log base e.

- Logarithms follow the five laws of logarithms. In particular, every log function is a multiple of $\ln(x)$.

- The derivatives of e^x and $\ln(x)$ are e^x and $1/x$, respectively.

- Sines, cosines, and the other trig functions appear:
    - As ratios of the length of sides of right triangles.
    - As locations of points on the unit circle.
    - As solutions to differential equations.

- The solution to $dy/dt = ry$ with $y(t_0) = y_0$ is $y(t) = y_0 e^{r(t-t_0)}$. That is, whenever something changes in proportion to how much is already there, you get exponential growth or decay.

- The law of 70: In exponential growth/decay, the doubling time or half-life is approximately 70 divided by the growth rate in percent.

- The logistic model describes population growth with limited resources, and also describes the interaction of warring populations (e.g., humans versus zombies).

- The Lotka-Volterra model describes interactions between predators and prey, and gives rise to oscillations. Modeling other forms of symbiosis is similar, only with some sign changes.
- Chemical reactions follow the law of mass action.

Expectations. You should be able to

- Turn a verbal description of a situation, such as a bank giving such-and-such interest and charging such-and-such fees, into a differential equation.
- Modify a differential equation to incorporate a new feature of a model, such as competition for resources.
- Solve differential equations numerically with Euler's method.
- Manipulate exponential functions, logs, and trig functions, using
  - The three laws of exponents.
  - The laws of logs.
  - The Pythagorean identities for trig functions.
- Sketch the graphs of e^x, $\ln(x)$, $\sin(x)$, and $\cos(x)$ accurately.
- Recite from memory the values of the sine, cosine and tangent of 0, $\pi/6$, $\pi/4$, $\pi/3$, and $\pi/2$.
- Solve the differential equations $dy/dt = ry$ and $dy/dt = ry + p$.
- Use the solutions of these differential equations in three ways:
  - Use exponentials to find $y(t)$ at some later time, given y_0 and r.
  - Use logs to find when $y(t)$ will reach a certain value.
  - Use logs to deduce r from data.
- Use sines and cosines to describe simple harmonic motion.
- Analyze a differential equation qualitatively by approximating it with something whose solution you already understand.
- Use differential equations to understand interactions between species.

6.8. Supplemental Material: A Crash Course in Trigonometry

Sines, Cosines, and Triangles. The trigonometric functions $\sin(x)$ and $\cos(x)$ show up in three different settings:

- Describing the shape of a right triangle.
- Describing the location of a point on a circle.
- As solutions to some very common differential equations, as we saw in Section 6.4.

Right Triangles. How can we parametrize the shape of a right triangle? We might give the lengths of the three sides, which certainly *determine* the shape, but those aren't properties *of* the shape, because two triangles can have the same shape without having the same size, as in Figure 6.11. To describe the shape, and only the shape, we either need to specify one of the angles, such as θ in Figure 6.11, or the ratio of two of the sides. Those are quantities that are the same for all triangles of the same shape.

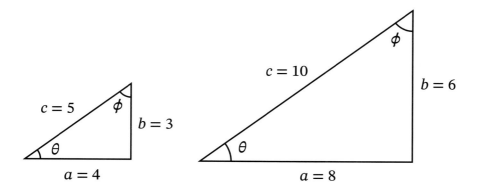

Figure 6.11. A 3-4-5 right triangle and a 6-8-10 right triangle have the same shape.

From the point of view of the angle θ, the side labeled a is called **adjacent** and the side labeled b is called **opposite**. The side labeled c is called the **hypotenuse**. Their lengths are related by the Pythagorean theorem:

$$(6.91) \qquad a^2 + b^2 = c^2.$$

The angles θ and ϕ are said to be **complementary**, which means that they add up to $\pi/2$ radians. From ϕ's point of view, b is adjacent, a is opposite, and c is still the hypotenuse.

There are six possible ratios of side lengths, and they are called the sine, cosine, tangent, cotangent, secant, and cosecant of the angle:

(6.92)

$$\sin(\theta) = \frac{\text{Opposite}}{\text{Hypotenuse}} = \frac{b}{c},$$

$$\cos(\theta) = \frac{\text{Adjacent}}{\text{Hypotenuse}} = \frac{a}{c},$$

$$\tan(\theta) = \frac{\text{Opposite}}{\text{Adjacent}} = \frac{b}{a},$$

$$\cot(\theta) = \frac{\text{Adjacent}}{\text{Opposite}} = \frac{a}{b},$$

$$\sec(\theta) = \frac{\text{Hypotenuse}}{\text{Adjacent}} = \frac{c}{a},$$

$$\csc(\theta) = \frac{\text{Hypotenuse}}{\text{Opposite}} = \frac{c}{b}.$$

The mnemonic SOH-CAH-TOA summarizes the first three. **S**ine equals **O**pposite over **H**ypotenuse. **C**osine equals **A**djacent over **H**ypotenuse. **T**angent equals **O**pposite over **A**djacent.

The functions tangent, cotangent, secant, and cosecant can all be expressed in terms of sine and cosine:

(6.93)

$$\tan(\theta) = \frac{\sin(\theta)}{\cos(\theta)},$$

$$\cot(\theta) = \frac{\cos(\theta)}{\sin(\theta)},$$

$$\sec(\theta) = \frac{1}{\cos(\theta)},$$

$$\csc(\theta) = \frac{1}{\sin(\theta)}.$$

Notice how the functions of the complementary angles θ and ϕ are related:

(6.94)

$$\cos(\phi) = \frac{a}{c} = \sin(\theta),$$

$$\cot(\phi) = \frac{b}{a} = \tan(\theta),$$

$$\csc(\phi) = \frac{c}{a} = \sec(\theta),$$

and similarly $\sin(\phi) = \cos(\theta)$, $\tan(\phi) = \cot(\theta)$ and $\sec(\phi) = \csc(\theta)$. The prefix "co" in cosine, cotangent, and cosecant stands for "complementary". The cofunctions of the angle θ are the regular functions of its complementary angle ϕ, and vice versa.

There are three special triangles associated with each angle θ, as shown in Figure 6.12. The first has $c = 1$ and is the key to understanding the sine and cosine functions. The second has $a = 1$ and is the key to understanding tangent and secant, while the third has $b = 1$ and is the key to understanding cotangent and cosecant. In the first

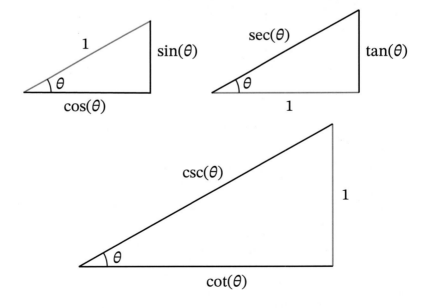

Figure 6.12. The three fundamental triangles for an angle θ

triangle, with $c = 1$, we have $a = c\cos(\theta) = \cos(\theta)$ and $b = c\sin(\theta) = \sin(\theta)$. The Pythagorean theorem then tells us that

$$(6.95) \qquad \sin^2(\theta) + \cos^2(\theta) = 1,$$

where the notation $\sin^2(\theta)$ means $(\sin(\theta))^2$. The second triangle has $a = 1$, $b = \tan(\theta)$ and $c = \sec(\theta)$, and the Pythagorean theorem tells us that

$$(6.96) \qquad \tan^2(\theta) + 1 = \sec^2(\theta).$$

Similarly, the third triangle shows that

$$(6.97) \qquad \cot^2(\theta) + 1 = \csc^2(\theta).$$

Equations (6.95), (6.96), (6.97) are called the **Pythagorean identities**.

The angles $\pi/6$ (also known as 30 degrees), $\pi/4$ (also known as 45 degrees), and $\pi/3$ (also known as 60 degrees) come up often. Let's compute the six functions of these angles.

In an equilateral triangle, say with side length 1, all three angles are $\pi/3$. If we cut an equilateral triangle in half, as in Figure 6.13, we get a right triangle with a base angle of $(\pi/3)/2 = \pi/6$ and with sides $c = 1$, $b = 1/2$ and $a = \sqrt{c^2 - b^2} = \sqrt{3}/2$. This tells us that

$$
\begin{aligned}
&\sin(\pi/6) = \cos(\pi/3) = \frac{1}{2}, &&\cos(\pi/6) = \sin(\pi/3) = \frac{\sqrt{3}}{2}, \\
(6.98) \quad &\tan(\pi/6) = \cot(\pi/3) = \frac{\sqrt{3}}{3}, &&\cot(\pi/6) = \tan(\pi/3) = \sqrt{3}, \\
&\sec(\pi/6) = \csc(\pi/3) = \frac{2\sqrt{3}}{3}, &&\csc(\pi/6) = \sec(\pi/3) = 2.
\end{aligned}
$$

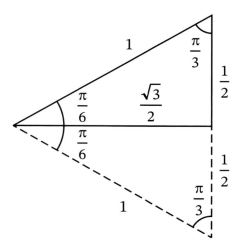

Figure 6.13. A 30-60-90 triangle is half of an equilateral triangle.

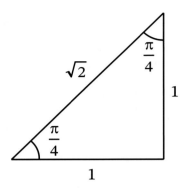

Figure 6.14. An isosceles right triangle.

An isosceles right triangle with $a = 1$ has $b = 1$ and $c = \sqrt{2}$ and $\theta = \phi = \pi/4$, as in Figure 6.14. This tells us that

$$\sin(\pi/4) = \cos(\pi/4) = \sqrt{2}/2,$$
(6.99)
$$\tan(\pi/4) = \cot(\pi/4) = 1,$$
$$\sec(\pi/4) = \csc(\pi/4) = \sqrt{2}.$$

If A and B are angles, then the sines and cosines of $A \pm B$ and $2A$ are given by the following formulas.

$$\sin(A + B) = \sin(A)\cos(B) + \cos(A)\sin(B),$$
$$\sin(A - B) = \sin(A)\cos(B) - \cos(A)\sin(B),$$
$$\cos(A + B) = \cos(A)\cos(B) - \sin(A)\sin(B),$$
(6.100)
$$\cos(A - B) = \cos(A)\cos(B) + \sin(A)\sin(B),$$
$$\sin(2A) = 2\sin(A)\cos(A),$$
$$\cos(2A) = \cos^2(A) - \sin^2(A) = 2\cos^2(A) - 1 = 1 - 2\sin^2(A).$$

These are called the **addition of angle** formulas and the **double angle** formulas.

The addition of angle formulas are explained in Figure 6.15.[8] The height of the triangle on the left is the sum of the heights of the triangles on the right, which gives the formula for $\sin(A + B)$. The width of the triangle on the left is the width of the bottom right triangle minus the width of the top right triangle, which gives the formula for $\cos(A+B)$. To get the formulas for $\sin(A-B)$ and $\cos(A-B)$, just replace B with $-B$ and use the facts that $\sin(-B) = -\sin(B)$ and $\cos(-B) = \cos(B)$. To get the formulas for $\sin(2A)$ and $\cos(2A)$, just set $B = A$.

---

[8]This way of seeing the proof of the formulas in a single figure was popularized by Blue the Trigonographer. You can find many more trigonometric proofs by picture, or "trigonographs", at Blue's blog: trigonography.com/toc.

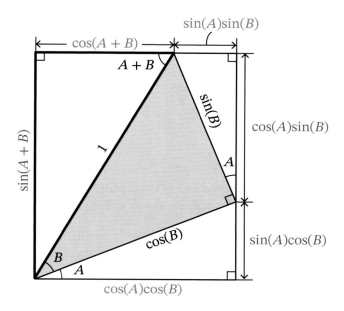

Figure 6.15. The sine and cosine of $A + B$

The Unit Circle. Next we look at the coordinates of points on the unit circle. If the line from the origin to a point on the unit circle makes an angle θ with the x-axis, then we have a triangle of hypotenuse 1 inscribed in the circle, as in Figure 6.16. The base and height of this triangle are $\cos(\theta)$ and $\sin(\theta)$, respectively, so the coordinates of our point must be $(\cos(\theta), \sin(\theta))$.

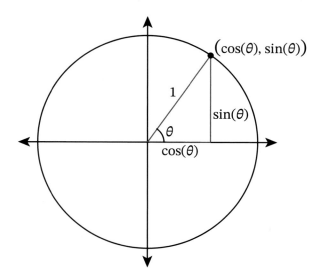

Figure 6.16. The coordinates of a point on the unit circle are $(\cos(\theta), \sin(\theta))$.

When dealing with right triangles, we only considered angles θ between 0 and $\pi/2$. However, the unit circle allows us to work with arbitrary angles. We simply *define* $\sin(\theta)$ and $\cos(\theta)$ to be the y and x coordinates (in that order!) of the corresponding point on the unit circle. The other trig functions are defined as ratios of sines and cosines as in (6.93), as long as this doesn't involve dividing by zero. For instance, $\sin(0) = 0$ and $\cos(0) = 1$, so

(6.101) $\tan(0) = 0/1 = 0$ and $\sec(0) = 1/1 = 1.$

Similarly, $\sin(\pi/2) = 1$ and $\cos(\pi/2) = 0$, so

(6.102) $\cot(\pi/2) = 0$ and $\csc(\pi/2) = 1.$

However, $\cot(0)$, $\csc(0)$, $\tan(\pi/2)$, and $\sec(\pi/2)$ are undefined, as they involve division by zero. The graphs of $\sin(\theta)$ and $\cos(\theta)$ are shown in Figure 6.17.

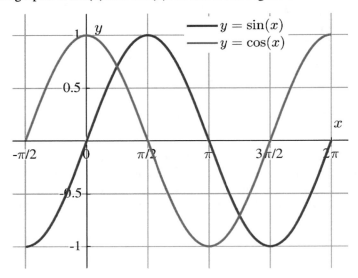

Figure 6.17. The graphs of $\sin(x)$ and $\cos(x)$

Because points on the unit circle have $x = \cos(\theta)$ and $y = \sin(\theta)$, symmetries of the unit circle give us relations among the sine and cosine functions. Rotating a circle by 2π sends (x, y) to itself, so the sine and cosine functions repeat every 2π:

(6.103) $\sin(\theta + 2\pi) = \sin(\theta)$ and $\cos(\theta + 2\pi) = \cos(\theta).$

Four other symmetries are shown in Figure 6.18.

Rotating by π sends θ to $\theta + \pi$ and sends (x, y) to $(-x, -y)$:

(6.104) $\sin(\theta + \pi) = -\sin(\theta)$ and $\cos(\theta + \pi) = -\cos(\theta).$

Reflecting about the x axis sends θ to $-\theta$ and sends (x, y) to $(x, -y)$:

(6.105) $\sin(-\theta) = -\sin(\theta)$ and $\cos(-\theta) = \cos(\theta).$

Reflecting about the y axis sends θ to $\pi - \theta$ and sends (x, y) to $(-x, y)$:

(6.106) $\sin(\pi - \theta) = \sin(\theta)$ and $\cos(\pi - \theta) = -\cos(\theta).$

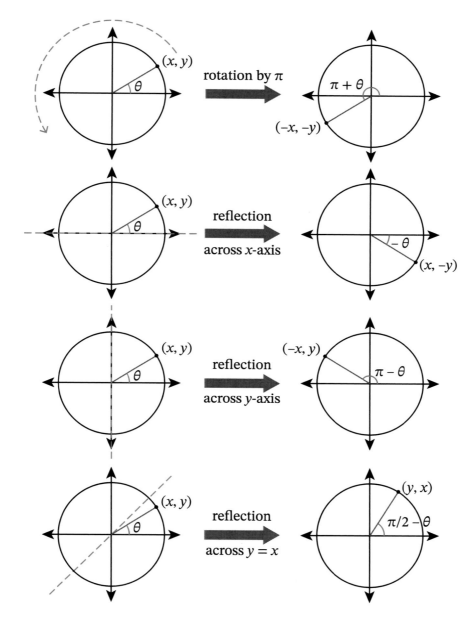

Figure 6.18. Symmetries of the unit circle

Finally, reflecting across the line $y = x$ sends θ to $\pi/2 - \theta$ and sends (x, y) to (y, x):

$$(6.107) \qquad \sin\left(\frac{\pi}{2} - \theta\right) = \cos(\theta) \qquad \text{and} \qquad \cos\left(\frac{\pi}{2} - \theta\right) = \sin(\theta).$$

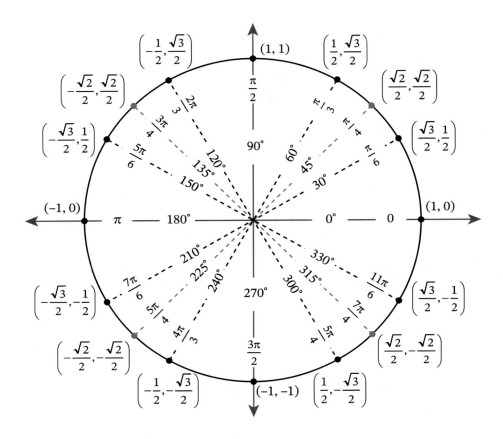

Figure 6.19. Angles and coordinates around the unit circle

With these identities, plus the known values of the sines and cosines of 0, $\pi/6$, $\pi/4$, $\pi/3$, and $\pi/2$, we can fill in the values of the sine and cosine of all the important angles between 0 and 2π, as in Figure 6.19.

If $-\pi/2 < \theta < \pi/2$, then $\tan(\theta)$ is the slope of the line that makes an angle of θ with the x axis. This is negative when $\theta < 0$ and is positive when $\theta > 0$. It is large and positive when θ is close to $\pi/2$, and is large and negative when θ is close to $-\pi/2$. It is not defined when $\theta = \pm\pi/2$, as that would involve dividing by zero. The graph of $\tan(\theta)$ is shown in Figure 6.20.

Note that

$$(6.108) \qquad \tan(\theta + \pi) = \frac{\sin(\theta + \pi)}{\cos(\theta + \pi)} = \frac{-\sin(\theta)}{-\cos(\theta)} = \frac{\sin(\theta)}{\cos(\theta)} = \tan(\theta),$$

so the graph repeats every π.

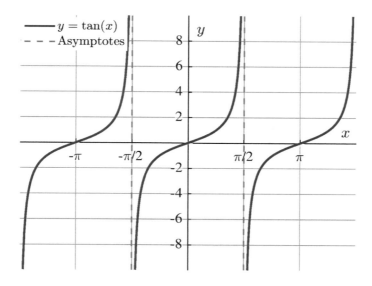

Figure 6.20. The graph of $\tan(x)$. Note the discontinuities at $x = \pm\pi/2$.

Derivatives and Inverse Trigonometric Functions. We previously computed the derivatives of the six basic trig functions in Sections 5.1 and 5.2:

$$\frac{d}{dx}\sin(x) = \cos(x) \qquad \frac{d}{dx}\cos(x) = -\sin(x)$$

(6.109)
$$\frac{d}{dx}\tan(x) = \sec^2(x) \qquad \frac{d}{dx}\cot(x) = -\csc^2(x)$$

$$\frac{d}{dx}\sec(x) = \sec(x)\tan(x) \qquad \frac{d}{dx}\csc(x) = -\csc(x)\cot(x).$$

The derivatives of the cofunctions all have minus signs, but otherwise look like "co" of the derivatives of the original functions. This is because

$$\text{cofunction}(x) = \text{function}\left(\frac{\pi}{2} - x\right)$$

(6.110)
$$(\text{cofunction})'(x) = \text{function}'\left(\frac{\pi}{2} - x\right) \times \left(\frac{\pi}{2} - x\right)'$$

$$= -\text{function}'\left(\frac{\pi}{2} - x\right)$$

$$= -\text{co-}(\text{function}')(x)$$

Also note that the derivative of $\tan(x)$ is always positive. This fits the fact that $\tan(x)$ increases on the entire interval from $x = -\pi/2$ to $x = \pi/2$, and then jumps down and starts all over again.

The **inverse sine** of x, also called the **arcsine** of x and written $\arcsin(x)$ or $\sin^{-1}(x)$, is the angle between $-\pi/2$ and $\pi/2$ whose sine is x. That is, $\theta = \sin^{-1}(x)$ means the same thing as $x = \sin(\theta)$. The **inverse tangent** or **arctangent** of x is the angle between $-\pi/2$ and $\pi/2$ whose tangent is x. The graphs of $\sin^{-1}(x)$ and $\tan^{-1}(x)$

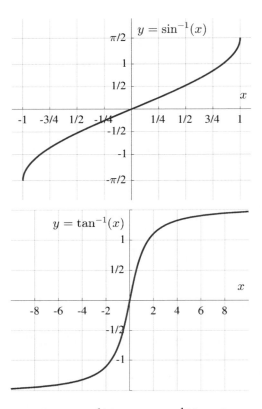

Figure 6.21. The graphs $y = \sin^{-1}(x)$ and $y = \tan^{-1}(x)$ are the same as $x = \sin(y)$ and $x = \tan(y)$, respectively, with y between $-\pi/2$ and $\pi/2$.

are obtained by taking the graphs of $\sin(x)$ and $\tan(x)$, restricting the input to be between $-\pi/2$ and $\pi/2$, and then swapping the input and output, as in Figure 6.21.

Warning: The notations $\tan^{-1}(x)$ and $\sin^{-1}(x)$ are standard but unfortunate, because they do not follow the same pattern as $\tan^2(x)$ or $\sin^2(x)$. $\tan^2(x)$ means $(\tan(x))^2$, but $\tan^{-1}(x)$ does *not* mean $(\tan(x))^{-1}$. It means the *inverse* tangent of x. The notation $\sin^n(x)$ is only used when n is a positive integer, in which case it means $(\sin(x))^n$, or when $n = -1$, in which case it means an inverse function. Expressions like $\sin^{-2}(x)$ don't mean anything at all.

The other inverse trig functions exist but are much less interesting. They are related to $\sin^{-1}(x)$ and $\tan^{-1}(x)$ by the following identities:

(6.111)
$$\cos^{-1}(x) = \frac{\pi}{2} - \sin^{-1}(x)$$
$$\cot^{-1}(x) = \tan^{-1}(1/x)$$
$$\sec^{-1}(x) = \frac{\pi}{2} - \sin^{-1}(1/x)$$
$$\csc^{-1}(x) = \sin^{-1}(1/x).$$

In practice, $\tan^{-1}(x)$ comes up frequently, $\sin^{-1}(x)$ comes up occasionally, $\cos^{-1}(x)$ and $\sec^{-1}(x)$ come up rarely, and $\csc^{-1}(x)$ and $\cot^{-1}(x)$ almost never come up.

We get the derivatives of the inverse trig functions from the chain rule:

$$
\begin{aligned}
x &= \tan(\tan^{-1}(x)) \\
x' &= \tan'(\tan^{-1}(x))(\tan^{-1}(x))' \\
1 &= \sec^2(\tan^{-1}(x))(\tan^{-1}(x))' \\
1 &= (1 + \tan^2(\tan^{-1}(x)))(\tan^{-1}(x))' \\
1 &= \left(1 + (\tan(\tan^{-1}(x)))^2\right)(\tan^{-1}(x))' \\
1 &= (1 + x^2)(\tan^{-1}(x))' \\
\frac{d}{dx}\tan^{-1}(x) &= \frac{1}{1+x^2}.
\end{aligned}
$$

(6.112)

$$
\begin{aligned}
x &= \sin(\sin^{-1}(x)) \\
x' &= \sin'(\sin^{-1}(x))(\sin^{-1}(x))' \\
1 &= \cos(\sin^{-1}(x))(\sin^{-1}(x))' \\
1 &= \sqrt{1 - \sin^2(\sin^{-1}(x))} \ \ (\sin^{-1}(x))' \\
1 &= \sqrt{1 - x^2}\,(\sin^{-1}(x))' \\
\frac{d}{dx}\sin^{-1}(x) &= \frac{1}{\sqrt{1-x^2}}.
\end{aligned}
$$

(6.113)

We can visualize inverse trig functions using the fundamental triangles of Figure 6.12. For instance, if $\theta = \tan^{-1}(1/2)$, then $\tan(\theta) = 1/2$. Replacing $\tan(\theta)$ with $1/2$ in the second fundamental triangle, and using the Pythagorean theorem to compute the hypotenuse, we get the triangle of Figure 6.22. If you prefer to avoid fractions, you can rescale this triangle to have sides 1 and 2 and hypotenuse $\sqrt{5}$, but that doesn't really change anything. Similar triangles have the same angles and the same ratios of side lengths.

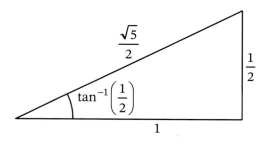

Figure 6.22. A fundamental triangle featuring $\tan^{-1}(1/2)$.

From either of these two triangles, we can compute all the trig functions of $\tan^{-1}(1/2)$ using SOH-CAH-TOA:

(6.114)

$$\sin(\tan^{-1}(1/2)) = 1/\sqrt{5}, \qquad \cos(\tan^{-1}(1/2)) = 2/\sqrt{5},$$
$$\tan(\tan^{-1}(1/2)) = 1/2, \qquad \cot(\tan^{-1}(1/2)) = 2,$$
$$\sec(\tan^{-1}(1/2)) = \sqrt{5}/2, \qquad \csc(\tan^{-1}(1/2)) = \sqrt{5}.$$

The same method words for the inverse tangent of any positive number x, not just $\tan^{-1}(1/2)$. Just draw a triangle with opposite side x, adjacent side 1, and hypotenuse $\sqrt{1+x^2}$. SOH-CAH-TOA does the rest.

Figure 6.23 shows two triangles featuring the angles $\sin^{-1}(x)$ and $\sec^{-1}(x)$. The remaining triangles are left to the exercises.

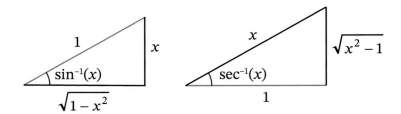

Figure 6.23. The fundamental triangles featuring $\sin^{-1}(x)$ and $\sec^{-1}(x)$.

6.9. Exercises

Differential Equations, Initial Conditions, and Solutions

6.1. For parts (a), (c), (e), and (g), answer in complete sentences.
   (a) Explain what a differential equation represents.
   (b) Write down an example of a differential equation.
   (c) Explain what an initial condition is.
   (d) Write down an example of an initial condition.
   (e) Explain what an Initial Value Problem (IVP) is.
   (f) Write down an example of an IVP.
   (g) What does it mean to say "$y = f(t)$ is a solution to an IVP"?

6.2. Verify that $y(t) = \tan\left(t + \dfrac{\pi}{4}\right)$ is a solution to the IVP $y' = y^2 + 1$, $y(0) = 1$.

6.3. Verify that $y(t) = \dfrac{1}{3}\left(4e^{3t} - 1\right)$ is a solution to the IVP $y' = 3y + 1$, $y(0) = 1$.

6.4. (a) Show that for any constant C, and for $t > -2C$, $y(t) = \left(\dfrac{t}{2} + C\right)^2$ is a solution to the differential equation $y' = \sqrt{y}$.
   (b) Why isn't this a solution when $t < -2C$?
   (c) By finding the correct C, solve the IVP $y' = \sqrt{y}$, $y(2) = 9$.

Food for Thought: The existence and uniqueness principle gives us a solution to every (reasonable) IVP for a short while. However, the solutions can blow up or have other problems, so the solutions may not work for all times. For instance, the solution to Exercise 6.2 only works for $\frac{-3\pi}{4} < t < \frac{\pi}{4}$ and the solution to Exercise 6.4 only works for $t > -2C$. Fortunately, the differential equations introduced at the beginning of the chapter, and discussed at length in Sections 6.3–6.6, have solutions that work for all values of t.

6.5.   Suppose that you take out a $200,000 mortgage at 2.5% annual interest (compounded continuously). You make payments of $2500/month. Write a differential equation that describes how the amount $P(t)$ that you owe the bank changes with time. What is the initial condition? (You are not expected to solve this initial value problem, just to write it down.)

6.6.   The vertical velocity of a skydiver jumping out of an airplane satisfies the IVP (6.18):

$$\frac{dv}{dt} = -32 + 0.001v^2; \qquad v(0) = 0.$$

(a) By modifying the SIREulers program from Chapter 3, find a numerical solution to this IVP and graph the solution for $0 \le t \le 15$.

(b) The velocity is approaching a certain value, called the **terminal velocity**. As the skydiver approaches that velocity, v' gets closer and closer to zero. By using the differential equation and setting $v' = 0$, compute the terminal velocity. How close is $v(15)$ to that value?

(c) The shape of the graph of $v(t)$ should remind you of one of the growth models in this chapter. Which one is it?

Exponentials and Logs

6.7.   Determine the *exact* value of each expression. You may leave your answers in terms of $\ln(2)$ and $\ln(10)$. For instance, if the question asked for $\ln(10e)$, you might write $\ln(10e) = \ln(10) + \ln(e) = 1 + \ln(10)$.

(a) $\ln(e/4)$ (b) $\ln\left(\sqrt{2e}\right)$ (c) $\ln(10)e^{4\ln(2)}$ (d) $\left(e^{\ln(3)}\right)^3$ (e) $e^{-\ln(5)}$

(f) $\ln(4/e)$ (g) $\ln(320)$ (h) $\ln(20^3)$ (i) $\ln(e^{500})$ (j) $e^{\ln(500)}$

6.8.   Find the derivative of each function with respect to x or t.

(a) $\ln(3x)$ (b) $\ln(2^t)$

(c) $\ln(2x^2 - 3x)$ (d) $\log_5(3t)$

(e) $5^{(x^2)}$ (f) $e^{\cos(x)+x^2}$

(g) $\sin^{-1}(3x)$ (h) $\tan^{-1}(t^2)$

(i) $e^{3x}\tan^{-1}(2x)$ (j) $2^x\sin^{-1}(x^2)$

Exponential Growth and Decay

6.9.   The rate of growth of invasive zebra mussels in a lake is proportional to the population. If the lake had 40,000 mussels in February and 45,000 in March, how many will there be in December?

6.10.   Suppose that a bacterial population grows so that its mass is

$$P(t) = 0.2e^{1.2t} \text{ grams}$$

after t hours. Its initial mass is $P(0) = 0.2$ grams.
(a) When will its mass double, to 0.4 grams?
(b) How much longer will it take to double again, to 0.8 grams?
(c) After the population reaches a mass of 0.8 grams, how long will it take for yet another doubling to happen?
(d) What is the doubling time of this population?
(e) At what time will there be 1 kg of bacteria? (Yuck!)

6.11.   Since the Civil War, the population of the Austin, Texas, metropolitan area has grown at a rate proportional to its population. In fact, it has grown at an amazingly steady rate of 3.5% per year. In the year 2000, the population was (approximately) 1,000,000. Assuming this growth pattern continues,
(a) What will the population of Austin be t years after the year 2000? (That is, let $t = 0$ correspond to the year 2000.)
(b) What will the population be in 2060?
(c) When will the population hit 5,000,000?

6.12.   Poaching of black rhinos in East Africa, and later in the rest of Africa, was a severe problem from the 1960s until the 1990s, nearly driving the rhinos to extinction. The population of black rhinos dropped from an estimated 100,000 individuals in 1960 to approximately 2400 in 1990. The good news is that the population has slowly rebounded since then, with the rhino population $R(t)$ increasing at a rate proportional to the number of rhinos present. In 1990 the population was increasing at approximately 100 rhinos per year.
(a) Write an initial value problem that governs the population of rhinos.
(b) Write the solution to this initial value problem.
(c) Use your model to predict the number of black rhinos in 2010. The actual number in the wild in 2010 was estimated to be 4900. Did your model provide a fairly close estimate?
(d) Use your model to predict the number of black rhinos in 2020. The actual number in the wild in 2020 was estimated to be 5630. Did your model provide a close estimate?
(e) Discuss the implications of your answer to part (d).

6.13.   In 1885, Hermann Ebbinghaus hypothesized that the speed of forgetting depends on a number of factors such as the effort spent trying to remember. If no effort is spent, we hypothesize that retention decreases at a rate proportional to how much is still retained. That is, if $m(t)$ is the percent of the memory retained after t days, then

$$\frac{dm}{dt} = -\frac{m}{s},$$

where s is a parameter that describes the stability of the memory. We also assume that $m(0) = 100$ (meaning that we recall 100% immediately after we learn something) and that $m(1) = 30$, meaning that we only retain only 30% of what we are taught one day later. (This is the actual rate found in research.)
  (a) Using the above information, write down a formula for $m(t)$.
  (b) What is the value of s?
  (c) Use your result to predict memory retention four days after learning something.

6.14. Suppose that sugar dissolves in water at a rate proportional to the amount left undissolved. Write a differential equation to describe this. If 50 lbs of sugar is added to a large quantity of water and 8 lbs are still undissolved after 6 hours, how long will it take until 98% of the sugar is dissolved?

6.15. In the early stages of an epidemic, almost everybody is a Susceptible and we can treat S as a constant. This means that I' is nearly proportional to I, and I grows exponentially. In March 2020, during the first wave of the Covid-19 pandemic, the number of confirmed cases in the United States rose from 541 on March 8 to 55,500 on March 24.
  (a) If we assume that the number of Infecteds was growing exponentially, on what date did the number of cases hit 5000? On what date did the number of cases hit 20,000?
  (b) How long did it take the number of cases to double?
  (c) If the number of cases continued to grow exponentially, how long would it have taken to reach 50 million cases?

Food for Thought: In fact, things weren't quite so bad. The entire world took drastic steps to limit the spread of the virus, reducing the transmission coefficient enough that infections started decreasing after reaching a peak of about 1 million in late May. Sadly, there were additional waves of infection after those emergency measures lapsed.

6.16. Suppose you open a saving account with an initial deposit of $5000, you continue to deposit money into the account at a rate of $2000 per year, and the account earns interest of 3% per year, compounded continuously. Write down an initial value problem for your balance $b(t)$. (*Hint:* "change in money = interest + savings".) In this exercise you do *not* need to find the solution to this IVP.

6.17. Solve the IVP of Exercise 6.16 as follows:
  (a) Define a variable z such that z' is proportional to z.
  (b) Write down the IVP that governs $z(t)$.
  (c) Solve for $z(t)$.
  (d) As you know how $b(t)$ is related to $z(t)$, write down the formula for $b(t)$.

Modeling Water Pollution. Between 1962 and 1970, the Dryden Chemical Company's pulp and paper mill dumped 11 tons of mercury into the rivers and lakes near Dryden, Ontario (Canada), with disastrous effects on human health and the environment. It is taking a very long time for the English-Wabigoon river system to recover, thanks in part to continued leaks, but mercury levels are finally dropping. In Exercises 6.18–6.21, we consider how a lake, or a chain of lakes, can gradually clean itself after being polluted.

6.18. Suppose that a lake is contaminated with a quantity $P(t)$ of pollutant, where $P_0 = P(0)$ is the size of the initial spill. The volume V of the lake (measured in liters) is constant, so the **density of pollutant** is $p(t) = P(t)/V$. Clean water flows into the lake at a constant rate r (measured in liters per unit of time), and is thoroughly mixed with the existing lake water. Meanwhile, water containing some pollutants flows out of the lake, also at rate r.

(a) In terms of $p(t)$ or $P(t)$, at what rate are pollutants leaving the lake?

(b) Write down an initial value problem for $P(t)$, the amount of pollution in the lake.

(c) Write down an initial value problem for $p(t)$, the *density* of pollution in the lake.

(d) Solve the IVPs from parts (b) and (c). That is, find formulas for $P(t)$ and $p(t)$. Your answer should depend on the parameters r and V and the initial value P_0.

6.19. The Great Lakes are connected to each other through a series of waterways, as shown in the following diagram. The numbers r_1 and r_2 represent the flow rates of water into Lake Superior and Lake Michigan, which then feed the other lakes. (For now, we are ignoring the various rivers that flow directly into Lake Huron, Lake Erie and Lake Ontario.)

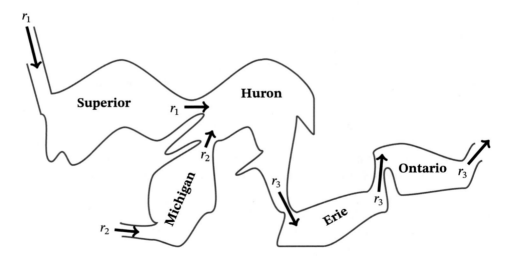

The lakes are polluted with plastic from littering and trash disposal. We denote the volume of water in the lakes by V_S, V_M, V_H, V_E, and V_O for Superior, Michigan, Huron, Erie, and Ontario, respectively, and similarly denote the density of plastic pollution in each lake by p_S, p_M, p_H, p_E, and p_O.

Suppose that all sources of plastic pollution are stopped. Water with no plastic in it flows into Lake Superior and Lake Michigan, gradually flushing the lakes. Write down a system of differential equations that models how the densities of plastic in the lakes changes. (You are *not* expected to solve these equations.)

6.20. When trying to model complex systems, we often start with a simple model, as in Exercise 6.19, and then add additional features, one at a time.

   (a) Based on the setup of Exercise 6.19, draw a diagram where another clean water source (the Maumee River) flows into Lake Erie. You will need to adjust your notation for the different flow rates, as the flow rate out of Lake Erie is no longer the same as the flow rate out of Lake Huron.

   (b) Use your diagram to write down a system of differential equations that models the amount of plastic in the lakes as they slowly become clean.

6.21. In the 1800s, Chicago's sewage was dumped into the Chicago River, which flowed into Lake Michigan, which was the source of Chicago's drinking water. To reduce the risks to public health, engineers built the Chicago Sanitary and Ship Canal, which opened in 1900. This canal reversed the flow of the Chicago River, bringing water from Lake Michigan through Chicago and ultimately to the Mississippi River basin. This **outflow** r_o of river water from Lake Michigan is less than the inflow r_2 from other rivers, so there is still some water flowing from Lake Michigan to Lake Huron.

   (a) Adjust your diagram and system of differential equations from Exercise 6.19 to account for this outflow from Lake Michigan.

   (b) Adjust your diagram and system of differential equations from Exercise 6.20 to account for both the (clean) Maumee River flowing into Lake Erie and the (polluted) Chicago River flowing out of Lake Michigan.

Growth and Decay with Offsets

6.22. Danielle just graduated from college. However, to help pay for her education she took out loans totaling $50,000. Now that she has graduated, the loans accrue interest at a rate of 6% per year (compounded continuously). Danielle has set up an online payment plan from her checking account which continuously pays off the loan at a rate of $6000 per year.

   (a) Write down an initial value problem that describes the balance of Danielle's loan at time t.

   (b) We saw how to solve initial value problems of the form

$$y' = ry + p, \quad y(0) = y_0$$

   in equation (6.62). Using this information, solve the IVP that you found in part (a).

   (c) By setting $y(t) = 0$, compute how long it takes Danielle to pay off her loan.

6.23. Bill is starting to save money for retirement. He sets aside $10,000/year to put into an investment account that grows at 7%/year. (*Note:* 7% is the historical average return for the stock market, adjusted for inflation.)

   (a) Write down an initial value problem that describes this situation.

   (b) Find the solution to this initial value problem. How much money will Bill have in t years?

   (c) Bill plans to retire when he has $1,000,000 in the account. When will that be?

6.24. The health department is using a radio advertising campaign to tell as many people as possible about a new vaccine. It is thought that the rate at which new people are exposed to the advertisement is proportional to the number of listeners who have not already heard it. No one is aware of the vaccine at the start of the campaign. After 10 days, 30% of the listeners are aware of the vaccine.

(a) Explain why this situation can be mathematically modeled by

$$\frac{dN}{dt} = -k(N - L), \qquad N(0) = 0, \qquad N(10) = 0.3L,$$

where $N(t)$ is the number of people aware of the vaccine and L is the total population of radio listeners.

(b) Solve the differential equation given in part (a).

(c) After how many days will 70% of the listeners be aware of the vaccine?

(d) Use MATLAB to graph the solution in part (b) for $0 \le t \le 100$.

6.25. The interior temperature $T(t)$ of a Thanksgiving turkey, t minutes after being removed from the oven, satisfies the initial value problem

$$\frac{dT}{dt} = -k(T - 70); \qquad T(0) = 165$$

for some unknown constant k, where we are measuring temperature in degrees Fahrenheit.

(a) If $T(10) = 140$, what is k?

(b) What will the temperature of the turkey be 30 minutes after it comes out of the oven?

(c) How long will the dinner guests need to wait before the turkey reaches 100 degrees?

6.26. An antibiotic is being injected into a hospital patient via a continuous IV drip, at a rate of 100 mg/hour. However, the antibiotic also breaks down (continuously) at a rate of 10%/hour. Let $A(t)$ be the amount of antibiotic after t hours.

(a) Write down an initial value problem for $A(t)$, assuming that there was no antibiotic in the patient's system when the injection started.

(b) $A(t)$ will eventually level off. At what level?

(c) Find a formula for $A(t)$.

(d) When will $A(t)$ reach 600 mg?

Rotations. In Section 6.4 we saw that $(x(t), y(t)) = (\cos(t), \sin(t))$ and $(x(t), y(t)) = (-\sin(t), \cos(t))$ are solutions to the differential equations

$$y' = x, \qquad x' = -y,$$

with initial conditions $(x(0), y(0)) = (1, 0)$ and $(x(0), y(0)) = (0, 1)$, respectively. These are said to be **basic** or **fundamental solutions** to the differential equations. All other solutions are obtained by taking a multiple of one basic solution plus a multiple of the other, also called a **linear combination** of the two solutions.

6.27. Find solutions the differential equations $y' = x$, $x' = -y$ with the following initial conditions.

(a) $x(0) = \sqrt{2}$ and $y(0) = 0$.

(b) $x(0) = 0$ and $y(0) = \pi^2$.

(c) $x(0) = \sqrt{2}$ and $y(0) = \pi^2$.

6.28. For the differential equations $y' = -x$, $x' = y$,

(a) Find a solution with $x(0) = 1$ and $y(0) = 0$.

(b) Find a solution with $x(0) = 0$ and $y(0) = 1$.

(c) Find a solution with $x(0) = \sqrt{2}$ and $y(0) = \pi^2$.

6.29. Let ω be an arbitrary number. For the differential equations $y' = \omega x$, $x' = -\omega y$,

(a) Find a solution with $x(0) = 1$ and $y(0) = 0$.

(b) Find a solution with $x(0) = 0$ and $y(0) = 1$.

(c) Find a solution with $x(0) = \sqrt{2}$ and $y(0) = \pi^2$.

Vibrations. In Exercises 6.30–6.32 we investigate solutions to the second-order differential equations

$$\frac{d^2x}{dt^2} = -\omega^2 x$$

that governs the vibration of a block of mass m on a spring with spring constant k, where $\omega = \sqrt{\frac{k}{m}}$.

6.30. Suppose that $\omega = 1$.

(a) Find a solution with $x(0) = 1$ and $\frac{dx}{dt}(0) = 0$.

(b) Find a solution with $x(0) = 0$ and $\frac{dx}{dt}(0) = 1$.

(c) Find a solution with $x(0) = \sqrt{2}$ and $\frac{dx}{dt}(0) = \pi^2$.

6.31. Now suppose that $\omega = 2$, so $\frac{d^2x}{dt^2} = -4x$.

(a) Find the solution with initial conditions $x(0) = 1$ and $\frac{dx}{dt}(0) = 0$.

(b) Find the solution with initial conditions $x(0) = 0$ and $\frac{dx}{dt}(0) = 1$.

(c) Find the solution with initial conditions $x(0) = 3$ and $\frac{dx}{dt}(0) = 0$.

(d) Find the solution with initial conditions $x(0) = 0$ and $\frac{dx}{dt}(0) = 2$.

(e) Find the solution with initial conditions $x(0) = 3$, $\frac{dx}{dt}(0) = 2$.

(f) In terms of the parameters x_0 and v_0, find the solution with initial conditions $x(0) = x_0$ and $\frac{dx}{dt}(0) = v_0$.

6.32. Now consider an arbitrary value of ω and arbitrary parameters x_0 and v_0.

(a) Find the solution with initial conditions $x(0) = 1$ and $\frac{dx}{dt}(0) = 0$.

(b) Find the solution with initial conditions $x(0) = 0$ and $\frac{dx}{dt}(0) = 1$.

(c) Find the solution with initial conditions $x(0) = x_0$ and $\frac{dx}{dt}(0) = v_0$.

Pendulums. Imagine a pendulum consisting of a heavy object (a "plumb bob") at the end of a string of length ℓ. The angle θ that the pendulum makes with the vertical axis is governed by the differential equation

$$\frac{d^2\theta}{dt^2} = -\frac{g}{\ell}\sin(\theta),$$

where $g \approx 32$ ft/s^2 ≈ 9.8 m/s^2 is the gravitational constant. Because $\sin(\theta) \approx \theta$ whenever θ is small,

$$\frac{d^2\theta}{dt^2} \approx -\frac{g}{\ell}\theta.$$

This is essentially the same as the equation that governs the position of a mass on a spring, only involving θ instead of x and g/ℓ instead of k/m. Note that the equation involves the length ℓ of the pendulum, but does not depend on the mass of the plumb bob.

6.33. (a) Compute the time it takes a pendulum with $\ell = 10$ cm to undergo a complete cycle, swinging from one side to the other and back again. This is called the **period of oscillation**.
        (b) Compute the period of a pendulum with $\ell = 40$ cm.
        (c) Compute the period of a pendulum with $\ell = 90$ cm.
        (d) Suppose that a pendulum has a period of 10 seconds. What is ℓ?

6.34. Make a pendulum by tying a small plumb bob to the end of a 90-cm (about 35.4-inch) piece of string.
        (a) Measure the period of oscillation. To avoid round-off errors, you might want to count how long it takes to make ten oscillations and divide by ten. If feasible, try the experiment with several different plumb bobs to check that the period really only depends on ℓ, and not on the mass of the bob. Compare your result to the answer you computed in part (c) of Exercise 6.33.
        (b) Now hold the string 40 cm from the middle of the bob and repeat the experiment. Compare your measured period to part (b) of Exercise 6.33.
        (c) Finally, hold the string 10 cm from the middle of the bob, repeat the experiment, and compare your measured period to part (a) of Exercise 6.33. (Be sure to use a bob that is much less than 10 cm across. If the mass is spread out, with one part of the bob being much closer to your fingers than another part, then the calculation becomes more complicated and involves the **moment of inertia** of the bob, a quantity we will learn about in Chapter 7.)

6.35. **Solving the Logistics Equation with MATLAB.** Modify the SIREulers program (see Section 3.6) to create a MATLAB program, called Logistics, to solve the logistic equation $y' = ry\left(1 - \frac{y}{L}\right)$. Your input parameters should be the time t, the number of steps N, the growth rate r, the carrying capacity L, and the initial value y_0. Your output should be arrays for t and y and a graph of y as a function of t.

6.36. **Deer Harvesting.** In many regions of the United States, deer populations have exploded due to a lack of predators. In those regions, harvesting of deer by hunters is considered a necessary management tool in order to maintain a sustainable and healthy herd. In the absence of predators, the population change for many species is modeled using the logistic growth model

$$\frac{dP}{dt} = rP\left(1 - \frac{P}{L}\right),$$

where $P(t)$ is the population at time t, r is the growth rate, and L is the carrying capacity of the environment.

(a) Suppose that $P(0) = 41$, $r = 0.2311$, and $L = 50^9$. Use MATLAB and the Logistics program that you wrote in Exercise 6.35 to plot the behavior of the deer population. On the same plot, graph what happens if $P(0) = 60$.

(b) To limit the population, the state allows harvesting a certain proportion of the deer population by hunters each year. How would you modify the model to include harvesting a certain proportion λ of the deer population each year?

(c) Modify the Logistics program of Exercise 6.35 to include the harvesting term from part (b). Use your modified program to make predictions about the deer population if harvesting is included. Using the parameter values given in part (a) of this exercise, explore using $\lambda = 0.05, 0.1, 0.15, 0.2$, and 0.25. Describe what is occurring in the model predictions. What you think is a sustainable harvest rate?

Dimensionless Models. When formulating a mathematical model for a biological process, we often recast the original model, which involves variables with units of volume or concentration or time, using dimensionless variables. There are several benefits to doing this, including reducing the number of parameters used in the model. Exercises 6.37–6.40 lead us through several examples of that approach.

6.37. (a) Consider the exponential population growth model

$$\frac{dy}{dt} = ry, \qquad y(0) = y_0.$$

The dimension for the independent variable t is time, and the dimension for the dependent variable y is individuals. What are the dimensions of the parameter r?

(b) Consider a new independent variable $\tau = rt$. What are the dimensions for τ?

(c) Consider a new dependent variable $Y = y/y_0$. What are the dimensions for Y?

(d) We can rewrite our original dimensionful differential equation in terms of our new dimensionless variables. Show that $\dfrac{dy}{dt} = y_0 r \dfrac{dY}{d\tau}$.

(e) Rewrite the right-hand side of the differential equation in part (a) in terms of Y.

(f) Combining the results of parts (d) and (e), show that $\dfrac{dY}{d\tau} = Y$. What happened to the parameter r?

(g) Restate the initial condition in terms of Y. What happened to the parameter y_0?

---

[9]$P(0)$ and r are the approximate number of deer per square mile, and the approximate growth rate, in Texas in 2020.

6.38.   In Exercise 6.37, we started with a dimensionful initial value problem

$$\frac{dy}{dt} = ry, \qquad y(0) = y_0,$$

and created an equivalent dimensionless initial value problem

$$\frac{dY}{d\tau} = Y, \qquad Y(0) = 1.$$

(a) If $r = 0.15$ and $y_0 = 20$, find the solution $y(t)$ to the dimensionful IVP.
(b) Find the solution $Y(\tau)$ to the dimensionless IVP.
(c) Using MATLAB, graph both $y(t)$ and $Y(\tau)$ on the same set of axes. What do you notice about the qualitative behavior of the two curves?

6.39.   Consider the logistic model

$$\frac{dy}{dt} = ry\left(1 - \frac{y}{L}\right), \qquad y(0) = y_0.$$

(a) List the three parameters in the model, and give the dimensions of each one.
(b) What are the dimensions of $\tau = rt$ and $Y = y/L$?
(c) Show that when you rewrite the differential equation in terms of τ and Y, you get

$$\frac{dY}{d\tau} = Y(1 - Y).$$

(d) Let $\alpha = y_0/L$. The initial condition $y(0) = y_0$ then becomes $Y(0) = \alpha$. How many parameters does your new dimensionless differential equation contain? How many parameters does your new dimensionless IVP contain?

6.40.   In Exercise 6.39, we used the dimensionless variable $Y = y/L$, as well as the dimensionless variable $\tau = rt$, to get rid of any dependence on L, but we wound up getting stuck with a dependence on y_0. Instead, consider the dimensionless variables $\tilde{Y} = Y/y_0$ and $\tau = rt$. Derive an IVP that governs $\tilde{Y}$ as a function of τ. When you compare the two dimensionless models, what differences do you see?

Competition for Limited Resources. In Exercises 6.41–6.53, we are going to study how natural selection causes certain genes to spread within a population. Imagine a situation where two species (one with a mutation and one without) share an ecosystem with a limited food supply. We could be talking about two kinds of rabbits invading Australia, or two kinds of trees repopulating a forest after a fire, or two kinds of Covid virus competing for a limited supply of humans to infect.

Let $x(t)$ be the population of the original ("wild type") version of the creature and let $y(t)$ be the population of the other ("mutant") version. Each creature has both wild type and mutant competitors, so the extent to which it is hurt by competition is proportional to $x + y$. The differential equations for x and y then take the general form

$$\frac{dx}{dt} = r_1 x - \alpha_1 x(x + y),$$

$$\frac{dy}{dt} = r_2 y - \alpha_2 y(x + y).$$

The parameters r_1 and r_2 refer to how fast each species can grow, while α_1 and α_2 have to do with how well they tolerate competition.

6.41.   First suppose that the mutation doesn't actually affect fitness. That is, suppose that $r_2 = r_1$ and that $\alpha_2 = \alpha_1$. Show that the ratio $y(t)/x(t)$ of mutants to wild type individuals is constant. (*Hint*: use the quotient rule.)

6.42.   Now suppose that the mutant strain breeds slightly faster than the wild type. That is, suppose that $\alpha_2 = \alpha_1$, but that $r_2 > r_1$. Show that

$$\frac{d}{dt}\left(\frac{y}{x}\right) = (r_2 - r_1)\frac{y}{x}.$$

That is, the ratio of the two populations will grow exponentially. This might involve the second population growing, the first population shrinking, or a combination of both.

6.43.   When there are two different populations, there are two (possibly different) carrying capacities. Let $L_1 = r_1/\alpha_1$, and let $L_2 = r_2/\alpha_2$.
   (a) Show that x keeps growing until the *total* population $x + y$ reaches L_1, while y keeps growing until the total population reaches L_2.
   (b) If $L_1 < L_2$, what will happen once the total population reaches L_1? Which population will dominate in the long run?

6.44.   In an alternate universe, Thomas Austin tested his plan to introduce rabbits to Australia before unleashing a continent-wide ecological disaster. In that universe, he introduced two kinds of rabbits to an island off the coast of Australia. He started with 100 long-eared rabbits and 10 short-eared rabbits. The short-eared rabbits grew at a rate of 11% per month, while the long-eared rabbits only grew at 10% per month. They had the same values of α, namely $\alpha_1 = \alpha_2 = 10^{-7}$. After how many years were the two populations equal? After how many years did the short-eared rabbits outnumber the long-eared rabbits 100:1?

6.45.   In the alternate universe of Exercise 6.44, what are the carrying capacities L_1 and L_2?

6.46.   We continue with the scenario of Exercise 6.44. After a while, the total rabbit population approached L_2 and most of the rabbits were short-eared. From then on, we can get a good estimate of the number of long-eared rabbits by multiplying the fraction of long-eared rabbits (which you can get from the ratio of long-eared to short-eared) by L_2.
   (a) How many long-eared rabbits were there after 100 years?
   (b) When did the long-eared population drop below 1? (That is, when did the long-eared rabbits go extinct?)

Using the Darwin Program. Exercises 6.47–6.53 use the following MATLAB program, called Darwin, which tracks the effect of the competition between rabbits. It is very similar to SIREulers, only with a different set of differential equations. The input parameters are the total time t_{max}, the number of steps N, the growth rates r_1 and r_2, the carrying capacities L_1 and L_2, and the initial populations $X0$ and $Y0$. As with SIREulers, you can either use the graphs generated by the program or make your own graphs using `plot`.

```
function [X,Y,t] = Darwin(tmax, N, r1, r2, L1, L2, X0, Y0)
dt = tmax/N;                                    % Compute Δt
a1 = r1/L1;                                      % Compute α₁
a2 = r2/L2;                                      % Compute α₂
X = zeros(1,N+1);                               % Start with empty
Y = zeros(1,N+1);                               % vectors for X and Y.
X(1) = X0;                                      % Initialize X
Y(1) = Y0;                                      % Initialize Y
for n = 1:N                                     % Start the loop
    X(n+1) = X(n) + dt*(r1*X(n) - a1*X(n)*(X(n)+Y(n)));   % Rate eq. for X
    Y(n+1) = Y(n) + dt*(r2*Y(n) - a2*Y(n)*(X(n)+Y(n)));   % Rate eq. for Y
end                                             % End the loop
t = linspace(0, tmax, N+1);                     % A vector of times
plot(t, X, 'blue', t,Y, 'red', t, X+Y,'green')  % Graph the results
title('Darwinian evolution')                    % Add a title
ylabel('Individuals'), xlabel('Time')           % Add labels
legend('Population X','Population Y', 'Total population')  % Add a legend.
```

In Exercises 6.47–6.49, use Darwin (or your own modification of Darwin) to track the populations of rabbits on Thomas Austin's hypothetical island, described in Exercise 6.44. Be careful when entering your parameters, as our growth rates are 10% and 11% per *month*, but we want to track the populations over a number of *years*.

6.47.   Plot the populations for the first 5 years. Describe what is happening. Are the long-eared rabbits thriving or dying out?

6.48.   Now plot the populations for the first 15 years. When did the long-eared population reach a peak? (Zoom into your graph, or look at the actual arrays, to determine this.) What was the total population at that time?

6.49.   Now plot the populations for the first 25 years. At what time did the short-eared population overtake the long-eared population? Does this agree with your answer to Exercise 6.44?

Food for Thought: The upshot of problems 6.44 through 6.49 is that, at first, the short-eared rabbits are no threat to the long-eared rabbits. There is plenty of food and plenty of room, so both populations grow, with the short-eared population growing a little faster than the long-eared population. However, sooner or later the total rabbit population levels out. Once it reaches that point, a greater ratio of short-ears to long-ears means more short-ears and fewer long-ears. Eventually, the long-ears are driven to extinction.

That's how natural selection works. Even a small evolutionary advantage can cause one population to completely replace another.

Differential Survival. One evolutionary strategy is to breed faster than the competition. A different strategy is to be hardier than the competition and to survive competition better. We explore this strategy in Exercises 6.50–6.53.

6.50. Suppose that two species of pine trees grow equally fast in a certain forest, but that the second species is better at competition and has a higher carrying capacity. Specifically, suppose that $r_1 = r_2 = 0.2$/year, $L_1 = 1,000,000$ trees, and $L_2 = 1,100,000$ trees. Describe *qualitatively* what will happen if the forest starts with 200 trees of the first type and 100 trees of the second type.

6.51. Check your answer to problem 6.50 numerically. Use the MATLAB program Darwin.m to simulate what happens in the forest for 150 years.

Coexistence. In the real world, it's rare for one species (or subspecies) to be flat-out better than another. If they were, the inferior species would have gone extinct long ago.[10] Instead, typically one species is better at one thing while its competitor is better at something else. Different local conditions might make one advantage more important that the other, so one species dominates in some ecosystems while the other dominates in other ecosystems. That is, the two species occupy different evolutionary **niches**.

6.52. Suppose that one species grows quickly, with $r_1 = 0.2$/year, while another species grows slowly, with $r_2 = 0.1$/year. However, suppose that the second species has a higher carrying capacity than the first species, say with $L_1 = 1$ million and $L_2 = 2$ million. If an ecosystem initially has a small number of each species, describe what happens.

6.53. Now use MATLAB to numerically plot what happens in Exercise 6.52 for 200 years, starting with 100 of each species. Do these numerical results agree with your prediction?

Food for Thought: The process in Exercises 6.52–6.53 is exactly what happens in a forest after logging or fire or some other disaster and is sometimes called the **meadow cycle** or **succession**. For a while, the burnt-out forest turns into a meadow that is dominated by species like grasses that spread very quickly. These are replaced by shrubs and by trees that grow fairly quickly but require direct sun, such as pines. These in turn are replaced by slower growing trees, such as oaks, that can thrive in the shade of other trees. An **old growth forest**, where this replacement process process has reached an equilibrium, is a very different (and much more diverse!) ecosystem than a **secondary growth forest** that is dominated by a few fast-growing species.

---

[10]The exception is when we're dealing with a new species. The Covid virus mutated quickly, with new variants developing considerable advantages over older variants, and the older variants quickly went extinct.

Bimolecular Chemical Reactions. In Exercises 6.54–6.56 we consider a bimolecular reaction $A + B \Leftrightarrow C$ with dynamics described by equation (6.86):

$$\frac{d[A]}{dt} = \frac{d[B]}{dt} = -k_1[A][B] + k_2[C]; \qquad \frac{d[C]}{dt} = k_1[A][B] - k_2[C].$$

6.54. (a) Use MATLAB and Euler's method to solve this rate equation for t running from 0 to 20, and to graph the solution. Use the parameters $k_1 = 0.02$ and $k_2 = 0.01$ and the initial conditions $[A](0) = 10$, $[B](0) = 8$, $[C](0) = 0$.
       (b) What is $\frac{d[A]}{dt} - \frac{d[B]}{dt}$ equal to? What does this imply about $[A](t) - [B](t)$?
       (c) Consider the general case with $[A](0) = A_0$, $[B](0) = B_0$, and $[C](0) = C_0$. What is $[A](t) - [B](t)$ equal to?

6.55. In Exercise 6.54, we explored the bimolecular reaction $A + B \Leftrightarrow C$ and discovered that $[A](t) - [B](t) = L$, where $L = [A](0) - [B](0)$. In this problem we will assume that k_2 is so small that we can ignore the rate at which the reaction runs backwards. That is, we are using the approximation $k_2 \approx 0$. This is often the case when C has a much lower free energy than A and B separately.
       (a) Why can we assume, without loss of generality, that $L > 0$?
       (b) Let $y = [A]$. By writing $[B](t) = [A](t) - L$ and substituting into the rate equation for $[A]$, show that

$$\frac{dy}{dt} = k_1 L y \left(1 - \frac{y}{L}\right).$$

This is our old friend the **logistic equation** with $r = k_1 L$. In studying Australian rabbits and humans vs. zombies, we mostly cared about what happens when $0 < y < L$. Here we are interested in what happens when $y \geq L$.
       (c) In words, what does $\frac{dy}{dt} = 0$ mean?
       (d) Eventually, the system will approach a steady state in which $dy/dt = 0$. What will $[A]$ and $[B]$ be then? In words, what causes the reaction to stop?
       (e) Use MATLAB and the Logistics program of Exercise 6.35 to make a plot of $y(t) = [A](t)$. Use $[A](0) = 10$, $[B](0) = 8$, $k_1 = 0.02$, and $t_{max} = 20$. Does this resemble your plot in part (a) of Exercise 6.54? How is it different?

6.56. Consider another bimolecular reaction $A + B \Leftrightarrow C$. In this exercise we do *not* assume that $k_2 \approx 0$. In fact, we will take $k_1 = k_2 = 1$. We also assume that we start with equal amounts of A and B. Let $y = [A]$.
       (a) Show that $[A] + [C]$ is a constant. Call this constant z.
       (b) Express the condition for equilibrium as a quadratic equation involving y and no other variables. (However, the constant z will appear.)
       (c) Solve for y in terms of z.
       (d) Suppose that $z = 0.01$. At equilibrium, which is bigger, $[A]$ or $[C]$?
       (e) Now suppose that $z = 100$. At equilibrium, which is bigger, $[A]$ or $[C]$?
       (f) Graph the ratio $[A]/[C]$ as a function of z, as z ranges from 0.01 to 100.

Food for Thought: Exercise 6.56 illustrates a general chemical principle. In dilute solutions, or even in interstellar space, chemicals and even atoms have a tendency to dissociate or ionize. But the higher the concentration, the more they tend to remain whole.

6.57.   **Enzymes.** An enzyme is a chemical, usually a protein, that increases the rate constants k_1 and k_2 of a chemical reaction. In this problem we'll examine how an enzyme can affect the trajectory of a bimolecular reaction $A + B \Leftrightarrow C$.

(a) In Exercise 6.54(a), you graphed the solution to the rate equations (6.86) with the parameters $k_1 = 0.02$ and $k_2 = 0.01$ and the initial conditions $[A](0) = 10$, $[B](0) = 8$, $[C](0) = 0$, and with t running from 0 to 20. Now suppose that $k_1 = 0.2$ and $k_1 = 0.1$. That is, suppose that both k_1 and k_2 are 10 times bigger than before. Graph the solution to the rate equations (6.86), with the same initial conditions, with t running from 0 to 2. How does this compare to the graph you found in Exercise 6.54(a)?

(b) Let $\tau = 10t$. In the situation of part (a), what are $d[A]/d\tau$, $d[B]/d\tau$, and $d[C]/d\tau$? How do those compare to $d[A]/dt$, $d[B]/dt$ and $d[C]/dt$ in the situation of Exercise 6.54(a)? What can you conclude about the trajectories of the two reactions and their equilibrium concentrations?

Food for Thought: In practice, enzymes can change the speed of a biochemical reaction by much more than a factor of ten. Although they don't change the equilibrium point, they can essentially turn reactions on and off. By controlling the amount of different enzymes it releases, your body regulates which reactions will happen, and when.

Trigonometry

6.58.   Draw a right triangle where one of the angles is $\tan^{-1}(3/4)$. Use this triangle to compute $\sin(\tan^{-1}(3/4))$ and $\sec(\tan^{-1}(3/4))$.

6.59.   Draw a right triangle where one of the angles is $\sin^{-1}(5/13)$. Use this triangle to compute $\cos(\sin^{-1}(5/13))$ and $\tan(\sin^{-1}(5/13))$.

6.60.   Draw a right triangle featuring $\cos^{-1}(x)$, where x is an unspecified number between 0 and 1. In terms of x, what are all six trig functions of $\cos^{-1}(x)$?

6.61.   Draw a right triangle featuring $\sec^{-1}(x)$, where x is an unspecified number greater than 1. In terms of x, what are all six trig functions of $\sec^{-1}(x)$?

The Whole
Is the Sum of the Parts

Chapters 2 through 6 have been about **differential calculus**. We developed the idea of a derivative, we developed rules for computing derivatives, and we used derivatives in a number of applications. Finally, we acknowledged the fact that most of the laws that govern the natural world are best described as differential equations, and we saw how the exponential, log and trig functions that come from solving the simplest of these differential equations appear over and over in the natural world.

In this chapter we change gears completely and introduce **integration**. The root meaning of *integral* is *whole*. An *integer* is a whole number. An airplane has *structural integrity* if it stays in one piece. *Personal integrity* means not speaking with a forked tongue. *Social integration* means bringing different social or racial groups into a unified community. *E Pluribus Unum*: Out of many, one.

Mathematically, an **integral** represents a bulk quantity, accumulated over a region in space or time. In all cases, we compute the whole by breaking it up into little pieces, estimating the contribution of each piece, and adding up the pieces. This is the 4th Pillar of Calculus:

> *The whole is the sum of the parts.*

If you have taken calculus before, you may be thinking that integration is the opposite of differentiation. *What in blazes do sums have to do with it?!* The truth is that anti-derivatives have nothing to do with what integrals *are*, and very little to do with that they are *good for*, but they have a tremendous amount to do with how integrals are *calculated*. Meanwhile, sums have everything to do with what integrals are and what they are good for.

In this chapter we're going to talk about what integrals are (sums!) and what they're good for. We'll start in Section 7.1 with a variety of problems that can be addressed with a slice-and-dice strategy. In Section 7.2 we'll develop notation for estimating all the parts of the whole and computing their sum. In Section 7.3 we will define the integral of an arbitrary function over an arbitrary interval. Finally, in Section 7.4 we will consider accumulation functions, which are running totals.

Once we've learned what integrals are and how they can be used, we can turn our attention to computing them. In Chapter 8, we'll learn how to do that using antiderivatives.

7.1. Slicing and Dicing

We start with three problems. Their real-world meanings are very different, but mathematically they are almost identical.

(1) A bear prepared for hibernation by gaining weight at a rate of $10\sqrt{16 - t^2}$ kilograms per month for four months, where t is the time in months since it started packing in the food. How much weight did it gain in all?

(2) A projectile was shot into a vat of viscous fluid (e.g., molasses) that slowed it down and stopped it after 4 seconds. The velocity of the projectile t seconds after hitting the fluid was $100\sqrt{16 - t^2}$ mm/second. How far did the projectile go before it stopped?

(3) What is the area under the curve $y = \sqrt{16 - t^2}$ between $t = 0$ and $t = 4$?

If the bear gained weight at a constant rate, or if the projectile moved at a constant speed, or if the graph was a horizontal line, these problems would be easy.

Weight gain	=	gain rate	×	time,
Distance	=	velocity	×	time,
Area	=	height	×	width.

Unfortunately, the function $\sqrt{16 - t^2}$ is *not* constant. However, it is *continuous*, which means that it only changes a little over a short period of time. We can get a good approximation to each total by breaking the interval $[0, 4]$ into smaller pieces, and then approximating $\sqrt{16 - t^2}$ by a constant on each piece.

The Bear. We can treat the bear's weight gain one month at a time. Let $f(t) = 10\sqrt{16 - t^2}$ kg/month be the rate at which the bear gained weight, as tabulated in Table 7.1.

In the first month, the bear gained weight at rate of $f(0) = 40$ kg/month at the beginning of the month, and gained weight at rate $f(1) \approx 38.7$ kg/month at the end of the month. All month long, it was gaining weight at a rate somewhere between 38.7 and 40 kg/month, so its weight gain for the month was somewhere between 38.7 kg/month × 1 month = 38.7 kg and 40 kg. Similarly, the gain in the second month was somewhere between $f(2) \times 1$ month and $f(1) \times 1$ month, the gain in the third month was somewhere between $f(3) \times 1$ month and $f(2) \times 1$ month, and the gain in the fourth

Table 7.1. Weight gain rate as a function of time

Time (in months)	Fattening rate (in kg/month)
0	40
1	38.7
2	34.6
3	26.5
4	0

month was somewhere between $f(4) \times 1$ month and $f(3) \times 1$ month. Adding these up, we get

$$
\begin{array}{rcccr}
38.7 \text{ kg} & < & \text{1st month gain} & < & 40 \text{ kg} \\
34.6 \text{ kg} & < & \text{2nd month gain} & < & 38.7 \text{ kg} \\
26.5 \text{ kg} & < & \text{3rd month gain} & < & 34.6 \text{ kg} \\
0 & < & \text{4th month gain} & < & 26.5 \text{ kg} \\
\hline
99.8 \text{ kg} & < & \text{Total gain} & < & 139.8 \text{ kg.}
\end{array}
$$

This calculation is very rough. All we know is that the bear gained between 99.8 kg and 139.8 kg. A reasonable guess is the average between the **lower bound** of 99.8 kg and the **upper bound** of 139.8 kg, namely 119.8 kg. But that's still just a rough guess.[1]

The trouble with this calculation is that the weight gain rate can change by a lot in the course of a single month. In the 4th month, it went all the way from 26.5 kg/month to zero! To do better, we can redo the calculation half a month at a time, or one week at a time, or one day at a time, or one hour at a time. The downside is that doing things by weeks requires 12 estimates and doing things by days requires 122 estimates. As with Euler's method, smaller time steps mean greater accuracy, but at the cost of more work.

The Projectile. We can estimate the distance traveled by our projectile the same way that we estimated the bear's weight gain. We make a table of the velocity $v(t) = 100\sqrt{16 - t^2}$ (mm/sec) as a function of time.

Time (in seconds)	Velocity (in mm/sec)
0	400
1	387
2	346
3	265
4	0

---

[1]120 kg is a fairly typical weight gain for a male black bear, but Alaskan grizzly bears have been observed to gain five times that much!

In the first second, the projectile was going between 387 mm/sec and 400 mm/sec, in the second second it was going between 346 mm/sec and 387 mm/sec, and so on.

387 mm	<	Distance traveled in 1st second	<	400 mm
346 mm	<	Distance traveled in 2nd second	<	387 mm
265 mm	<	Distance traveled in 3rd second	<	346 mm
0 mm	<	Distance traveled in 4th second	<	265 mm
998 mm	<	Total distance traveled	<	1398 mm

Averaging the upper and lower bounds we get an estimated distance of 1198 mm, or just over a meter.

The Area Under an Arc. The area under the curve is similar. We break the interval $[0, 4]$ into four pieces of width $\Delta t = 1$, which breaks the area under the curve into four vertical strips. See Figure 7.1. The area of each strip is greater than the area of the darkly shaded rectangle that sits inside it, and is less than what you get by including the area of the lightly shaded rectangle. That is,

3.87	<	Area of first strip	<	4
3.46	<	Area of second strip	<	3.87
2.65	<	Area of third strip	<	3.46
0	<	Area of fourth strip	<	2.65
9.98	<	Total area	<	13.98

Averaging the upper and lower bounds gives an estimated area of 11.98.

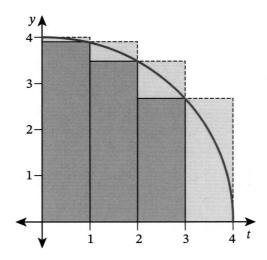

Figure 7.1. The area under $y = \sqrt{16 - t^2}$

As with the bear and projectile problems, we can get greater accuracy by using more and more strips, as seen in Figure 7.2.

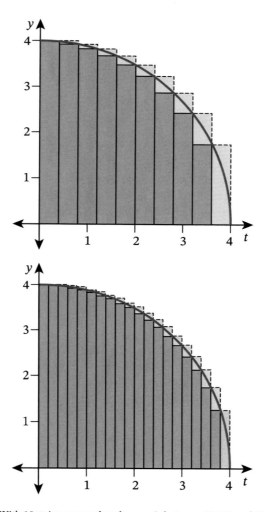

Figure 7.2. With 10 strips we see that the area is between 11.618 and 13.218. With 20 we see that it is between 12.114 and 12.914.

Geometry to the Rescue. The calculations in the three problems are *exactly* the same, except for factors of 10 kg and 100 mm. The amount of weight gained by the bear, in kilograms, is equal to the area under the curve $y = 10\sqrt{16 - t^2}$ between $t = 0$ and $t = 4$, which is 10 (kg) times the area under the curve $y = \sqrt{16 - t^2}$. The distance (in millimeters) traveled by the projectile is equal to the area under the curve $y = 100\sqrt{16 - t^2}$, which is 100 (mm) times the area under $y = \sqrt{16 - t^2}$.

This is a general pattern. The total accumulation of a quantity (the **integral**) can often be visualized as the area under the graph of that quantity, just as the rate of change of that quantity (the **derivative**) can be visualized as the slope of the graph. In calculus, we do a lot of slope and area-under-the-curve problems, but it isn't because we're obsessed with geometry. It's because problems that don't seem to have anything to do with geometry can often be solved with a little geometrical insight.

In this case, we can solve the bear and projectile problems exactly with the help of some high school geometry. If $y = \sqrt{16 - t^2}$, then $y^2 = 16 - t^2$ and $t^2 + y^2 = 16 = 4^2$, so our curve is part of a circle of radius 4 with its center at the origin. The area we are looking for is the region in the first quadrant enclosed by this circle, and therefore equals

$$(7.1) \qquad \frac{1}{4}\pi r^2 = \frac{1}{4}16\pi = 4\pi \approx 12.57.$$

This means that the projectile traveled a distance of $400\pi \approx 1257$ mm, and that the bear gained $40\pi \approx 125.7$ kg.

Even with only four pieces, the estimates we got by averaging the upper and lower bounds (namely a weight gain of 119.8 kg, a distance of 1198 mm and an area of 11.98) turned out to be accurate to within 5%. This approximation is called the **trapezoidal rule**, and it is pictured in Figure 7.3. We will revisit this method, with more details about its accuracy, in Chapter 9.

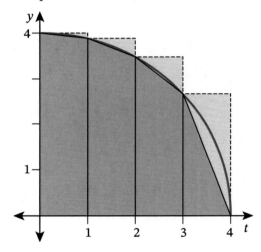

Figure 7.3. The area of each trapezoid is the average of the areas of the outer and inner rectangles.

And now for something completely different. Or maybe not. At first glance, the next three problems look like they have nothing to do with bears gaining weight or bullets moving through molasses. However, they are solved with the exact same slice-and-dice algorithm. That's the heart of integration.

(1) The **moment of inertia** of a particle of mass m at distance r from the origin is mr^2. This quantity measures how much effort (torque) it takes to rotate the particle around the origin. The moment of inertia of a larger object is the sum of the moments of inertia of its pieces. What is the moment of inertia of a thin uniform rod of mass 1 kg and length 1 m, sitting on the x-axis between $x = 0$ and $x = 1$ (meter)?

(2) What is the volume of a pyramid with a 1 meter $\times$ 1 meter square base and a height of 1 meter?

(3) What is the area under the parabola $y = x^2$ between $x = 0$ and $x = 1$?

I claim that all three problems have the same answer, so solving any one of them is tantamount to solving all three. To see why, let's break each of the three quantities—the moment of inertia of the rod, the volume of the pyramid, and the area under the parabola—into 10 pieces and estimate the size of each piece.

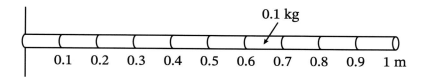

Figure 7.4. A rod broken into 10 pieces

The Rod. If we break the rod into 10 pieces, then each piece has a mass of 1 kg/10 = 0.1 kg (see Figure 7.4). The points in the first piece have a distance to the origin that ranges from 0 to 0.1 m, so the moment of inertia of the first piece is somewhere between $0.1 \times (0)^2$ kg-m^2 and $0.1(0.1)^2$ kg-m^2. The second piece has mass 0.1 kg, and every point in the second piece is at distance at least 0.1 m from the origin and at most 0.2 m from the origin, so the moment of inertia of the second piece is between $0.1(0.1)^2$ kg-m^2 and $0.1(0.2)^2$ kg-m^2. Continuing in this way we get

0 kg-m^2	<	Moment of inertia of 1st piece	< 0.001 kg-m^2
0.001 kg-m^2	<	Moment of inertia of 2nd piece	< 0.004 kg-m^2
0.004 kg-m^2	<	Moment of inertia of 3rd piece	< 0.009 kg-m^2
0.009 kg-m^2	<	Moment of inertia of 4th piece	< 0.016 kg-m^2
0.016 kg-m^2	<	Moment of inertia of 5th piece	< 0.025 kg-m^2
0.025 kg-m^2	<	Moment of inertia of 6th piece	< 0.036 kg-m^2
0.036 kg-m^2	<	Moment of inertia of 7th piece	< 0.049 kg-m^2
0.049 kg-m^2	<	Moment of inertia of 8th piece	< 0.064 kg-m^2
0.064 kg-m^2	<	Moment of inertia of 9th piece	< 0.081 kg-m^2
0.081 kg-m^2	<	Moment of inertia of 10th piece	< 0.100 kg-m^2
0.285 kg-m^2	<	Total moment of inertia	< 0.385 kg-m^2

We can also use the trapezoidal rule, averaging the upper and lower bounds to get an approximate moment of inertia of 0.335 kg-m^2.

The Pyramid. To find the volume of the pyramid, we lay it on its side and slice it into 10 slabs (see Figure 7.5). Each of these slabs is a square tile with beveled edges, as shown in Figure 7.6. The ith tile has thickness 0.1 m. On the right its cross-section is a square of side length $\frac{i}{10}$ meters and area $\frac{i^2}{100}$ square meters, and on the left its cross-

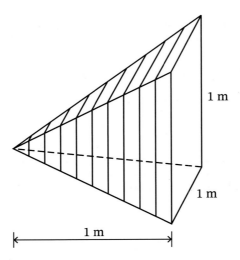

Figure 7.5. A pyramid sliced into 10 pieces

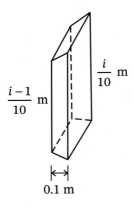

Figure 7.6. The ith slice of the pyramid

section is a square of side length $\frac{i-1}{10}$ meters and area $\frac{(i-1)^2}{100}$ square meters. This means that its volume is somewhere between $\frac{(i-1)^2}{1000}$ and $\frac{i^2}{1000}$ cubic meters. That is,

0 m^3	$<$	Volume of 1st slab	$<$	0.001 m^3
0.001 m^3	$<$	Volume of 2nd slab	$<$	0.004 m^3
0.004 m^3	$<$	Volume of 3rd slab	$<$	0.009 m^3
0.009 m^3	$<$	Volume of 4th slab	$<$	0.016 m^3
0.016 m^3	$<$	Volume of 5th slab	$<$	0.025 m^3
0.025 m^3	$<$	Volume of 6th slab	$<$	0.036 m^3
0.036 m^3	$<$	Volume of 7th slab	$<$	0.049 m^3
0.049 m^3	$<$	Volume of 8th slab	$<$	0.064 m^3
0.064 m^3	$<$	Volume of 9th slab	$<$	0.081 m^3
0.081 m^3	$<$	Volume of 10th slab	$<$	0.100 m^3
0.285 m^3	$<$	Total Volume	$<$	0.385 m^3

The Parabola. Finally, we look at the area under the parabola $y = x^2$ and cut the region into 10 strips. The area of the each strip is at least the area of biggest rectangle that fits inside it and at most the area of the smallest rectangle that surrounds it (see Figure 7.7).

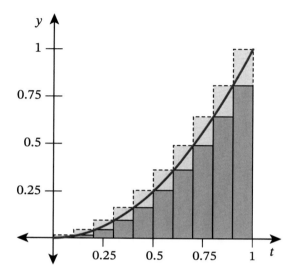

Figure 7.7. The area under a parabola sliced into 10 pieces

0	<	Area of 1st strip	<	0.001
0.001	<	Area of 2nd strip	<	0.004
0.004	<	Area of 3rd strip	<	0.009
0.009	<	Area of 4th strip	<	0.016
0.016	<	Area of 5th strip	<	0.025
0.025	<	Area of 6th strip	<	0.036
0.036	<	Area of 7th strip	<	0.049
0.049	<	Area of 8th strip	<	0.064
0.064	<	Area of 9th strip	<	0.081
0.081	<	Area of 10th strip	<	0.100
0.285	<	Total Area	<	0.385

The upper and lower bounds for the volume of the pyramid (in cubic meters) are exactly the same as the upper and lower bounds for the moment of inertia of the rod (in kg-m^2), which are exactly the same as the upper and lower bounds for the area under the parabola. If we want more accuracy, we can cut our rod and our pyramid and the region under the parabola into more pieces. The more pieces we use, the closer our upper and lower bounds will be to the true answers. But no matter how many pieces we use, the volume (in m^3) will still equal the moment of inertia (in kg-m^2) and the area. To a mathematician, the three problems are the same.

In Section 8.2 we will see that the exact answer to all three problems is $1/3 \approx 0.333$. The trapezoidal rule estimate of $(0.285 + 0.385)/2 = 0.335$ is only off by half a percent.

Summary. We have looked at six different problems in this section. One involved eating, one involved distance, one involved a moment of inertia, one involved volume, and two involved areas. However, all six were solved with the same slice-and-dice strategy, following the slogan *the whole is the sum of the parts*.

---

(1) Break the quantity we are trying to study into little pieces.

(2) Estimate the size of each piece. In our examples were were able to say "at least this much and at most that much", which is ideal, but estimates of the form "approximately this much" are also useful. If we have upper and lower bounds, then the average of the two (the **trapezoidal rule**) is a good approximate value.

(3) Add up the estimated sizes of the pieces to get an estimated total. We can also add up the lower bounds for the pieces to get a lower bound for the total, and add up the upper bounds for the pieces to get an upper bound for the total.

(4) For more accuracy, repeat the process with more pieces. As the number of pieces goes to infinity, the lower bound, the upper bound and any estimate in between will all home in on the exact answer.

In each case, the estimated contribution of each piece was of the form $f(x)\Delta x$ or $f(t)\Delta t$, where Δx (or Δt) was the width of each piece. This is the same thing that we would get if we studied the area under the curve $y = f(x)$ (or $y = f(t)$). In other words, *most bulk quantities can be visualized as the area under some curve*. For the hungry bear problem, $f(t)$ was the rate of weight gain. For the projectile problem, it was the velocity. For the moment of inertia of the rod it was the square of the distance from the origin, and for volume of the pyramid it was the cross-sectional area.

There no single magic formula that tells us what $f(x)$ or $f(t)$ is supposed to be. Depending on the application, we have to reason that out using our understanding of biology or physics or three-dimensional geometry. Calculus doesn't tell us what functions to study! It just tells us what we can do with those functions once we've found them.

Finally, the phrase "area under the curve" needs to be taken with a grain of salt. When $f(x)$ is negative, the negative contribution $f(x)\Delta x$ isn't the approximate area of a strip lying below the curve $y = f(x)$. Instead, it is *minus* the area of a strip lying *above* the curve. For instance, suppose that $f(x) = \cos(x)$ and that we are working on the interval $[0, 2\pi/3]$, as in Figure 7.8. When we slice it into pieces and add up terms of the form $f(x)\Delta x$ we are actually estimating the area of the blue region *minus* the area of the red region. But we still refer to the result, somewhat inaccurately, as "the area under $y = \cos(x)$ as x goes from 0 to $2\pi/3$".

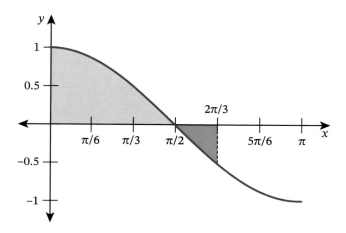

Figure 7.8. The area "under" $y = \cos(x)$ is actually a difference of two areas.

7.2. Riemann Sums

In the last section we saw that all sorts of bulk quantities can be represented as areas under (or minus areas over) a curve $y = f(x)$. In this section we develop the formalism for expressing that area as the limit of a sum.

Because we are always adding many terms, we need a compact notation for sums. We use an oversized capital Greek letter sigma, written $\sum$. The formula

$$(7.2) \qquad z = \sum_{j=1}^{5} j^2$$

is a shorthand way of saying "add up five numbers, namely j^2 when j equals 1, 2, 3, 4, and 5, and call the result z". Underneath the $\sum$ we write the name of the counter and its starting value, and on top we write its ending value. You might think of it as a loop in a computer program:

sum = 0;	% Start your sum at zero.
for j=1:5	% Do a loop with j running from 1 to 5.
sum = sum + j*j;	% Add j^2 to our running total.
end;	% Finish the loop.
z = sum	% Set z equal to the total.

We can call the counter in the sum i, j, m, n, or any other letter. That is,

$$(7.3) \qquad \sum_{i=1}^{5} i^2 = \sum_{j=1}^{5} j^2 = \sum_{m=1}^{5} m^2 = \sum_{n=1}^{5} n^2 = 1^2 + 2^2 + 3^2 + 4^2 + 5^2 = 55.$$

The counter usually starts at 1, but it doesn't have to. We can also write

$$(7.4) \quad \sum_{k=9}^{13} (k - 10)^3 = (9 - 10)^3 + (10 - 10)^3 + (11 - 10)^3 + (12 - 10)^3 + (13 - 10)^3.$$

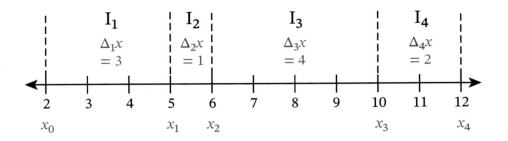

Figure 7.9. The interval [2,12] broken into four unequal pieces

Next we develop some notation for what happens when we break the interval $[a, b]$ into pieces. We usually call the number of pieces N. Most of the time we break $[a, b]$ into equal sized pieces, but that isn't strictly necessary. For instance, we *might* choose to break the interval $[2, 12]$ into $[2, 5]$, $[5, 6]$, $[6, 10]$, and $[10, 12]$, as in Figure 7.9.

If our input variable is called x, then the endpoints of these intervals are called x_0, x_1, ..., x_N, so in this case $a = x_0 = 2$, $x_1 = 5$, $x_2 = 6$, $x_3 = 10$ and $b = x_4 = 12$. (If the input variable is called t then we call the endpoints t_0, t_1, etc.) We call the ith interval I_i (or call the jth interval I_j or call the nth interval I_n). I_i runs from x_{i-1} to x_i, and its length is $\Delta_i x = x_i - x_{i-1}$. In this case $\Delta_1 x = 3$, $\Delta_2 x = 1$, $\Delta_3 x = 4$ and $\Delta_4 x = 2$. Breaking the interval $I = [a, b]$ into subintervals I_i is equivalent to cutting the region under the curve $y = f(x)$ between $x = a$ and $x = b$ into vertical strips, each one sitting over an interval I_i.

We need to estimate the area of the strip sitting above I_i. This strip is approximately a rectangle of width $\Delta_i x$ and height $f(x)$, so its area is approximately $f(x)\Delta_i x$. But which value of x do we plug into the function $f(x)$?

There are many ways to pick a point x_i^* to represent the interval I_i. We could use the **left endpoint** x_{i-1}, the **right endpoint** x_i, or any point in between. Sometimes it's useful to pick the point with the smallest value of $f(x)$, in which case $f(x_i^*)\Delta x$ will be a lower bound for the area of the strip. (If $f(x)$ is an increasing function on I_i, this is the same as picking the left endpoint, while if $f(x)$ is decreasing it's the same as picking the right endpoint.) Other times it's useful to pick the point with the largest value of $f(x)$, in which case $f(x_i^*)\Delta x$ will be an upper bound for the area.

There is no single right answer for which point to choose. When computing derivatives, we could use forward differences, backward differences, or centered differences. Centered differences were the most accurate, but they all gave the same answer in the limit of smaller and smaller intervals. Likewise, we can compute areas using left endpoints, right endpoints, midpoints, or a host of other schemes. Some are more accurate than others, but as $N \to \infty$ (and the sizes of the intervals go to zero) they all approach the correct area.

Now that we have our notation settled, we can summarize the procedure for finding the area under the curve $y = f(x)$ between $x = a$ and $x = b$ (see Figure 7.10).

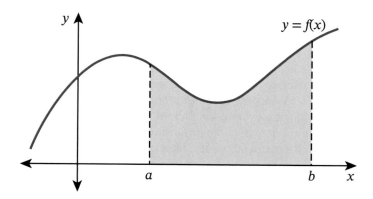

Figure 7.10. What is the area of the shaded region?

(1) Decide on the number N of pieces we are going to break the interval $I = [a, b]$ into. This is the same as the number of strips that we are going to break the region under the curve $y = f(x)$ from $x = a$ to $x = b$ into.

(2) Let $x_0 = a$, $x_N = b$, and pick points $x_1 < x_2 < \cdots < x_{N-1}$ in between. Most of the time we want these points to be spaced an equal distance $\Delta x = (b-a)/N$ apart, in which case we have $x_i = a + i\Delta x = a + \frac{i(b-a)}{N}$. However, other choices are possible, in which case the various widths $\Delta_i x = x_i - x_{i-1}$ may not all be the same.

(3) In each interval $I_i = [x_{i-1}, x_i]$, pick a representative point x_i^*.

(4) Approximate the ith strip (that is, the region over I_i) as a rectangle of height $f(x_i^*)$ and width $\Delta_i x$, so it has area $f(x_i^*)\Delta_i x$.

(5) Add up the pieces. The total area under the curve is approximately

(7.5)
$$\sum_{i=1}^{N} f(x_i^*)\Delta_i x.$$

This expression is called a **Riemann sum**.[2]

(6) For more accuracy, increase N. The exact area under the curve is

(7.6)
$$\lim_{N\to\infty} \sum_{i=1}^{N} f(x_i^*)\Delta_i x.$$

This limit does not depend on the choices you made. As long as $f(x)$ is continuous and the lengths of all the intervals approach 0, the Riemann sum (7.5) will get closer and closer to the true area as $N \to \infty$.

We are using area problems to develop intuition for Riemann sums, but Riemann sums apply to *much* more than just area problems. The same ideas apply to distance,

---

[2]Bernhard Riemann (1826–1866) was one of the greatest mathematicians in history. He died of tuberculosis at age 39 and only published about a dozen papers (some posthumously), but every one of his papers was groundbreaking. Besides being the first person to define integrals rigorously, he invented much of differential geometry (sometimes called Riemannian Geometry), complex analysis (Riemann surfaces), and number theory (the Riemann zeta function).

volume, arc-length, accumulated income, present value, and many other applications. You'll see some of these in the exercises, but the list of possible applications is way too long to fit in one book. The common feature of all of these applications is that there is a big quantity that can be broken into many little pieces, and that the contribution of the ith piece can be approximated as $f(x_i^*)\Delta x$ for some function $f(x)$ and some variable x (or as $f(y_i^*)\Delta y$ or $f(t_i^*)\Delta t$ or $f(r_i^*)\Delta r$ or whatnot). To understand what the function $f(x)$ is, we need to understand something about distance or volume or arc-length or income or present value. *There is no magic formula for that step!* But once we've done that, the machinery of Riemann sums, and later of definite integrals, takes care of the rest.

Area Problems. Let's apply our procedure to the two area problems of Section 7.1.

First, we want the area under $y = f(x) = \sqrt{16 - x^2}$ between $x = 0$ and $x = 4$. We set $N = 4$ and break $[0, 4]$ into equal sized intervals, each of length $\Delta x = \frac{4-0}{4} = 1$. We then have

$$
\begin{aligned}
x_0 &= a & &= 0 \\
x_1 &= a + \Delta x & &= 1 \\
(7.7) \qquad x_2 &= a + 2\Delta x & &= 2 \\
x_3 &= a + 3\Delta x & &= 3 \\
x_4 &= a + 4\Delta x & &= 4 = b
\end{aligned}
$$

To do the calculation with left endpoints we pick $x_1^* = x_0 = 0$, $x_2^* = x_1 = 1$, $x_3^* = x_2 = 2$, and $x_4^* = x_3 = 3$. In other words, $x_i^* = i - 1$. Our estimated area is then

$$
\begin{aligned}
\sum_{i=1}^{4} f(x_i^*)\Delta x &= \sum_{i=1}^{4} f(i-1)\Delta x \\
(7.8) \qquad &= \sum_{i=1}^{4} \sqrt{16 - (i-1)^2} \\
&= 4 + \sqrt{15} + \sqrt{12} + \sqrt{7} \approx 13.98.
\end{aligned}
$$

In general, equal-sized intervals and left endpoints always give us

$$
\begin{aligned}
(7.9) \qquad \sum_{i=1}^{N} f(x_i^*)\Delta x &= \sum_{i=1}^{N} f(x_{i-1})\Delta x \\
&= \Delta x\,(f(x_0) + \cdots + f(x_{N-1})).
\end{aligned}
$$

If instead we used right endpoints, then each x_i^* would be x_i instead of x_{i-1}. Our estimated area would then be

$$
\begin{aligned}
\sum_{i=1}^{4} f(x_i^*)\Delta x &= \sum_{i=1}^{4} f(i)\Delta x \\
(7.10) \qquad &= \sum_{i=1}^{4} \sqrt{16 - (i)^2} \\
&= \sqrt{15} + \sqrt{12} + \sqrt{7} + \sqrt{0} \approx 9.98.
\end{aligned}
$$

In general, equal-sized intervals and right endpoints always give us

(7.11)
$$\sum_{i=1}^{N} f(x_i^*)\Delta x = \sum_{i=1}^{N} f(x_i)\Delta x$$
$$= \Delta x \left(f(x_1) + \cdots + f(x_N) \right).$$

If the function is decreasing, as in this case, then using left endpoints gives too big an answer and using right endpoints gives too small an answer. If the function is increasing, then it's the other way around, with left endpoints giving an underestimate and right endpoints giving an overestimate. The **trapezoidal rule** says to take the average of the answers. It gives

(7.12)
$$\text{Area} \approx \frac{1}{2}\left(\sum_{i=1}^{N} f(x_{i-1})\Delta x + \sum_{i=1}^{N} f(x_i)\Delta x \right)$$
$$= \Delta x \left(\frac{1}{2}f(x_0) + f(x_1) + f(x_2) + \cdots + f(x_{N-1}) + \frac{1}{2}f(x_N) \right)$$
$$= \frac{1}{2}4 + \sqrt{15} + \sqrt{12} + \sqrt{7} + \frac{1}{2}\sqrt{0} \approx 11.98.$$

Another obvious compromise between left and right endpoints is to use midpoints. Because $I_i = [x_{i-1}, x_i]$, the **midpoint rule** has

(7.13) $\quad x_i^* = \frac{1}{2}(x_{i-1} + x_i) = \frac{1}{2}(a + (i-1)\Delta x + a + i\Delta x) = a + \left(i - \frac{1}{2}\right)\Delta x.$

In our example, that gives an estimated area of

(7.14)
$$\sum_{i=1}^{4} f(x_i^*)\Delta x = \sum_{i=1}^{4} f\left(i - \frac{1}{2}\right)\Delta x$$
$$= \sum_{i=1}^{4} \sqrt{16 - \left(i - \frac{1}{2}\right)^2}$$
$$= \sqrt{15.75} + \sqrt{13.75} + \sqrt{9.75} + \sqrt{3.75} \approx 12.74.$$

The midpoint rule is slightly more accurate than the trapezoidal rule and is a *lot* more accurate than either the left endpoint rule or the right endpoint rule.

For the second area problem of Section 7.1, namely computing the area under the parabola $y = x^2$, we will program a computer to do our calculations for us using left endpoints. Because the algorithm is the same for all choices of starting point a, endpoint b, an number of intervals N, and even for all functions $f(x)$, we will leave these as input parameters for our program. In the exercises you will be asked to modify the program to work with right endpoints, midpoints, or the trapezoidal rule and with functions other than $y = x^2$. The following program is written for MATLAB, but it can be easily adapted to work on other platforms. Note that the counter in this program is called k instead of i. That is because MATLAB also works with complex numbers, where i already means $\sqrt{-1}$.

```
function v = Riemann(f,a,b,N)    % f, a, b and N are
                                 % input parameters
v = 0;                           % Initialize our sum at 0
deltax = (b - a)/N;              % Compute Δx
for k = 1:N
    xstar = a + (k-1)*deltax;    % Compute x*_k = x_{k-1}.
    v = v + f(xstar);            % Add f(x*_k) to v
end                              % Finish the loop.
v = deltax*v;                    % Multiply Σ^N_{k=1} f(x*_k) by Δx.
```

Instructions for using this program can be found just above Exercise 7.21.

7.3. The Definite Integral

The Definition of an Integral. In Section 7.1 we saw that a lot of bulk quantities can be expressed as limits of sums. The procedure was the same whether we were talking about a bear's weight gain, the distance traveled by a projectile, the moment of inertia of a bar, the volume of a pyramid, or the area under a curve. In Section 7.2 we looked more closely at the area under a curve and developed notation and terminology to describe what we were doing in Section 7.1. With that notation and terminology in hand, we are finally ready to define the **definite integral**.

Definition. Suppose that $f(x)$ is a function defined on the interval $[a, b]$. The quantity

$$(7.15) \qquad \lim_{N \to \infty} \sum_{i=1}^{N} f(x_i^*)\Delta_i x$$

is called the *definite integral of $f(x)$ with respect to x from a to b* and is denoted

$$\int_a^b f(x)dx.$$

The **integral sign** $\int$, sometimes written $\smallint$, is designed to look like a stretched out letter S. The S stands for "sum", as did the Greek letter Σ. The expression dx in the integral is analogous to the Δx in the Riemann sum. It tells us that our variable is x, and that we are adding up contributions from all values of x from a to b. *Do not leave it out!* The expression $\int_2^3 x^2$, without a dx in it, is meaningless.

There are a lot of details, and a hidden assumption, built into equation (7.15). To begin with, we need to build the Riemann sum $\sum_{i=1}^{N} f(x_i^*)\Delta_i x$ as in Section 7.2. That is, we have to break the interval $[a, b]$ into N subintervals of length $\Delta_1 x$, $\Delta_2 x$, etc., pick a representative point x_i^* in each subinterval, and add up all the values of $f(x_i^*)\Delta_i x$. As we have seen, this process involves choices. For the integral to make sense, we need the Riemann sum to approach the same limit as $N \to \infty$, regardless of those choices. If that happens, we say that the function $f(x)$ is **integrable** on the interval $[a, b]$.

Fortunately, *all continuous functions are integrable*, and functions with a finite number of jumps are also integrable. The only functions you're likely to encounter that aren't integrable are artificial examples cooked up for that purpose.

In case you are curious, here is one of those artificial examples. Let

(7.16)
$$f(x) = \begin{cases} 0 & \text{if } x \text{ is rational,} \\ 1 & \text{if } x \text{ is irrational.} \end{cases}$$

If we try to compute $\int_0^1 f(x)dx$ by using N equal-sized intervals and left endpoints, then each x_i^* will be rational, each $f(x_i^*)$ will equal zero, and our Riemann sum $\sum_{i=1}^N f(x_i^*)\Delta x$ will be zero. However, if we pick each x_i^* to be an irrational number between x_{i-1} and x_i, then $f(x_i^*)$ will be 1 and our Riemann sum will be 1. This is true no matter how big N is. As the limiting answer depends on how we set up our Riemann sums, there is no single limiting value. In this example, $f(x)$ is not integrable and $\int_0^1 f(x)dx$ is not defined.

Most of the time that we write $\int_a^b f(x)dx$, we imagine that $a < b$. However, expressions like $\int_3^2 x^2dx$ make sense, too. We just have to break the interval from 3 to 2 into pieces, add up the $f(x_i^*)\Delta_i x$'s, and take a limit. This is exactly the same as breaking the interval from 2 to 3 into pieces, adding up the $f(x_i^*)\Delta_i x$'s, and taking a limit, except for one thing. When going from a bigger number to a smaller number, the quantity $\Delta_i x = x_i - x_{i-1}$ is *minus* the distance from x_i to x_{i-1}. This means that $\int_3^2 f(x)dx = -\int_2^3 f(x)dx$, and more generally

(7.17)
$$\int_b^a f(x)dx = -\int_a^b f(x)dx.$$

Properties of Integrals.

- If $f(x) = c$ is a constant function, then $\int_a^b f(x)dx = c(b - a)$. No matter how we break up our interval $[a, b]$ and no matter how we pick the points x_i^*, we will have $f(x_i^*) = c$ for every i, so

(7.18)
$$\sum_{i=1}^N f(x_i^*)\Delta_i x = \sum_{i=1}^N c\Delta_i x = c(b - a).$$

  Because this is true for every N, the limit as $N \to \infty$ is also $c(b - a)$.

- If $f(x)$ and $g(x)$ are integrable functions on the interval $[a, b]$ and c_1 and c_2 are constants, then $c_1 f(x) + c_2 g(x)$ is integrable on $[a, b]$, and

(7.19)
$$\int_a^b (c_1 f(x) + c_2 g(x))dx = c_1 \int_a^b f(x)dx + c_2 \int_a^b g(x)dx.$$

In particular,

(7.20)
$$\int_a^b (f(x) + g(x))dx = \int_a^b f(x)dx + \int_a^b g(x)dx, \quad \text{and}$$
$$\int_a^b (f(x) - g(x))dx = \int_a^b f(x)dx - \int_a^b g(x)dx.$$

Equation (7.19) comes straight from the corresponding property of sums. If we break the interval $[a, b]$ into N pieces and pick our representative points x_i^*, then

(7.21)
$$\int_a^b (c_1 f(x) + c_2 g(x))dx \approx \sum_{i=1}^N (c_1 f(x_i^*) + c_2 g(x_i^*))\Delta_i x$$
$$= c_1 \sum_{i=1}^N f(x_i^*)\Delta_i x + c_2 \sum_{i=1}^N g(x_i^*)\Delta_i x$$
$$\approx c_1 \int_a^b f(x)dx + c_2 \int_a^b g(x)dx.$$

The bigger N is, the better the approximations in the first and third lines. In the limit all the expressions are equal.

- If $a < b$ and if $f(x) > 0$ on the interval $[a, b]$, then $\int_a^b f(x)$ equals the area under the curve $y = f(x)$ and above the x axis between $x = a$ and $x = b$. However, if $f(x)$ is sometimes negative, then $\int_a^b f(x)dx$ is a **signed area** (see Figure 7.11). That is, it is the area of the region above the x-axis and below the curve $y = f(x)$, minus the area of the region below the x-axis and above the curve.

  In one sense this is obvious, as we already know that areas can be computed as integrals. However, it tells us that integrals can *always* be visualized as signed areas, even if they come from completely different problems, like distance or volume or weight gain.

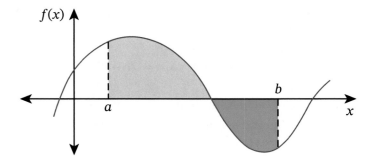

Figure 7.11. The integral $\int_a^b f(x)\,dx$ is a signed area.

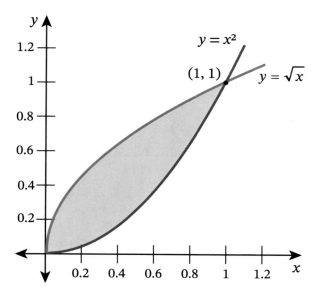

Figure 7.12. The area between the vertical parabola $y = x^2$ and the horizontal parabola $y = \sqrt{x}$ is $\int_0^1 (\sqrt{x} - x^2)\, dx$.

- Likewise, if $f(x)$ and $g(x)$ are integrable functions, then $\int_a^b f(x) - g(x)dx$ is the signed area between $y = f(x)$ and $y = g(x)$, as shown in Figure 7.12. If we cut the region between the two curves into vertical strips, each strip will have height $f(x) - g(x)$ and width Δx. Adding up the approximate areas $(f(x) - g(x))\Delta x$ of these strips and taking a limit gives us the integral $\int_a^b (f(x) - g(x))dx$.

- If a, b, and c are real numbers and the function $f(x)$ is integrable on the relevant intervals, then

(7.22)
$$\int_a^c f(x)dx = \int_a^b f(x)dx + \int_b^c f(x)dx.$$

This is easy to see in terms of area. If $a < b < c$, as in Figure 7.13, then the area from a to c is the same as the area from a to b plus the area from b to c.

If $a < c < b$, as in Figure 7.14, then we also need a little algebra:

$$\int_a^b f(x)dx = \int_a^c f(x)dx + \int_c^b f(x)dx$$

(7.23)
$$\int_a^b f(x)dx - \int_c^b f(x)dx = \int_a^c f(x)dx$$

$$\int_a^b f(x)dx + \int_b^c f(x)dx = \int_a^c f(x)dx$$

That is, the area from a to c is the area from a to b minus the area from c to b, which is the integral from a to b plus the integral *from b to c*.

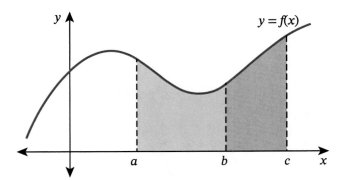

Figure 7.13. The area from a to c is the area from a to b plus the area from b to c.

- All continuous functions are integrable. The reason is that if the function $f(x)$ is continuous, then the value of $f(x)$ can only change a tiny amount when x changes a tiny amount. This means that when we try to integrate $f(x)$ from a to b using a Riemann sum with N terms, with N large, then $f(x)$ can only change by a small amount in each interval I_i. (Mathematicians use the Greek letter ε to denote such small quantities.)

  As all of the choices of x_i^* give values of $f(x_i^*)$ that are within ε of one another, $f(x_i^*)\Delta_i x$ is within $\varepsilon \Delta_i x$ of the true area of the ith strip. We then have

$$\text{Error computing area of } I_i \leq \varepsilon \Delta_i x$$

$$\text{Total error} \leq \sum_{i=1}^{N} \text{Error in } I_i$$

(7.24)

$$\leq \sum_{i=1}^{N} \varepsilon \Delta_i x$$

$$= \varepsilon(b - a).$$

As $N \to \infty$, ε goes to zero and our total error disappears, regardless of how we pick the points x_i^*.

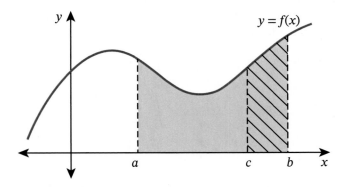

Figure 7.14. The area from a to c is the area from a to b minus the area from c to b.

7.4. The Accumulation Function

So far we have talked a lot about what a definite integral *is*, and relatively little about how to compute one. In this section we introduce the **accumulation function**, and we use it to compute some definite integrals.

As a warmup, consider $\int_2^5 x\,dx$, shown in Figure 7.15. This is the area of the trapezoid underneath the line $y = x$ between $x = 2$ and $x = 5$. The area of this trapezoid is the area $\frac{5^2}{2}$ of the right triangle of width 5 and height 5 minus the area $\frac{2^2}{2}$ of the right triangle of width 2 and height 2, or $\dfrac{5^2 - 2^2}{2} = \dfrac{21}{2}$.

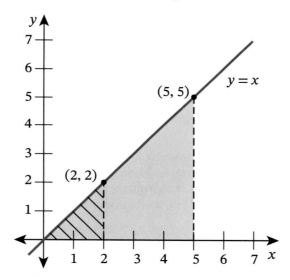

Figure 7.15. The area from 2 to 5 is the difference of the areas of two triangles.

More generally, $\int_a^b x\,dx$ is the area of a trapezoid of width $b - a$, as shown in Figure 7.16. That is, it equals the area of the triangle with vertices $(0,0)$, $(b,0)$, and (b,b) minus the area of the triangle with vertices $(0,0)$, $(a,0)$, and (a,a). In other words,

$$(7.25) \qquad \int_a^b x\,dx = \int_0^b x\,dx - \int_0^a x\,dx = \frac{b^2}{2} - \frac{a^2}{2} = \frac{b^2 - a^2}{2}.$$

The function $\frac{b^2}{2} = \int_0^b x\,dx$ is an example of an **accumulation function**. If $f(x)$ is an integrable function, we define the accumulation function $A(X)$ to be

$$(7.26) \qquad \int_0^X f(x)\,dx.$$

Geometrically, this represents a running total of all the area under the curve $y = f(x)$ from 0 to X. Note the difference between X and x! The variable of the A function is capital X, while the variable in the integral $\int_0^X f(x)\,dx$ is lower case x. An integral is a kind of sum, so x takes on *all* values between 0 and X, while X is a particular number.

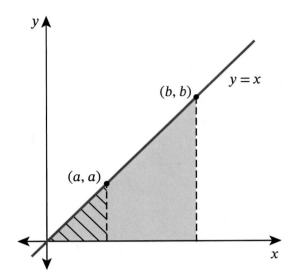

Figure 7.16. The area from a to b is the difference of two accumulations.

An accumulation function is a running total, like an odometer or the reading on a water meter. It measures the total accumulation of some quantity, whether area under a curve or distance driven or water usage, from the starting point (when your car rolled off the assembly line, or when your house was hooked up to the water main) until the point indicated by the variable X.

If we want to call the variable of the A function x instead of X, we have to give the variable in the integral a different name, like s or t, as shown in Figure 7.17:

$$(7.27) \qquad A(x) = \int_0^x f(s)ds = \int_0^x f(t)dt.$$

s and t are called **dummy variables**. They only serve to label the values of the f function that we are adding up to make our integral. Just as the expressions

$$(7.28) \qquad \sum_{s=1}^{100} s^2, \qquad \sum_{t=1}^{100} t^2, \quad \text{and} \quad \sum_{w=1}^{100} w^2$$

all represent the same sum $1^2 + 2^2 + \cdots + 100^2$, the expressions

$$(7.29) \qquad \int_0^x f(s)\,ds, \qquad \int_0^x f(t)\,dt, \quad \text{and} \quad \int_0^x f(w)\,dw$$

all represent limits of the same Riemann sums, and give the area under the f curve from 0 to x.

If we can figure out the accumulation function $A(x)$, then it's easy to figure out any definite integral involving $f(x)$:

$$(7.30) \qquad \int_a^b f(x)\,dx = \int_0^b f(x)\,dx - \int_0^a f(x)\,dx = A(b) - A(a).$$

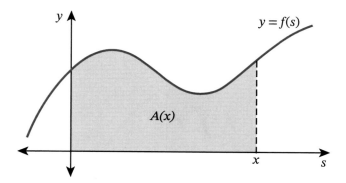

Figure 7.17. The accumulation function $A(x)$ is all the area from 0 up to x. Because x has a fixed value, the variable of integration must have a different name, in this case s.

The total distance traveled between time a and time b is the odometer reading at time b minus the odometer reading at time a. The total water usage in March is the meter reading on April 1 minus the meter reading on March 1. Of course, this approach is only useful if we know $A(x)$!

In the example where $f(x) = x$, $A(x) = \frac{x^2}{2}$, as this is the area of a triangle with base x and height x. We could also write $A(b) = \frac{b^2}{2}$ or $A(z) = \frac{z^2}{2}$ or $A(X) = \frac{X^2}{2}$; the name of the variable doesn't matter. Eventually we will learn a lot of tricks for figuring out accumulation functions. Here are two more that we can work out using geometry to identify areas.

(1) If $f(x) = 1$, then $A(x) = \int_0^x f(s)ds = x$. This is the area of a rectangle of height 1 and width x.

(2) If $f(x) = \sqrt{1 - x^2}$, then $A(x) = \int_0^x f(s)ds$ is the area of the shaded region in Figure 7.18. This region consists of a triangle and a pie wedge. The sine of the angle of the pie wedge is x, so the angle itself is $\sin^{-1}(x)$. Because this is a fraction $\frac{\sin^{-1}(x)}{2\pi}$ of a circle of area π, the pie wedge has area $\sin^{-1}(x)/2$. Meanwhile, the triangle has base x and height $\sqrt{1 - x^2}$, and so has area $x\sqrt{1 - x^2}/2$. Putting this together, we have

(7.31)
$$A(x) = \int_0^x \sqrt{1 - s^2}ds = \frac{\sin^{-1}(x) + x\sqrt{1 - x^2}}{2}.$$

We can also add accumulation functions. If $f(x) = 13 + 6x + 8\sqrt{1 - x^2}$, then

$$S(x) = \int_0^x \left(13 + 6s + 8\sqrt{1 - s^2}\right) ds$$

(7.32)
$$= 13 \int_0^x 1 dx + 6 \int_0^x s \, ds + 8 \int_0^x \sqrt{1 - s^2} \, ds$$

$$= 13x + 3x^2 + 4\sin^{-1}(x) + 4x\sqrt{1 - x^2}.$$

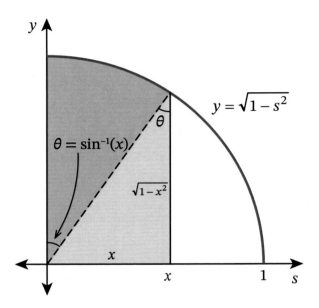

Figure 7.18. The accumulation function $A(x)$ is the area of a triangle plus the area of a pie wedge.

Do you notice anything about the derivatives of the accumulation functions that we have found so far? We'll have a lot more to say about them in the next chapter.

7.5. Chapter Summary

The Main Ideas.

- Integration means adding up the pieces to make a whole. It applies to any quantity that can be broken into pieces, not just to the area under a curve.

- Our strategy is to break the quantity into pieces, estimate the size of each piece, and add up the pieces. *The whole is the sum of the parts.*

- Most of the time, adding up the pieces results in an expression of the form $\sum_{i=1}^{N} f(x_i^*)\Delta_i x$. This is called a Riemann sum and can be visualized as the area under the curve $y = f(x)$.

- Setting up Riemann sums involves choices. We need to decide on the number N of pieces, on the sizes of the intervals I_i, and the choices of the representative points x_i^*.

- Some choices give much more accurate approximations than others. Both the trapezoidal rule and the midpoint rule are much more accurate than using left endpoints or right endpoints.

- A definite integral is the limit of a Riemann Sum as the number of pieces goes to ∞. That is,

(7.33)
$$\int_a^b f(x)dx = \lim_{N \to \infty} \sum_{i=1}^{N} f(x_i^*)\Delta_i x.$$

- If $f(x)$ is integrable then all choices give the same limit as $N \to \infty$. (That's the definition of integrability!) All continuous functions, and most other functions we will encounter, are integrable.

- While integrals come from a variety of problems, all integrals can be *visualized* as signed areas. This allows us to apply geometric arguments to integrals.

- The accumulation function $A(X)$ is the running total of the integral of $f(x)$, namely $\int_0^X f(x)dx$. We often write $A(x) = \int_0^x f(s)\,ds$, where s is a dummy variable. So far we only know how to compute a few accumulation functions, but they will eventually turn out to be very useful.

- If $A(x)$ is the accumulation function of $f(x)$, then $\int_a^b f(x)dx = A(b) - A(a)$.

Expectations. You should be able to

- Apply the strategy of breaking things into pieces, estimating the pieces, and adding up the pieces to approximate a variety of bulk quantities as Riemann sums.

- Understand and use $\sum$ notation for sums.

- Approximate bulk quantities numerically via Riemann sums, using either left endpoints, right endpoints, midpoints, or the trapezoidal rule.

- Take the limit as $N \to \infty$ to describe bulk quantities exactly as integrals.

- Convert back and forth between a Riemann sum and an integral that it approximates.

- Relate integrals to areas under, over, and between curves.

- Understand and use the main properties of integrals.

- Use accumulation functions to compute integrals of any linear function $f(x) = c_0 + c_1 x$.

7.6. Exercises

Note: The exercises for this chapter are about understanding bulk quantities by slicing and dicing them into pieces that we can estimate, adding up the pieces, and taking a limit to get an integral. Some exercises are about setting up those integrals, some are about evaluating Riemann sums numerically, and some are about accumulation functions. A few are about evaluating integrals that can be visualized as the area under a curve, but mostly the point is to understand how any bulk quantity, not just area, can be understood as an integral.

Riemann Sums

7.1.   Express each sum using $\sum$ notation.
        (a) $3 + 6 + 9 + \cdots + 300$
        (b) $1 + \dfrac{1}{2} + \dfrac{1}{3} + \cdots + \dfrac{1}{N}$
        (c) $\tan(x_1^*)\Delta x + \tan(x_2^*)\Delta x + \cdots + \tan(x_8^*)\Delta x$
        (d) $\sqrt{x_1}\Delta x + \sqrt{x_2}\Delta x + \cdots + \sqrt{x_N}\Delta x$

7.2.   Write out each sum explicitly. Then evaluate it.
        (a) $\displaystyle\sum_{i=1}^{10}(i - 1)$

        (b) $\displaystyle\sum_{j=1}^{5} e^{-j}$

        (c) $\displaystyle\sum_{k=2}^{5}\left(\dfrac{1}{k-1} - \dfrac{1}{k}\right)$. Simplify your final answer as much as possible.

        (d) $\displaystyle\sum_{n=2}^{1000}\left(\dfrac{1}{n-1} - \dfrac{1}{n}\right)$. In this part you only have to evaluate the sum. You do
        not have to write out the 999 terms!

7.3.   A CAT scan of a football-shaped tumor shows slices that are spaced 2.5 cen-
        timeters apart. The areas of the slices are 0, 20, 24, 38, 56, 40, 30, 23, and 0
        square centimeters.
        (a) Write down a Riemann sum that approximates the volume of the tumor.
        (b) Evaluate that Riemann sum.

Drug Titers. When you take an antibiotic, the concentration of the drug in your body
(the **drug titer**) shoots up at first but then decays exponentially. If you take in a quantity A of the drug, all at once, there will be Ae^{-rt} left after a time t, where r is the rate at
which your body breaks down or otherwise eliminates the antibiotic. (The same idea
applies to alcohol and other poisons in your system, but we'll stick to antibiotics.) In
Exercises 7.4–7.7 we consider the drug titer of an antibiotic that you take continuously,
as with an IV drip.

7.4.   Suppose that you receive an antibiotic from an IV drip at a steady rate of 100
        mg/hour for 10 hours. We will compute the amount of antibiotic in your body
        at the end of that period, assuming that your body eliminates the antibiotic at
        a rate of 4% per hour.
             For this problem, it is convenient to let t represent how long ago you re-
        ceived the drug. That is, $t = 0$ is the current time (at the end of the 10 hours),
        $t = 1$ is one hour ago, $t = 2$ is two hours ago, and the medication started at
        $t = 10$. That way, we only have to consider positive values of t, and the effect
        of the medicine administered at time t is reduced by a factor of e^{-rt}.
        (a) Consider the portion of the drug that was administered in the past hour,
            that is between $t = 0$ and $t = 1$. Find upper and lower bounds for how
            much of that drug remains in your body at time 0.

(b) Repeat part (a) for the portions administered between $t = 1$ and $t = 2$, between $t = 2$ and $t = 3$, and so on up to the portion administered between $t = 9$ and $t = 10$.

(c) Add up the pieces from parts (a) and (b) to get upper and lower bounds for how much antibiotic remains in your body now.

(d) Average the upper and lower bounds from part (c) to get a good estimate for how much antibiotic is currently in your body.

7.5. Repeat Exercise 7.4, only now breaking the 10 hours into N pieces instead of 10. (For instance, if you were calculating things ten minutes at a time, N would be 60.) Using $\sum$ notation, write down expressions for an upper bound on the amount of antibiotic currently in your body, for a lower bound, and for the average.

7.6. Now take the limit as $N \to \infty$ of the sum you wrote in Exercise 7.5. Write down a definite integral that gives the exact amount of antibiotic in your body. You are not expected to evaluate this integral.

7.7. Generalizing the scenario of Exercises 7.4–7.6, suppose that a drug is administered at rate A per hour for T hours, and that it is broken down in the body at rate r, usually measured in percent per hour. Write down an integral involving A, T and r that gives the drug titer at the end of the period.

Area and Volume Problems. In Exercises 7.8–7.12, we use our "slice and dice" procedure to compute areas and volumes in unusual ways.

7.8. We are going to compute the area of the region between the parabola $y = 4 - x^2$ and the horizontal line $y = 2$ by slicing it into *horizontal* strips, estimating the area of each strip, adding them up, and taking a limit to get an integral *over* y. (Yes, there are ways to get this area by slicing the region into vertical strips, but we aren't doing that!) We imagine cutting the region into N strips of equal height. The figure shows how that looks when $N = 5$, but you should think of N as an unspecified parameter. We number the strips from the bottom of the region to the top.

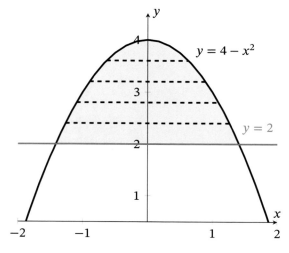

(a) What is the vertical thickness of each strip?

(b) Let y_i be the y-value at the top of the ith strip. Find a formula for y_i in terms of i and N.

(c) What is the *width* of the top of the ith strip? You can leave your answer in terms of y_i.

(d) In terms of i and N, what is the approximate area of the ith strip?

(e) Express the (approximate) total area of the region as a Riemann sum.

(f) Express the limit of the Riemann sum, as $N \to \infty$, as a definite integral over y. You do not have to evaluate this integral.

7.9.   Now we are going to compute the volume of the bullet-shaped solid obtained by rotating the region of Exercise 7.8 around the y axis. We will slice the solid horizontally into roughly cylindrical slabs that we'll call "pancakes", estimate the volume of each pancake, and add them up.

(a) If there are N pancakes, each of equal thickness, what is the thickness of each pancake? Call this distance Δy.

(b) Let y_i be the height above the ground of the top of the ith pancake, where we're numbering pancakes starting at the bottom. Find an explicit formula for y_i.

(c) Find the cross-sectional area of the solid at height y from the ground.

(d) Based on your result in part (c), what is the approximate volume of the ith pancake? You can write your answer either in terms of i and N or in terms of y_i and Δy.

(e) Express the approximate total volume of the solid as a Riemann sum.

(f) Taking a limit of that Riemann sum, express the total volume of the solid as a definite integral over y. You do not have to evaluate that integral.

7.10.   Imagine that we cut down a 4-year-old tree, leaving a circular stump with a radius of 10 cm. The tree has four equally spaced growth rings. We can estimate the area of each ring by multiplying its circumference by its width. However,

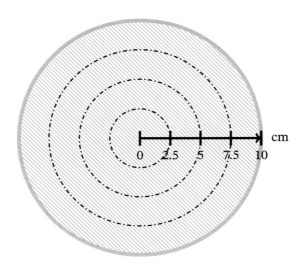

we have to decide which circumference to use: the outside of the ring or the inside of the ring.

(a) Use the outer circumference of each ring to estimate its area. Add up these estimates to get an estimated area for the entire stump surface. Is your result an underestimate or an overestimate of the true total area?

(b) Now imagine having N rings instead of just 4, with the radius of the stump still being 10 cm. Write an expression for the area of the ith ring, counting outwards, using outer circumferences.

(c) Add up the results from part (b) to get a Riemann sum that approximates the area of the stump surface. You do not have to evaluate this sum.

(d) Taking a limit of the Riemann sum as $N \rightarrow \infty$, write down a definite integral that computes the exact area of the stump surface.

(e) Evaluate the integral from part (d) using what you know about areas of triangles. Does the result agree with the usual formula for the area of a circle of radius 10 cm?

7.11. Repeat steps (b)–(e) of Exercise 7.10 for a stump of arbitrary radius R instead of radius 10 cm. The final answer should look very familiar.

7.12. Sometimes, an oyster ingests a speck of sand that gets stuck on its inside. To protect itself, the oyster coats the speck with a material that hardens, much like a coat of paint. Then it applies another coat, and another coat, and so on, until the speck of sand grows into a pearl. The volume of the pearl is then the sum of the volumes of all the coats of "paint".

(a) Suppose that a pearl of radius 5 mm consists of five coats, each a millimeter thick. Estimate the volume of the outermost coat as surface area times thickness, using the fact that the surface area of a sphere of radius r is $4\pi r^2$. (You have to decide whether to use the inner or outer radius of each coat. One gives an underestimate of the volume, while the other gives an overestimate.)

(b) Add up the estimated volumes of the five coats to get an approximate volume of the pearl.

(c) Now imagine that the pearl consists of N coats. Write down a Riemann sum that approximates the volume of the sphere. You do not need to evaluate this sum.

(d) Finally, take a limit of the Riemann sum as $N \rightarrow \infty$ to get a definite integral that gives the exact volume of the pearl. You do not need to evaluate the integral.

Distance. In the projectile problem of Section 7.1, we saw that distance traveled is the integral of velocity. We explore this idea further in Exercises 7.13–7.14.

7.13. You are going on a road trip to Big Bend National Park. At noon, you start the trip and set your trip odometer to zero. You drive at 60 miles/hour (mph) for two hours, until you reach the open road. You then drive at 75 mph for four hours. At 6:00 p.m. you turn onto a park road and drive one more hour at 40 mph, reaching your destination at 7:00 p.m.

Let $v(t)$ be your speed at time t, measured in miles per hour, where t is the number of hours since noon. Let $D(t)$ be the reading of your trip odometer at time t, indicating the total distance you have traveled since noon. This is the same as the accumulation function of V: $D(t) = \int_0^t v(s)ds$.

(a) Graph your velocity $v(t)$ (measured in mph) as a function of the time t (measured in hours since noon).

(b) What will your trip odometer read at 1:00 p.m.? At 2:00 p.m.? At 3:00 p.m.? At 4:00 p.m.? At 5:00 p.m.? At 6:00 p.m.? At 7:00 p.m.?

(c) Find a formula for $D(t)$ that applies for $0 \leq t \leq 2$. Make sure that it agrees with your answers to part (b).

(d) Find formulas for $D(t)$ when $2 \leq t \leq 6$ and when $6 \leq t \leq 7$. Again, make sure that these formulas agree with your answers to part (b).

(e) Graph $D(t)$ for $0 \leq t \leq 7$.

7.14. After having fun at Big Bend, you decide to go bungee jumping from the Royal Gorge Bridge in Colorado, almost 1000 feet above the canyon below.[3] You jump off the bridge, attached to a 484 foot bungee cord. Your downward velocity $v(t)$ for the first part of your fall is $32t$, where the velocity $v(t)$ is measured in feet/second and t is the time in seconds since you jumped. This continues until you have fallen 484 feet, at which point the cord starts to stretch, slowing you down and eventually pulling you back up.

(a) Graph $v(t)$ as a function of t for the first part of your fall.

(b) Express the distance traveled between $t = 1$ and $t = 4$ as the area of a region under your graph. Using what you know about triangles, compute that area.

(c) Write an expression for how far you fall in the first T seconds that applies until your bungee cord starts to stretch.

(d) After how many seconds will your bungee cord start to stretch?

7.15. **Burning Calories.** When exercising, it is better to do a long but gentle workout, or a short but intense workout? When it comes to burning calories, neither duration nor intensity are enough. What really matters is the product of the intensity and the duration.

Let $i(t)$ be the intensity of a workout, measured in calories per minute. A half hour workout might involve 10 minutes of warm-up, 15 minutes of hard exercise, and 5 minutes of cool-down, with

$$i(t) = \begin{cases} 6, & 0 \leq t < 10, \\ 12, & 10 \leq t < 25, \\ 3, & 25 \leq t \leq 30. \end{cases}$$

(a) Draw the graph of the step function defined here, and compute the total number of calories burned during the workout.

(b) Let $C(t)$ be the total number of calories burned up to time t. Draw the graph of $C(t)$.

---

[3]From 2003 to 2008, bungee jumping at Royal Gorge was part of the annual Go Fast! games, but to the best of my knowledge it has not been allowed since 2008.

7.16. One weekend you decide to skip the gym and go for a 7-hour hike instead. It's a hot day, so you bring along 2.4 liters of water. For the first 3.5 hours, the trail is downhill and shaded and you consume water at a rate of 200 ml/hour. After that, the trail is uphill and exposed and you start drinking at a rate of 500 ml/hour.
    (a) For how many hours will your water last?
    (b) Let $d(t)$ (d for "drinking") be the rate at which you drink water. Draw a graph of $d(t)$. End your graph at the time that you run out of water.
    (c) Let $W(t)$ be the total amount of water drunk up to time t. Draw the graph of $W(t)$.

7.17. (a) What is the average rate of burning calories in Exercise 7.15?
    (b) What is the average rate of water consumption in Exercise 7.16?

Electrical Energy. Exercises 7.18–7.20 deal with electricity. The work done by electricity is usually referred to as **electrical energy**, and the rate at which this work is done is called **electric power**. Electric *power* is measured in **watts**. Electrical *energy* is measured in **watt-hours** (or kilowatt-hours or megawatt-hours, which are thousands and millions of watt-hours, respectively). Running a 20-watt light bulb for five hours consumes 100 watt-hours of electrical energy, as does running a 100-watt light bulb for one hour.

7.18. On a summer day, a 2000-watt air conditioner runs from 10:00 a.m. until 10:00 p.m. How many watt-hours of energy does it consume? How many kilowatt-hours does it consume?

7.19. A certain air conditioner has three settings: "economy" runs at 1200 watts, "normal" runs at 1600 watts, and "maximum" runs at 2000 watts. A homeowner programs the air conditioner to run on economy from midnight to 7:00 a.m., on normal from 7:00 a.m. until noon, on maximum from noon until 7:00 p.m., and on normal again from 7:00 p.m. until midnight.
    (a) Let $p(t)$ be the rate of power consumption, where t is the number of hours since midnight. Sketch the graph of $p(t)$.
    (b) Express the total energy consumption during the day (in watt-hours, or in kilowatt-hours if you prefer) as a definite integral. Evaluate that integral.
    (c) Let $P(t)$ be the accumulated energy consumption in the first t hours of the day. Sketch the graph of $P(t)$.

7.20. Most air conditioners are controlled by thermostats. The hotter it is outside, the more they run and the more energy they consume. On a certain day, it got hot enough to trigger the air conditioner at 9:00 a.m., and it kept getting hotter until 4:00 p.m. The rate of power consumption was $p(t) = 200t$ watts, where t is the number of hours since 9:00 a.m.
    (a) How much energy was used in the first t hours? That is, compute an accumulation function.
    (b) How much energy was used in the entire 7-hour period?
    (c) How much energy was used between noon and 3:00 p.m.?

Using and Modifying the Riemann Program. Exercises 7.21–7.25 use the Riemann program, shown at the end of Section 7.2. This program tells MATLAB how to approximate $\int_a^b f(x)dx$ as a Riemann sum

$$\sum_{i=1}^{N} f(x_i^*)\Delta x$$

using N intervals, all of size $\Delta x = (b-a)/N$, with t_i^* being the left endpoint of the ith interval. Riemann is not a built-in part of MATLAB, but needs to be added by hand. If your instructor has not already provided you with an electronic copy of this file, you can simply type it in yourself to a file called Riemann.m. The ".m" ending tells MATLAB that it is a program called Riemann. For MATLAB to find this program, you must be working in the directory that contains Riemann.m.

To use the program, you first have to specify what the function $f(x)$ is. For instance, if you want to compute $\int_1^4 x^3 dx$ with $N = 100$, you would first type
f = @(x) x^3;
at the command prompt. You then type
Riemann(f,1,4,100)
and MATLAB returns the answer 62.8084.

7.21.   (a) Using the Riemann program, calculate left-endpoint Riemann sums for the function e^{-x^2} on the interval $[0, 3]$ using $N = 50, 500, 5000$, and $50{,}000$. (Type f = @(x) exp(-x^2); at the command prompt to tell MATLAB what $f(x)$ is.)

        (b) The left-endpoint Riemann sums that you calculated in part (a) seem to be approaching a limit as the number N of pieces increases without bound. Estimate that limit to four decimal places.

7.22.   (a) Modify the Riemann program so it calculates the Riemann sum using right endpoints instead of left endpoints. Describe exactly how you changed the program to do this. Call your new function RiemannRit and save it as RiemannRit.m.

        (b) Use your RiemannRit program to calculate right-endpoint Riemann sums for the function e^{-x^2} on the interval $[0, 3]$ using $N = 50, 500, 5000$, and $50{,}000$.

7.23.   (a) Modify the Riemann program so it calculates the Riemann sum by taking t_i^* to be the midpoint of the ith interval, instead of the left or right endpoint. Describe exactly how you changed the program to do this. Call your new function RiemannMid and save it as RiemannMid.m.

        (b) Use RiemannMid to calculate midpoint Riemann sums for the function e^{-x^2} on the interval $[0, 3]$ using $N = 50, 500, 5000$, and $50{,}000$.

        (c) Compare your results from using RiemannMid to the results of using Riemann, the results of using RiemannRit, and the results of taking the average of Riemann and RiemannRit. Which method do you think gives the most accurate estimate of $\int_0^3 e^{-x^2} dx$? Which gives the least accurate estimate?

(d) For each method, estimate how big N needs to be to get within 0.0001 of that actual integral. How can you tell? (We will have a lot more to say about the accuracy of different numerical methods in Chapter 9.)

7.24. Suppose we want to compute the (signed) area "under" the curve $y = \sin(x^2)$ between $x = 0$ and $x = 4$.
  (a) Use Riemann to estimate this area with Riemann sums using left endpoints and $N = 100, 1000$, and $10,000$.
  (b) Use the RiemannRit program from Exercise 7.22 to estimate this area with Riemann sums using right endpoints and $N = 100, 1000$, and $10,000$.
  (c) Use the RiemannMid program from Exercise 7.23 to estimate this area with Riemann sums using midpoints and $N = 100, 1000$, and $10,000$.

7.25. Suppose we want to compute the (signed) area "under" the curve $y = x^5 + x^3$ between $x = -2$ and $x = 2$.
  (a) Use the RiemannMid program from Exercise 7.23 to estimate this area with Riemann sums using midpoints and $N = 10, 100$, and 1000. Explain why each Riemann sum is zero.
  (b) Repeat part (a) using Riemann sums with left endpoints. Are the sums still zero? Explain how these estimates are (or aren't) different from those in part (a).

Probability. Most random quantities follow a bell-shaped distribution, also called a **normal distribution** or a **Gaussian distribution**. The most widely accepted measure of the width of the distribution is called the **standard deviation**. For instance, if you flip 100 coins many times, the number of heads in each set of 100 flips follows a normal distribution with an average (or **mean**) of 50 heads and a standard deviation of 5. If you flip 10,000 coins, the number of heads is described by a normal distribution with a mean of 5000 and a standard deviation of 50. Exercises 7.26 and 7.27 explore this distribution and Exercise 7.28 explores another common distribution, the exponential distribution.

7.26. Estimate the area under the curve $y = e^{-x^2}$ between $x = 0$ and $x = 1$ to four decimal places. Sketch the curve and shade the area.

7.27. For a random quantity following a normal distribution, the probability of being between a standard deviations and b standard deviations above the mean is the area under the graph of

$$f(x) = \frac{1}{\sqrt{2\pi}} e^{-x^2/2}$$

between $x = a$ and $x = b$. That is, the probability is $\int_a^b f(x)\, dx$. Note that a and b don't have to be positive, although we do want $a \le b$. The probability of being between 3 standard deviations below the mean and 2 standard deviations above the mean is $\int_{-3}^{2} f(x)\, dx$. This is almost (but not exactly) the same function as in Exercise 7.26.

(a) With the help of MATLAB, compute to 4 decimal places the probability of being within 1 standard deviation of the mean (that is, the probability of being between 1 standard deviation below the mean and one standard deviation above the mean).

(b) Compute the probability, to four decimal place accuracy, of being within 2 standard deviations of the mean.

(c) Compute the probability, to four decimal place accuracy, of being between 0.5 standard deviations below the mean and 1.5 standard deviations above the mean.

7.28. The **exponential distribution** describes how long a piece of machinery lasts. The probability of lasting between a and b times the average lifetime is $\int_a^b e^{-x}\, dx$.

(a) A certain brand of light bulb has an average lifetime of 100 hours. What is the probability of its lasting less than 75 hours? Feel free to use MATLAB (and specifically the Riemann program) to compute this.

(b) What is the probability that the light bulb lasts less than 150 hours?

(c) What is the probability that the light bulb lasts between 75 and 150 hours? How is that related to your answers to parts (a) and (b)?

7.29. An oil field is currently producing oil at a rate of 3 million barrels of oil per year. The production rate typically falls at a rate proportional to the current rate of production, with rate constant 0.15/year. That is, if $P(t)$ (P for petroleum) represents the production rate in barrels/year after t years, then $P(t)$ solves the initial value problem

$$P'(t) = -0.15P(t); \qquad P(0) = 3{,}000{,}000.$$

(a) What is the solution to this initial value problem?

(b) If a and b are specific times, what real-world quantity does $\int_a^b P(t)\, dt$ represent?

(c) Using $\Delta t = 1$ year and right endpoints, estimate how much oil is produced between $t = 0$ and $t = 5$.

(d) Write down a definite integral that gives the exact total oil production over the 5-year period of part (c).

7.30. Evaluate each integral.

(a) $\int_2^{10} 3\, dx$

(b) $\int_{10}^2 3\, dx$

(c) $\int_{-4}^9 -2\, dz$

(d) $\int_9^{-4} -2\, dz$

7.31. This exercise uses the function $f(x) = \begin{cases} -2 & 0 \le x < 4, \\ 5 & 4 \le x \le 9. \end{cases}$

(a) Sketch the graph of $f(x)$.

(b) Evaluate $\int_0^6 f(x)\, dx$, $\int_6^9 f(x)\, dx$, and $\int_0^9 f(x)\, dx$.

7.32. A pyramid is 20 feet tall. The area of a horizontal cross-section x feet from the top of the pyramid is $4x^2$ square feet.
  (a) What is the area of the base?
  (b) What is the volume of the pyramid, to the nearest cubic foot? (*Hint:* Use Riemann or RiemannRit or RiemannMid.)

7.33. Find a formula for the accumulation function $A(x) = \int_0^x 2s + 5\,ds$. Then use that accumulation function to compute $\int_{-1}^4 2x + 5\,dx$.

7.34. **Comparing Cattle Breeds.** The rate of weight gain in two calves of two different breeds of cattle (in pounds per month) is given in the figure below. Assume the calves had the same initial weight.
  (a) Which calf weighs more after 3 months? Explain.
  (b) Which weighs more after 6 months? Explain.
  (c) Which weighs more after 12 months? Explain.
  (d) Sketch the accumulation functions that represent the weight gain of each calf.
  (e) Write down an integral that represents the difference in the weight of the two calves after 6 months.
  (f) Write down an integral that represents the difference in the weight of the two calves after t months.

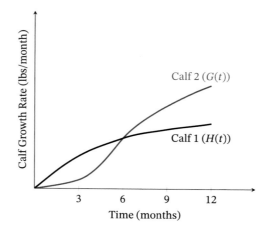

7.35. **Yeast and Alcohol.** Brewing alcohol begins by producing grain mash from grain mixed with water; enzymes in the mixture convert starch in the grain to sugar. A colony of living yeast cells is then added to a vat of grain mash. The yeast consumes the sugar and produces alcohol and carbon dioxide as waste products. It is reasonable to expect that the *rate* of alcohol production is proportional to the size of the colony, so the *amount* of alcohol produced is proportional to both the size of the colony and the amount of time that passes. That is, if the colony has constant size P and produces alcohol for a time Δt, then it will produce a quantity $kP\Delta t$ of alcohol, where k is a proportionality constant.

  Suppose the amount $P(t)$ of yeast is not constant. (The yeast will grow at first, but will eventually die from the toxic effects of its own waste.) In this exercise, you should estimate the amount of alcohol produced between time a and b with our usual slice-and-dice procedure:

(a) Divide the time period $[a, b]$ into N pieces. What are the endpoints of the ith piece?

(b) Estimate the alcohol production in the ith piece in terms of k, $P(t)$, and Δt.

(c) Add up the pieces to get a Riemann sum. (Write down the sum.)

(d) Take the limit as $N \to \infty$ to get a definite integral. (Write down the integral.)

(e) Suppose that we measure P and the alcohol produced in grams, that we measure t in hours, that $P(t) = 300e^{0.2t}$, and that $k = 0.1$. Use Riemann or RiemannRit or RiemannMid to estimate the amount of alcohol produced in the first four hours.

Chapter 8

The Fundamental Theorem of Calculus (One Step at a Time)

As noted at the beginning of Chapter 5, there are three questions you should ask about every mathematical concept that you encounter.

- What is it?

- How do you compute it?

- What is it good for?

When it comes to integrals, we tackled the first and third questions in Chapter 7, but we barely touched the second. At this point, we haven't even learned how to evaluate $\int_0^1 x^2\,dx$. In this chapter, we're going to see how anti-derivatives can help us evaluate integrals. At the end of this chapter, and into Chapter 9, we'll develop techniques for finding those anti-derivatives.

The **Fundamental Theorem of Calculus**, or FTC, relates three quantities that come up a lot in integral calculus: the definite integral $\int_a^b f(x)dx$, the accumulation function $\int_0^x f(s)ds$, and the anti-derivative of $f(x)$. We already saw the first two in Chapter 7, and we learned how they are related. We will introduce the third in Section 8.1. The full situation is illustrated in Figure 8.1.

The FTC actually has two parts, called the first and second Fundamental Theorems of Calculus, or FTC1 and FTC2 for short.[1] In Section 8.2 we will go over FTC2, which relates definite integrals and anti-derivatives. This gives us our most powerful tool for computing definite integrals, and with it we will see why, for instance, $\int_0^1 x^2 dx = \frac{1}{3}$.

---

[1]We follow the practice of most contemporary authors in calling the two parts FTC1 and FTC2, although we discuss them in the opposite order.

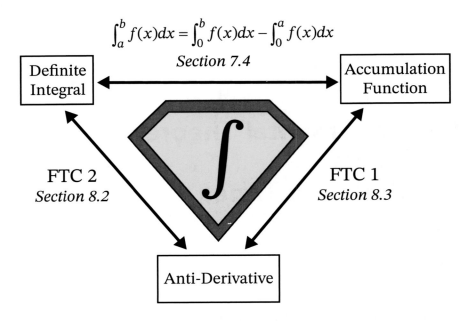

Figure 8.1. The main topics of this chapter

In Section 8.3 we will go over FTC1, which relates accumulation functions and anti-derivatives. We will also see how the two halves of the FTC fit together. In Section 8.4 we will treat anti-derivatives in their own right, and we'll see how they solve many problems involving basic physics. These physics problems will give us an additional perspective on why the FTC is true. Finally, in Section 8.5 we will develop some tools for computing anti-derivatives.

8.1. Three Different Quantities

Let's start by defining our three related concepts: definite integrals, accumulation functions, and anti-derivatives.

Definite Integrals.

The **definite integral** $\int_a^b f(x)dx$ is the limit of a Riemann sum:

$$(8.1) \qquad \int_a^b f(x)dx = \lim_{N \to \infty} \sum_{i=1}^{N} f(x_i^*)\Delta_i x.$$

A definite integral can always be visualized as the (signed) area under the curve $y = f(x)$ between $x = a$ and $x = b$, but it applies to lots of different things as well. For instance, if $v(t)$ is the velocity of a particle at time t, then $\int_a^b v(t)dt$ is the net distance traveled between time a and time b.

Defining a definite integral as the limit of a Riemann sum is more precise than just saying "the area under the curve", but making sense of that definition in Chapter 7 took a lot of machinery. We needed $\sum$ notation, we needed to understand what $\Delta_i x$ means, we needed to pick the representative points x_i^*, and we needed to understand how to take a limit as $N \to \infty$. Most of all, we had to learn how to estimate bulk quantities with Riemann sums, and how to get them exactly as limits of Riemann sums. The upshot of Chapter 7 is that lots and lots of real-world quantities can be represented as definite integrals.

Remember that a definite integral is a *number*, not a function. Between 1:00 p.m. and 3:00 p.m. we might travel 105 miles (by car), or we might travel 5 miles (on foot), but we don't travel $f(t)$ miles. Also, there isn't any single value of x (or t) to associate with the integral. The integral $\int_a^b f(x)dx$ adds up the contributions of *all* values of x between a and b.

Accumulation Functions. The **accumulation function** $\int_0^x f(s)ds$ is a running total. It is a function of the endpoint x. For instance, if the speed of your car is $v(t)$, then $\int_0^t v(s)ds$ is the reading of your odometer at time t. It gives the total distance that the car drove between time 0 (the day the car was made) and time t. The dummy variable s in the integral runs from 0 to t, as the odometer reading today is the result of all of the driving that you did between time 0 and time t. Remember that the accumulation function is a *function*, not a number. It takes x or t as the input and outputs the number

$$(8.2) \qquad A(x) = \int_0^x f(s)ds \quad \text{or} \quad A(t) = \int_0^t f(s)ds.$$

Accumulation functions are closely related to definite integrals. Partly this is because the output of the accumulation function *is* a definite integral. Partly it is because we can use accumulation functions to compute definite integrals. As we saw in Chapter 7,

$$(8.3) \qquad \int_a^b f(x)dx = \int_0^b f(x)dx - \int_0^a f(x)dx = A(b) - A(a).$$

So far, we have defined our accumulation functions starting at 0, but we could just as well have used a different starting point or starting time. For instance, a *trip odometer* is an accumulation function that starts each time you press the reset button. To figure out how far we drove in a certain period of time (that is, $\int_a^b v(t)dt$) we can compare our car's odometer readings at times a and b, or we can compare the *trip* odometer readings at times a and b.

In particular, if $A_1(x)$ and $A_2(x)$ are accumulation functions for the same function $f(x)$ with different starting points, then $A_1(x)$ and $A_2(x)$ just differ by a constant. For instance, if A_1 starts at $x = 0$ and A_2 starts at $x = 3$, then

$$(8.4) \qquad \begin{aligned} A_1(x) &= \int_0^x f(s)ds = \int_0^3 f(s)ds + \int_3^x f(s)ds \\ &= A_2(x) + \int_0^3 f(s)ds. \end{aligned}$$

The *function* $A_1(x)$ equals the *function* $A_2(x)$ plus the *constant* $\int_0^3 f(s)ds$. No matter what a and b are, we can compute $\int_a^b f(x)dx$ using either function:

$$\int_a^b f(x)dx = A_1(b) - A_1(a)$$

(8.5)
$$= \left(A_2(b) + \int_0^3 f(s)ds \right) - \left(A_2(a) + \int_0^3 f(s)ds \right)$$

$$= A_2(b) - A_2(a).$$

The only thing wrong with accumulation functions is that we don't know how to actually compute them. That will change once we learn FTC1.

Anti-Derivatives.

An **anti-derivative** of $f(x)$ is a function $F(x)$ whose derivative (with respect to x) is $f(x)$. For instance, the derivative of x^3 is $3x^2$, so we say that x^3 is an anti-derivative of $3x^2$.

We already know a lot about derivatives, so we know lots of pairs of functions where the second is the derivative of the first. For instance, $5x^4$ is the derivative of x^5, $2e^{2x}$ is the derivative of e^{2x}, and $\sec^2(x)$ is the derivative of $\tan(x)$. That means we know lots of pairs of functions where the first is the anti-derivative of the second! x^5 is an anti-derivative of $5x^4$, e^{2x} is an anti-derivative of $2e^{2x}$, and $\tan(x)$ is an anti-derivative of $\sec^2(x)$.

Another example comes from motion. The derivative of the position $x(t)$ is the velocity $v(t)$, so the position is an anti-derivative of the velocity. The derivative of the velocity $v(t)$ is the acceleration $a(t)$, so the velocity is an anti-derivative of the acceleration.

In the last paragraph I wrote *an* anti-derivative, rather than *the* anti-derivative, for a reason. The anti-derivative of a function isn't unique. The derivative of $x^3 + 7$ is $3x^2$, so $x^3 + 7$ and x^3 are both anti-derivatives of $3x^2$. In general, adding a constant to any anti-derivative gives another anti-derivative. However, on connected regions, that is the only ambiguity.

Here's why: If $F_1(x)$ and $F_2(x)$ are both anti-derivatives of $f(x)$, then

(8.6)
$$\frac{d}{dx}(F_1(x) - F_2(x)) = F_1'(x) - F_2'(x) = f(x) - f(x) = 0.$$

Because the derivative of $F_1(x) - F_2(x)$ is always zero, $F_1(x) - F_2(x)$ never changes. That is, $F_1(x) - F_2(x)$ is a *constant*, and we can write

(8.7)
$$F_1(x) = F_2(x) + C,$$

where C is a constant. In other words, every anti-derivative of $3x^2$ is x^3 plus a constant, and we say that *the* anti-derivative of $3x^2$ is

(8.8)
$$x^3 + C,$$

where now C is an *arbitrary* constant.

To summarize,

(1) Checking whether $F(x)$ is an anti-derivative of $f(x)$ is easy. Just take the derivative of $F(x)$ and see if it equals $f(x)$.

(2) Once we have one anti-derivative $F(x)$ of $f(x)$, perhaps by guessing and checking, we know that every other anti-derivative of $f(x)$ is of the form $F(x) + C$, and we might say that $F(x) + C$ is *the* anti-derivative of $f(x)$.

Unfortunately, we still don't know what anti-derivatives are good for, so there isn't anything for us to do with this knowledge. That is about to change.

As a hint, consider the way that position and velocity are related. The position $x(t)$ is an *anti-derivative* of velocity. The change in position from time a to time b, namely $x(b) - x(a)$, is the distance traveled, which is the *definite integral* of the velocity from time a to time b. This suggests that definite integrals and anti-derivatives have something to do with one another. FTC2 will tell us what.

8.2. FTC2: The Integral of the Derivative

The second Fundamental Theorem of Calculus relates anti-derivatives and definite integrals.

Theorem 8.1 (FTC2). *Suppose that $f(x)$ and $F(x)$ are functions defined on all of $[a, b]$ and that $F(x)$ is an anti-derivative of $f(x)$. Then $f(x)$ is integrable on $[a, b]$ and*

(8.9)
$$\int_a^b f(x)dx = F(b) - F(a).$$

In other words, computing the definite integral $\int_a^b f(x)dx$ boils down to finding an anti-derivative $F(x)$, evaluating it at the endpoints and taking the difference. We often use the notation $F(x)\Big|_a^b$ to mean $F(b) - F(a)$. For instance, $x^3/3$ is an anti-derivative of x^2, so

(8.10)
$$\int_0^1 x^2 \, dx = \frac{x^3}{3}\Big|_0^1 = \frac{1^3}{3} - \frac{0^3}{3} = \frac{1}{3}.$$

Similarly, $-e^{-x}$ is an anti-derivative of e^{-x} (as you can check by taking the derivative of $-e^{-x}$ with the help of the chain rule), so

(8.11)
$$\int_0^3 e^{-x} \, dx = -e^{-x}\Big|_0^3 = -e^{-3} - (-e^0) = 1 - e^{-3}.$$

That's it! No messy sums, no worrying about left vs. right endpoints, and no limits. Those sums were *extremely* important for understanding why $\int_0^1 x^2 \, dx$ is the volume of a pyramid and the moment of inertia of a rod, but when it comes to *computing* $\int_0^1 x^2 \, dx$ (or $\int_0^3 e^{-x} \, dx$), using FTC2 is much, much easier.

This begs two questions:

(1) Why is FTC2 true? That's the content of the rest of this section.

(2) How do we go about finding anti-derivatives? We will start on that in Section 8.5, and follow it up more seriously in Sections 9.1 and 9.2.

The reason that FTC2 works can be summed up in the 5th Pillar of Calculus:

> ### One step at a time.

If we want to know how much something changes in a calendar year, we can figure out how much it changes on January 1, how much it changes on January 2, all the way through December 31, and add up those little changes. Likewise, if we have points

$$(8.12) \qquad a = x_0 < x_1 < x_2 < \cdots < x_{N-1} < x_N = b$$

and we ask how much a function $F(x)$ changes between $x = a$ and $x = b$, then the whole change is the change from x_0 to x_1, plus the change from x_1 to x_2, all the way up to the change from x_{N-1} to x_N. That is,

$$
\begin{aligned}
\sum_{i=1}^{N} F(x_i) - F(x_{i-1}) &= -F(x_0) + F(x_1) \\
&\quad - F(x_1) + F(x_2) \\
&\quad\quad - F(x_2) + F(x_3) \\
&\quad\quad\quad \ddots \\
&\quad\quad\quad\quad - F(x_{N-1}) + F(x_N) \\
&= -F(x_0) + F(x_N) \\
&= F(b) - F(a),
\end{aligned}
$$

(8.13)

as all the terms in the middle cancel in pairs.

Next we look at the change $F(x_i) - F(x_{i-1})$ that occurs in each step of the sum (8.13). By the microscope equation, this is approximately $F'(x_{i-1})(x_i - x_{i-1}) = f(x_{i-1})\Delta_i x$. That is,

$$(8.14) \qquad F(b) - F(a) \approx \sum_{i=1}^{N} f(x_{i-1})\Delta_i x.$$

However, the right-hand side is exactly the Riemann sum we would get if we tried to estimate $\int_a^b f(x)dx$ using left endpoints! As $N \to \infty$, the approximation gets better and better, and the right-hand side gets closer and closer to that integral. In the limit, we have

$$(8.15) \qquad F(b) - F(a) = \int_a^b f(x)dx,$$

which is exactly FTC2.

For another perspective on FTC2, imagine that a rock moves upward with velocity $v(t) = 160 - 32t$ feet/second, where t is the time measured in seconds. That is, at time $t = 0$ it is shot upward at 160 feet/second, and then it slows down thanks to gravity. How much does the rock rise between time $t = 1$ and $t = 4$?

On the one hand, the distance traveled is the integral of velocity, so our answer must be $\int_1^4 160 - 32t\, dt$. On the other hand, the distance traveled is the change in position, so our answer must be $z(4) - z(1)$, where $z(t)$ is the height at time t. Equating our two answers tells us that

$$(8.16) \qquad \int_1^4 v(t)dt = z(t)\Big|_1^4.$$

The derivative of $z(t)$ is $v(t)$, so $z(t)$ is an anti-derivative of $v(t)$ and this equation is *exactly* what FTC2 says.

We can go further with anti-derivatives. Because the derivative of $160t - 16t^2$ is $160 - 32t$, our position function must be

$$(8.17) \qquad z(t) = 160t - 16t^2 + C,$$

for some constant C. The problem doesn't give us enough information to figure out what C is, but that doesn't stop us, because the *change* in z doesn't depend on C.

$$
\begin{aligned}
(8.18) \qquad \int_1^4 160 - 32t\, dt = \int_1^4 v(t)dt &= z(t)\Big|_1^4 \\
&= \left(160t - 16t^2 + C\right)\Big|_1^4 \\
&= \left(160t - 16t^2\right)\Big|_1^4 + C\Big|_1^4 \\
&= \left(160t - 16t^2\right)\Big|_1^4.
\end{aligned}
$$

To find the integral of $v(t)dt$ from 1 to 4, just take *any* anti-derivative of $v(t)$, plug in the values $t = 4$ and $t = 1$, and take the difference to get a final answer of 240 feet.

There are three ways of thinking about FTC2, and all three are useful.

(1) From the perspective of $f(x)$, FTC2 is a statement about *integrals*, and in particular it's a recipe for computing them. To compute $\int_a^b f(x)\, dx$, just find an anti-derivative of $f(x)$, plug in the beginning and ending values of x, and take the difference.

(2) From the perspective of $F(x)$, FTC2 is a statement about *changes*. The change in the quantity $F(x)$ is the integral of the derivative of $F(x)$. The change in position is the integral of velocity. The change in velocity is the integral of acceleration. The change in atmospheric carbon over a period of time is the integral of the net rate of carbon production over that period of time. In fact, some texts treat this perspective as a theorem in its own right and call it the **net change theorem**.

(3) If we start with a function $F(x)$, take its derivative, and then integrate, we get something involving $F(x)$ again. (Specifically, we get $F(x)\Big|_a^b$.) So in some sense differentiation and integration are inverse operations. We will explore this further when we get to FTC1.

8.3. FTC1: The Derivative of the Accumulation

Suppose that $f(x)$ is a function, that $A(x)$ is its accumulation function, and that $F(x)$ is an anti-derivative of $f(x)$. The second Fundamental Theorem of Calculus tells us that

$$(8.19) \qquad \int_a^b f(x)\,dx = F(x)\Big|_a^b.$$

However, we already know that

$$(8.20) \qquad \int_a^b f(x)\,dx = A(x)\Big|_a^b.$$

This suggests that $A(x)$ and $F(x)$ must have something to do with each other. The first Fundamental Theorem of Calculus tells us what.

> **Theorem 8.2** (FTC1). *If $f(x)$ is an integrable function and $A(x)$ is its accumulation function, then $A'(x) = f(x)$. In equations,*
>
> $$(8.21) \qquad \frac{d}{dx}\int_0^x f(s)\,ds = f(x).$$

In other words, accumulation functions are anti-derivatives. If $F(x)$ is any other anti-derivative, then we must have $A(x) = F(x) + C$ for some constant C.

We have already seen this in a couple of examples. The accumulation function of $f(x) = 1$ is $A(x) = x$, whose derivative is 1. The accumulation function of $f(x) = x$ is $A(x) = x^2/2$, whose derivative is x. The accumulation function of $\sqrt{1-x^2}$ is complicated (specifically, we worked it out to be $\frac{1}{2}(\sin^{-1}(x) + x\sqrt{1-x^2})$), and if we take the derivative of *that* we get $\sqrt{1-x^2}$.

To see why this works for an arbitrary function f, think about the graph of $y = f(s)$, shown in Figure 8.2. (Note that we are using s as a dummy variable, as with all

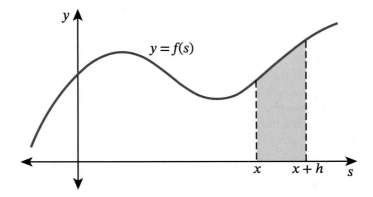

Figure 8.2. $A(x+h) - A(x)$ is the area of the shaded region.

accumulation functions.) We have

$$A(x) = \quad \text{Area under the curve from 0 to } x,$$

(8.22)
$$A(x + h) = \quad \text{Area under the curve from 0 to } x + h,$$

$$A(x + h) - A(x) = \quad \text{Area under the curve from } x \text{ to } x + h.$$

When h is small, the region under the curve $y = f(s)$ between $s = x$ and $s = x + h$ is a strip of width h and approximate height $f(x)$, so its area is approximately $hf(x)$. That is, $A(x + h) - A(x) \approx hf(x)$. Dividing by h, we get that

(8.23)
$$\frac{A(x + h) - A(x)}{h} \approx f(x).$$

By definition, $A'(x)$ is the limit of the left-hand side of (8.23) as $h \to 0$. But the smaller h is, the better our approximation becomes, and the closer the left-hand side gets to $f(x)$. In other words,

(8.24)
$$\frac{d}{dx} \int_0^x f(s)\, ds = \frac{d}{dx} A(x) = f(x).$$

This completes the picture of how integration and differentiation can be viewed as inverse operations. If you start with $f(x)$, integrate to get its accumulation function $A(x)$, and then take the derivative of $A(x)$, you get exactly $f(x)$. If you start with $F(x)$, compute its derivative $f(x)$, and then integrate $f(x)$ to get its accumulation function $A(x)$, you wind up with $F(x)$ plus a constant. In words,

(1) The derivative of the accumulation function (integral) is the original function.

(2) The accumulation function (integral) of the derivative is the original function plus a constant.

The Graph of $A(x)$. Armed with FTC1, we can sketch accumulation functions. For instance, consider the function

(8.25)
$$f(x) = \begin{cases} -1 & x < -1, \\ x & -1 \leq x \leq 1, \\ 1 & x > 1, \end{cases}$$

shown in Figure 8.3. At each point on the graph of $A(x)$, the slope of $A(x)$ is $f(x)$. This tells us that the graph of the accumulation function $A(x)$ must:

- Have slope -1 when $x < -1$. This part of the graph of $A(x)$ must be a straight line of slope -1.

- Have a slope that gradually increases from -1 to 1 as x goes from -1 to 1. This part of the graph must be concave up, and there must be a local minimum at $x = 0$, as that is where $f(x) = A'(x)$ goes from negative to positive.

- Have slope 1 when $x > 1$. This part of the graph of A is a straight line of slope 1.

- Be differentiable everywhere (as $A'(x) = f(x)$, which is defined everywhere) and in particular not have any jumps.

- Go through $(0,0)$, as $A(0) = \int_0^0 f(s)ds = 0$.

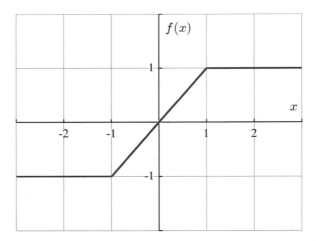

Figure 8.3. A function $f(x)$ defined in pieces

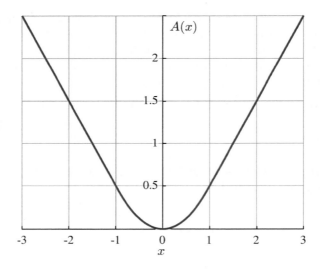

Figure 8.4. The accumulation function $A(x)$

Based on these clues, we see that the graph of $A(x)$ has to more-or-less look like Figure 8.4.

Notation for Integrals and Anti-Derivatives. Integrals are so closely tied to anti-derivatives that we use the same notation for both. The expression

(8.26)
$$\int_a^b f(x)\,dx$$

denotes a definite integral, which as we have seen is the limit of a sum. The expression

$$(8.27) \qquad \int f(x)\,dx$$

without limits of integration, means the anti-derivative of $f(x)$. This notation makes FTC2 especially easy to write down:

$$(8.28) \qquad \int_a^b f(x)\,dx = \left(\int f(x)\,dx \right) \Big|_a^b.$$

Not only is the notation for anti-derivatives and integrals the same, but so is the language. The process of finding anti-derivatives is usually called **integration**. In fact, most people call $\int f(x)\,dx$ an **indefinite integral**.[2] In this language, equation (8.28) says that a definite integral (the left-hand side) is the difference of the values of the indefinite integral at the endpoints. For both definite and indefinite integrals, we call the quantity inside the integral sign the **integrand**. Note that the integrand of $\int f(x)\,dx$ is $f(x)\,dx$, not just $f(x)$!

Unfortunately, this unified terminology *completely* obscures the fact that integration is all about adding up little parts to make a whole! However, resisting it is hopeless. For the rest of this book, I will raise the white flag and write things like "the indefinite integral of $x^2\,dx$ is $x^3/3$ plus a constant".

Relating FTC1 and FTC2. We proved the two halves of the Fundamental Theorem of Calculus separately, but it's easy to get each half from the other. Suppose that $f(x)$ is a function, $F(x)$ is an anti-derivative, and $A(x)$ is the accumulation function.

- If FTC1 is true, then $A'(x) = f(x) = F'(x)$, so $A(x) = F(x) + C$ for some constant C. But then

$$
\begin{aligned}
(8.29) \qquad \int_a^b f(x)\,dx &= A(b) - A(a) \\
&= F(b) + C - (F(a) + C) \\
&= F(b) - F(a),
\end{aligned}
$$

  which is FTC2.

- If FTC2 is true, then

$$(8.30) \qquad A(x) = \int_0^x f(s)\,ds = F(x) - F(0).$$

Because $F(0)$ is a constant, this means that $A'(x) = F'(x) = f(x)$, which is FTC1.

---

[2]A few curmudgeons, including the author, believe that the term "indefinite integral" should only be applied to accumulation functions, as an accumulation function is an integral whose upper limit is unspecified, hence indefinite. But it doesn't really matter, as FTC1 says that accumulation functions are anti-derivatives.

Evaluating Derivatives of More Complicated Integrals. Students' knowledge of FTC1 is often tested with computations like

$$(8.31) \qquad \frac{d}{dx}\int_0^x \cos(s^2)\,ds, \qquad \frac{d}{dx}\int_x^7 e^s \tan^{-1}(s)\,ds, \text{ and } \frac{d}{dx}\int_{\ln(x)}^{x^2} \frac{e^s\,ds}{s+1}.$$

FTC1 applies directly to the first problem, and there are tricks for applying it to the other two. However, all three of these problems can be solved easily with FTC2.

In the first problem, we are studying the derivative of $\int_0^x f(s)\,ds$, where $f(s) = \cos(s^2)$. But $\int_0^x f(s)\,ds = F(x) - F(0)$, so the derivative of this with respect to x is $F'(x) = f(x) = \cos(x^2)$. Notice that we never had to figure out what $F(x)$ actually was! All we needed was its derivative, which we already knew.

The second problem is similar. If we let $f(x) = e^x \tan^{-1}(x)$ and let $F(x)$ be an anti-derivative, then

$$\int_x^7 f(s)ds = F(7) - F(x)$$

$$(8.32) \qquad \frac{d}{dx}\int_x^7 f(s)ds = \frac{d}{dx}F(7) - \frac{d}{dx}F(x)$$

$$= 0 - F'(x)$$

$$= -f(x)$$

$$= -e^x \tan^{-1}(x).$$

The first line is FTC2. In the third line, the derivative of $F(7)$ is *not* $F'(7)$! $F(7)$ is a number that doesn't change as we vary x, so its derivative with respect to x is zero. Meanwhile, $F'(x) = f(x)$, as $F(x)$ is an anti-derivative of $f(x)$. Once again, we never needed to actually figure out what $F(x)$ was.

The third problem is the same story, plus the chain rule. Let $f(x) = e^x/(x+1)$, and let $F(x)$ be its anti-derivative, which we don't need to ever compute. Then

$$\int_{\ln(x)}^{x^2} f(s)ds = F(x^2) - F(\ln(x)),$$

$$\frac{d}{dx}\int_{\ln(x)}^{x^2} f(s)ds = \frac{d}{dx}F(x^2) - \frac{d}{dx}F(\ln(x))$$

$$(8.33) \qquad\qquad = F'(x^2)\frac{d(x^2)}{dx} - F'(\ln(x))\frac{d(\ln(x))}{dx}$$

$$= 2xf(x^2) - \frac{1}{x}f(\ln(x))$$

$$= \frac{2xe^{x^2}}{1+x^2} - \frac{1}{x}\frac{e^{\ln(x)}}{1+\ln(x)}$$

$$= \frac{2xe^{x^2}}{1+x^2} - \frac{1}{1+\ln(x)}.$$

The first four lines apply to any function $f(x)$. In the fifth line we entered the values of $f(x^2)$ and $f(\ln(x))$, and on the last line we used the fact that $e^{\ln(x)} = x$. We never needed to know what $F(x)$ or $F(x^2)$ or $F(\ln(x))$ were. We just needed their derivatives, and those came from the chain rule.

In general, whenever we're faced with a problem of the form $\frac{d}{dx} \int_{g(x)}^{h(x)} f(s)\,ds$, we can write

$$\int_{g(x)}^{h(x)} f(s)\,ds = F(h(x)) - F(g(x))$$

(8.34)
$$\frac{d}{dx} \int_{g(x)}^{h(x)} f(s)\,ds = \frac{d}{dx}F(h(x)) - \frac{d}{dx}F(g(x))$$
$$= F'(h(x))h'(x) - F'(g(x))g'(x)$$
$$= f(h(x))h'(x) - f(g(x))g'(x).$$

Please do not attempt to memorize equation (8.34). Remember the *method* instead.

8.4. Anti-Derivatives and Ballistics

Anti-derivatives (a.k.a. indefinite integrals) are essential for using the Fundamental Theorem of Calculus to compute definite integrals. However, they are also very important in their own right. In particular, Newton's second law says that the acceleration of a particle is proportional to the force applied to it. If we know the force, we know the acceleration. To get from the acceleration to the velocity, we just need to integrate. To get from the velocity to the position, we just need to integrate a second time.

For instance, in Section 6.1 we considered the problem of a falling rock. At time $t = 1$ the rock was 41 feet above the ground and falling at 12 feet/second. What is the rock's elevation $z(t)$ as a function of time? When will it hit the ground?

Under gravity (and ignoring air resistance), all objects near the surface of the Earth accelerate downward at 32 feet/second^2. That is, if the velocity is $v(t) = z'(t)$ and the acceleration is $a(t) = v'(t) = z''(t)$, and if we measure z in feet, v in feet/second and a in feet/second^2, then

(8.35)
$$a(t) = -32,$$

where the minus sign indicates that the acceleration is downward. Because $a(t) = v'(t)$, we must have $v(t) = \int a(t)dt = \int -32dt$. We already know that an anti-derivative of 1 is t, so we must have

(8.36)
$$v(t) = -32t + c_1,$$

where c_1 is a constant.

Every time we compute an indefinite integral, there is an unknown **constant of integration** like c_1. To figure out what that constant is, we have to compare our formulas to actual data. In this case, we know that $v(1) = -12$, as the rock was falling at

12 feet/second at time $t = 1$. This means that

$$v(1) = -12$$

$$-32(1) + c_1 = -12$$

(8.37)

$$c_1 = 20$$

$$v(t) = 20 - 32t.$$

Once we have $v(t)$, we can integrate again to get $z(t)$. An anti-derivative of t is $t^2/2$, so we must have

(8.38) $$z(t) = 20t - 16t^2 + c_2,$$

where once again we have a constant of integration c_2. We fix the constant by using the known value of $z(1) = 41$:

$$z(1) = 41$$

$$20(1) - 16(1)^2 + c_2 = 41$$

(8.39)

$$c_2 = 37$$

$$z(t) = 37 + 20t - 16t^2.$$

The position, velocity, and acceleration are shown in Figure 8.5.

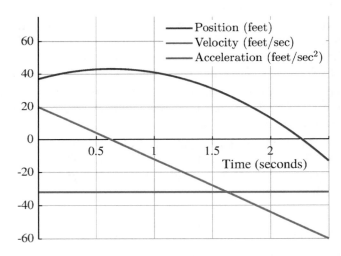

Figure 8.5. The position $z(t) = -16t^2 + 20t + 37$, velocity $v(t) = -32t + 20$, and acceleration $a(t) = -32$ of a falling rock; note that the slope of $z(t)$ is always equal to $v(t)$, and the slope of $v(t)$ is always equal to $a(t)$.

Finally, setting $z(t) = 0$, we get the time when the rock hits the ground. By the quadratic formula, the solutions to $37 + 20t - 16t^2 = 0$ are approximately $t = 2.2691$ and $t = -1.0191$, so the rock will hit the ground at time $t = 2.2691$. (If the rock was thrown from the ground, then the other solution represents the time that it was thrown. However, it's also possible that the rock was thrown or dropped from a different height at a different time.)

Even if we don't know the position and velocity at time $t = 1$, we can use integration to find the general form of a rock's trajectory:

$$a(t) = -32$$

(8.40)
$$v(t) = -32t + c_1$$

$$z(t) = -16t^2 + c_1 t + c_2.$$

The height of a projectile is always a quadratic function of time. The coefficients of 1 and t depend on the initial position and velocity, but the coefficient of t^2 is always -16. Similarly, in any situation where a particle has constant acceleration a, its position will be of the form $\frac{1}{2}at^2 + c_1 t + c_2$, while its velocity will be $at + c_1$. That formula applies equally well to falling rocks, to drag racing (where cars undergo more-or-less constant acceleration for the first few seconds), and to slamming on the brakes (where the braking force is more-or-less constant).

In the last problem we considered height as a function of time. We can use the same sort of reasoning to figure out the arc of a thrown object, such as a baseball thrown from center field to home plate, or a cannon ball fired at a castle.

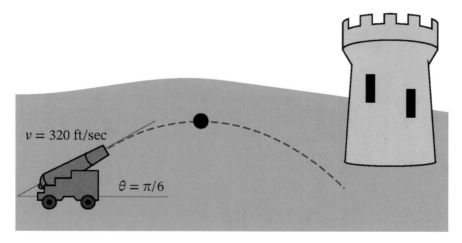

Figure 8.6. A cannon shooting at a castle

Suppose that a cannon is pointed at an angle of $\pi/6$ (30 degrees) above the horizontal, and it fires a cannon ball at an initial speed of 320 feet/second, as in Figure 8.6.[3] How far does the cannon ball fly before it hits the ground?

This is actually two problems rolled into one, as we have to separately keep track of the height $z(t)$ of the cannon ball and its horizontal distance $x(t)$ downrange. Our strategy is

(1) Figure out the initial height $z(0)$ and the initial vertical velocity $v_z(t)$.

(2) Use that to compute the height $z(t)$ as a function of time.

---

[3]Technically, an artillery piece that shoots at a high angle is called a *mortar* rather than a cannon. A velocity of 320 feet/second is realistic for 17th and 18th-century mortars. Modern artillery pieces are much more powerful.

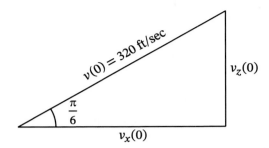

Figure 8.7. The vertical and horizontal components of the initial velocity

(3) Compute the time t_f at which the cannon ball hits the ground.

(4) Compute the horizontal position $x(t)$ as a function of time.

(5) Plug t_f into the formula for $x(t)$.

Because the cannon is fired at an angle of $\pi/6$, the initial horizontal velocity $v_x(0)$, the initial vertical velocity $v_z(0)$ and the speed form a triangle, as shown in Figure 8.7. We then have

$$(8.41) \qquad v_z(0) = 320\sin(\pi/6) = 160; \qquad v_x(0) = 320\cos(\pi/6) \approx 277,$$

where both v_x and v_z are measured in feet/second.

Because $a_z(t) = -32$, we must have $v_z(t) = c_1 - 32t$ for some constant c_1. Comparing to our initial velocity $v_z(0) = 160$, we see that $c_1 = 160$, so $v_z(t) = 160 - 32t$. $z(t)$ is the anti-derivative of $v_z(t)$, so $z(t) = c_2 + 160t - 16t^2$. Because the cannon ball essentially started at the ground, we have $z(0) = 0$, so we must have $c_2 = 0$. In other words,

$$(8.42) \qquad\qquad\qquad z(t) = 160t - 16t^2.$$

To figure out when the cannon ball hits the ground, we just solve $z(t) = 0$. The two solutions are $t = 0$ and $t = 10$. The first is when the cannon is fired. The second is when the cannon ball hits the ground. In other words, the cannon ball hits the ground 10 seconds after it is fired.

Next we consider the horizontal motion. As there is no horizontal force on the cannon ball, the horizontal acceleration is zero, so the horizontal velocity is constant. We already found $v_x(0) = 277$, so $v_x(t) = 277$. We then integrate $v_x(t)\,dt$ to get $x(t) = c_3 + 277t$. As $x(0) = 0$, we have $c_3 = 0$, so

$$(8.43) \qquad\qquad\qquad x(t) = 277t.$$

Finally, we plug in $t_f = 10$ to get $x(10) = 2770$. The cannon ball will land at a spot about 2770 feet downrange, or a little over half a mile away.

8.5. Computing Anti-Derivatives

We already know the derivatives of a lot of functions. To get a table of indefinite integrals, we just take a table of derivatives, swap the columns, and write $+ C$ a few times; see Tables 8.1 and 8.2.

Table 8.1. A table of derivatives

Function	Derivative
x^n	nx^{n-1}
e^x	e^x
$\ln(x)$	$1/x$
$\sin(x)$	$\cos(x)$
$\cos(x)$	$-\sin(x)$
$\tan(x)$	$\sec^2(x)$
etc.	etc.

Table 8.2. A table of integrals

Function	Integral
nx^{n-1}	$x^n + C$
e^x	$e^x + C$
$1/x$	$\ln(x) + C$
$\cos(x)$	$\sin(x) + C$
$-\sin(x)$	$\cos(x) + C$
$\sec^2(x)$	$\tan(x) + C$
etc.	etc.

Table 8.2 is correct, but it isn't as useful as it could be. We usually aren't interested in the integral of $5x^4$. We'd rather just know the integral of x^4, and we'd rather describe that as $n = 4$ than as $n = 5$. Likewise, we'd rather know the integral of $\sin(x)$ than of $-\sin(x)$. By tweaking this table and adding a few more rows, we get the list of common integrals shown in Table 8.3.

Those aren't all the functions we will need, but they're a good start. The last two lines are just the chain rule and the product rule. The two most powerful techniques for finding additional anti-derivatives, called **integration by substitution** and **integration by parts**, are based on these two rules. We will learn these techniques, and use them to expand our table further, in Chapter 9.

Table 8.3. A better table of integrals

Function	Integral	Comment
x^n	$x^{n+1}/(n+1) + C$	for $n \neq -1$
$1/x$	$\ln(x) + C$	which is the $n = -1$ case
e^x	$e^x + C$	world's easiest integral
e^{kx}	$e^{kx}/k + C$	
$\sin(x)$	$-\cos(x) + C$	note the minus sign
$\cos(x)$	$\sin(x) + C$	
$\sec^2(x)$	$\tan(x) + C$	
$\sec(x)\tan(x)$	$\sec(x)$	
$f'(g(x))g'(x)$	$f(g(x)) + C$	substitution
$f(x)g'(x) + f'(x)g(x)$	$f(x)g(x) + C$	integration by parts

8.6. Chapter Summary

The Main Ideas.

- Definite integrals, accumulation functions, and anti-derivatives are different things:
  - A definite integral is the limit of a sum.

$$(8.44) \qquad \int_a^b f(x)\,dx = \lim_{N\to\infty} \sum_{i=1}^N f(x_i^*)\Delta_i x.$$

    It is a number, not a function.
  - The accumulation function $A(x) = \int_0^x f(s)\,ds$ is a running total, like an odometer or the reading on a water meter. It is a function, not a number.
  - An anti-derivative $F(x)$ is a function with $F'(x) = f(x)$.
    Anti-derivatives are unique, up to adding a constant.
- Accumulation functions and definite integrals are related by

$$(8.45) \qquad \int_a^b f(x)\,dx = A(b) - A(a).$$

- Anti-derivatives and definite integrals are related by FTC2:

$$(8.46) \qquad \int_a^b f(x)\,dx = F(b) - F(a).$$

  Depending on your perspective, this is either a great way to compute the definite integral $\int_a^b f(x)\,dx$, or a great way to compute the change in the function $F(x)$.
- FTC1 says that the accumulation function is an anti-derivative:

$$(8.47) \qquad \frac{d}{dx} \int_0^x f(s)\,ds = f(x).$$

- Together, FTC1 and FTC2 say that differentiation (that is, computing rates of change) and integration (getting whole quantities by adding up the pieces) are in some sense inverse operations.

- Because of this, the process of finding anti-derivatives is usually called **integration**. Anti-derivatives are often called **indefinite integrals** and are denoted $\int f(x)dx$.

- Anti-derivatives are important in their own right. If you know the acceleration of an object, you can get its velocity by integration. If you know the velocity, you can get the position by integration.

- Every time you compute an indefinite integral, there is an arbitrary additive constant that must be included in the answer. The value of this constant can be determined from the initial conditions.

- Every fact about derivatives is also a fact about anti-derivatives. From this, we already know the anti-derivatives of x^n, e^x, $\cos(x)$, $\sin(x)$, and $\sec^2(x)$.

Expectations. You should be able to:

- Explain the difference between definite integrals, accumulation functions and anti-derivatives.

- Explain the key ideas behind FTC1 and FTC2.

- Use your knowledge of derivatives to sketch an accumulation function.

- Use correct notation for definite and indefinite integrals. This involves the use of the integral sign $\int$, the limits of integration, using correct variable names, and including dx (or dt or ds, etc.) inside the integral.

- Find the anti-derivatives of functions such as $x^3 - 3x + \frac{2}{x} + \cos(x) - 5e^x$.

- Use FTC2 to compute definite integrals of such functions.

- Use FTC2 and the chain rule to compute quantities such as $\frac{d}{dx} \int_{x^2}^{x^3} \sin(s^2)\,ds$.

8.7. Exercises

8.1. Clearly and concisely explain the difference between an anti-derivative, an accumulation function, and a definite integral.

Drug titers. In the Chapter 7 exercises we computed the amount of an antibiotic in the body after continuous delivery, as from an IV drip, over a period of time. If a drug is being administered at a constant rate A per hour, and if the body eliminates the drug at a fractional rate r per hour, then the drug titer DT after T hours can be expressed as an integral

(8.48)
$$DT = \int_0^T Ae^{-rt}\,dt.$$

In Exercises 8.2 and 8.3, we evaluate that integral with the help of the Fundamental Theorem of Calculus.

8.2. First we need to compute an anti-derivative.

(a) What is the derivative of $f(t) = e^{-.02t}$?

(b) Find a function whose derivative is $f(t)$.

(c) Using FTC2, find the accumulation function $\int_0^t f(s)ds$. Although FTC1 says that this is an anti-derivative of $f(t)$, it probably isn't the same as what you found in part (b).

8.3. Using what you learned in Exercise 8.2, find the accumulation function of Ae^{-rt}, where r is an arbitrary constant.

Mortgages. The answer to Exercise 8.3 is important to doctors who are writing prescriptions. It is especially important to anesthesiologists, who have to make sure that the titers of anesthetics remain in the desired range throughout an operation.

As we will explore in Exercises 8.4–8.6, the exact same accumulation function is important to bankers and homeowners. When figuring out mortgages, banks assume that a dollar paid t years from now is only as good as e^{-rt} dollars today, as e^{-rt} dollars today, invested for t years at annual interest rate r, will turn into one dollar t years from now. This is called the **present value** of the future payment of $1.

To compute a mortgage, a bank figures out the present value of all of your future mortgage payments. That (minus some profit, of course) is what they're willing to give you today to help you buy your house. Specifically, if you make payments at a rate of A dollars/year for T years, and if the prevailing interest rate is r, then the present value of all those payments is $PV = \int_0^T Ae^{-rt}\, dt$. If we know three of the four quantities PV, A, r and T, we can compute the fourth. In practical terms, PV is the size of the mortgage, A is the yearly payment (usually broken up into 12 monthly payments), T is the length of the mortgage (usually 20 or 30 years), and r is the interest rate.

8.4. Rina is buying a house. If she takes out a 30-year $200,000 mortgage at 3% interest, how much will her monthly payments be?

8.5. Rina decides that she can only afford to spend $1500/month on mortgage payments. If she can find a bank that offers 30-year mortgages at 2.5% interest, how big a mortgage can she afford?

8.6. In the end, Rina finds a house that she can buy with a $150,000 mortgage. If the interest rate is 2.5% and she makes payments of $1500/month, how long will it take her to pay off the mortgage?

Accumulation Functions and Their Derivatives.

8.7. Let $f(x) = 1 + 2x$.

(a) Sketch the graph of $f(x)$ on the interval $-2 \le x \le 4$.

(b) Using the methods we learned in Chapter 7, find a formula for the accumulation function

$$A(x) = \int_0^x f(s)\, ds.$$

Sketch the graph of $A(x)$ on the interval $-2 \le x \le 4$.

(c) Verify that $A'(x) = f(x)$. This implies that, at any point x, the *slope* of the graph of $A(x)$ is the same as the *height* of the graph of $f(x)$.

8.8. Consider the accumulation function

$$A(x) = \int_0^x s^4 \, ds.$$

(a) Using the fact that $A'(x) = x^4$, write a formula that expresses A in terms of x.

(b) Now consider a modified accumulation function $\tilde{A}$ that begins at $x = 2$ instead of at $x = 0$. That is,

$$\tilde{A}(x) = \int_2^x s^4 \, ds.$$

It is still true that $\tilde{A}'(x) = x^4$, but now we have $\tilde{A}(2) = 0$ instead of $A(0) = 0$. Write an explicit formula for $\tilde{A}(x)$ in terms of x. How do the formulas for $\tilde{A}(x)$ and $A(x)$ differ?

8.9. Consider the accumulation function $A(x) = \int_0^x g(t) \, dt$, where $g(t)$ is the function whose graph is given below.

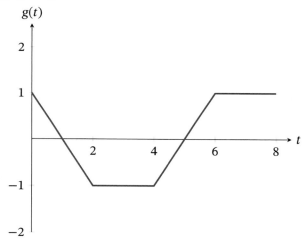

(a) Evaluate $A(0)$, $A(1)$, $A(4)$, and $A(8)$.

(b) On what interval(s) is $A(x)$ increasing?

(c) On what interval(s) is $A(x)$ decreasing?

(d) On what interval(s) is the graph of $A(x)$ concave up?

(e) On what interval(s) is the graph of $A(x)$ concave down?

(f) Where does $A(x)$ have a local maximum? Where does $A(x)$ have a local minimum?

(g) Sketch a graph of $A(x)$.

8.10. In each part, use the Fundamental Theorem of Calculus to find $A'(x)$.

(a) $A(x) = \int_1^x \sin^3(s) \, ds$

(b) $A(x) = \int_x^1 \sin^3(s) \, ds$

(c) $A(x) = \displaystyle\int_e^x \sqrt{t}\ln(t)\,dt$

(d) $A(x) = \displaystyle\int_e^{x^3} \sqrt{t}\ln(t)\,dt$

(e) $A(x) = \displaystyle\int_{x^2}^{x^4} e^s \sin^3(s)\,ds$

8.11.   Find all critical points of the function $A(x) = \displaystyle\int_1^x s^3 - 16s^5\,ds$.

8.12.   Find an anti-derivative $F(t)$ of $f(t) = t + \cos(t)$ for which $F(0) = 3$.

8.13.   Find a formula for the solution of each initial value problem.
   (a) $y' = 2x^3 - 5x^4$, $y(1) = 3$
   (b) $y' = \sin(5x)$, $y(\pi/2) = -2$

8.14.   Express the solution to each initial value problem using an accumulation function. That is, each solution should include an integral with a variable upper limit of integration.
   (a) $y' = \sin(x^4)$, $y(0) = 3$
   (b) $y' = e^{x^4}$, $y(1) = 0$
   (c) $y' = f(x)$, $f(x_0) = y_0$. Leave your answer in terms of the (unknown) function $f(x)$ and the unspecified constants x_0 and y_0.

8.15.   Determine the exact value of each definite integral. See Exercise 8.16 for the anti-derivative of $\cos(4x)$.

(a) $\displaystyle\int_0^2 1 + 2x + 6x^2\,dx$

(b) $\displaystyle\int_2^3 \frac{4}{x^2}\,dx$

(c) $\displaystyle\int_0^{3\pi} \sin(x) + 3\,dx$

(d) $\displaystyle\int_0^{\pi/8} \cos(4x)\,dx$

(e) $\displaystyle\int_0^2 4^t\,dt$

(f) $\displaystyle\int_{\ln(3)}^5 1 + e^t\,dt$

(g) $\displaystyle\int_{-2}^2 t^4\,dt$

(h) $\displaystyle\int_{-2}^2 t^5\,dt$

8.16.   Evaluate each indefinite integral in terms of the given parameters. Parts (b), (c), and (d) can all be done using integration by substitution, which we'll cover in Section 9.1, but that's overkill. You should be able to just guess an anti-derivative, and then adjust your guess if it isn't quite right. *Guess and check* is an important method of integration!

(a) $\displaystyle\int rx^3\,dx$

(b) $\displaystyle\int 2^{kt}\,dt$

(c) $\displaystyle\int \cos(\alpha x)\,dx$

(d) $\displaystyle\int 3 + \sec^2(Cz)\,dz$

(e) $\displaystyle\int \alpha x^n\,dx$, $n \neq -1$

(f) $\displaystyle\int \lambda x^4 - x^5\,dx$

8.17.   Suppose that bacteria are growing in a Petri dish at rate $b(t)$, where t is measured in hours and b is measured in colonies/hour. The sample begins with $b(0) = b_0$ colonies.

(a) Provide a clear and concise interpretation for the quantity $\int_1^2 b(t)\,dt$.

(b) Provide a clear and concise interpretation for the quantity $b_0 + \int_0^T b(t)\,dt$.

8.18.   A medical engineering company has designed a high resolution ultrasound device. Records indicate that 10,000 staff-hours were used for the design and assembly of the first 8 units. As production continued, the production team found that the function

$$L'(x) = 1200x^{-1/3}$$

worked well to describe the staff-hours per item needed to produce additional devices. That is, after the production of 8 units of this device, it takes approximately $1200(8)^{-1/3} = 600$ staff-hours to produce the next device.

(a) How many staff-hours $L(x)$ are required for the design and assembly of a total of x units? (*Note*: $L(0)$ is not zero, as it takes many staff-hours of design work to even begin assembly.)

(b) How many staff-hours are required for the design and assembly of the first 27 units?

Describing Motion. Exercises 8.19 and 8.20 explore the relationship between velocity, position, and distance traveled.

8.19.   The following graphs represent the velocity of a particle moving on the x-axis. The particle starts at $x = 2$ when $t = 0$. Time is measured in seconds, velocity is measured in feet/second and distance is measured in feet.

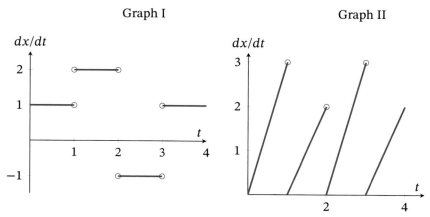

(a) For each graph, find where the particle is at the end of the trip 4 seconds later.

(b) For each graph, find the total (as opposed to net) distance traveled.

Food for Thought: Real-world objects can't change velocity instantaneously, as that would require infinite acceleration and infinite force. However, a baseball bat can apply a huge force to a ball, changing the ball's velocity *very* quickly! Physicists refer to such sudden intense forces as **impulses** and treat the resulting changes in velocity as if they were instantaneous. In other words, the graphs in Exercise 8.19 can't be taken literally, but they are exactly how a physicist would approximate a situation with impulsive forces.

8.20.   The following graph represents the velocity of a particle moving on the x-axis. The particle starts at $x = 2$ when $t = 0$. Time is measured in seconds and distance is measured in meters.

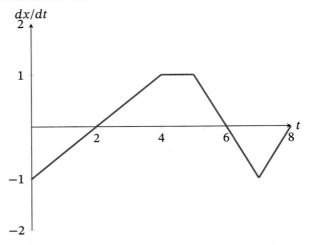

(a) Find where the particle is at the end of the trip 8 seconds later.
(b) Find the total (as opposed to net) distance traveled.

8.21.   **Baby Boom.** After World War II ended in 1945, the birth rate in the United States increased dramatically. Suppose the birth rate $b(t)$ (in millions of babies per year) for the first 20 years after the end of the war was

$$b(t) = 4 + 0.025t,$$

where t is the number of years since the end of the war.
(a) In all, how many babies were born in the 20 years after the end of World War II?
(b) What was the average number of babies born each year during these 20 years?
(c) Find an accumulation function $B(t)$ that gives a running total number of babies born since the end of the war.
(d) Starting at the end of the war, how long did it take for 20 million babies to be born?

Insulin and Diabetes. Insulin is a hormone produced by the pancreas to help regulate glucose levels in the human body. The digestive system converts the carbohydrates that are eaten into glucose and other simple sugars, which then enter the bloodstream. Insulin signals the body's tissues to absorb that glucose (glucose uptake) and convert it to glycogen or fats, which are then stored for later use. Type-1 diabetes occurs when the pancreas can't produce enough insulin. Type-2 diabetes occurs when the body doesn't respond strongly enough to the insulin that the pancreas does produce. Type-1 diabetics depend on injections of synthetic insulin to keep their glucose levels under control. In Exercise 8.22, we compare the effects of two kinds of insulin: Insulin A, which acts faster, and Insulin B, which lasts longer.

8.22. Let $g_A(t)$ and $g_B(t)$ be the rate at which the body absorbs glucose from the bloodstream after an injection of Insulin A or Insulin B (not counting the uptake that would happen without any injection), as shown in the figure.

(a) Let $G_A(t)$ and $G_B(t)$ represent the total amount of glucose uptake caused by Insulin A and Insulin B, respectively, up to time t. At the time when the two insulin curves in the figure intersect, which is bigger, G_A or G_B? Explain your answer.

(b) At what time is the difference between $G_A(t)$ and $G_B(t)$ greatest?

(c) Use a Riemann sum with $\Delta t = 1$ and midpoints to estimate $G_A(4)$ and $G_B(4)$. You can obtain the (approximate) relevant values of $g_A(t)$ and $g_B(t)$ by using a ruler to read the graph.

(d) At approximately what time(s) is $G_A(t) = G_B(t)$?

(e) Write down a definite integral involving $g_A(t)$ and $g_B(t)$ that equals $G_A(5) - G_B(5)$. Do you think the value of this definite integral is positive, negative, or zero? Explain.

8.23. **Ballistics.** A projectile is fired straight up from a 60-foot platform at an initial velocity of 28 feet per second. Ignoring air resistance,

(a) Find a formula for $z(t)$, the height (in feet) of the projectile t seconds after launch.

(b) What is the maximum height the projectile attains?

(c) What is the velocity of the projectile when it reaches the ground?

8.24.   Imagine that you are an NFL kicker. With 5 seconds left in the Super Bowl, your team is behind by 2 points. You line up to attempt a 60 yard field goal. The center snaps the ball and you give the ball a mighty kick!

   (a) Once kicked, the ball has an initial velocity of 80 ft/sec and an initial angle of 30 degrees, and is aimed straight toward the middle of the goal posts. Ignoring air resistance, do you win? (*Note*: The crossbar that the ball must clear is 10 feet high, and a successful kick is worth 3 points.)

   (b) Wait! The opposing team called a time out just before the snap, so your previous attempt doesn't count! You redo the kick, but this time you kick the ball (with perfect aim) at an angle of 40 degrees, with an initial velocity of 80 ft/sec. Again ignoring air resistance, do you win?

Chapter 9

Methods of Integration

In this chapter we complete our survey of one-dimensional integral calculus. In Chapter 7 we learned about representing bulk quantities as definite integrals. In Chapter 8 we learned how anti-derivatives, also known as indefinite integrals, can help us compute definite integrals. In this chapter we'll go over some of the main techniques for computing both definite and indefinite integrals. The two most powerful techniques for finding indefinite integrals are called **integration by substitution** and **integration by parts**. These are covered in Sections 9.1 and 9.2. If we can find an anti-derivative, then the Fundamental Theorem of Calculus makes computing a definite integral easy.

But what if we can't find an anti-derivative? There are some functions, like e^{-x^2}, whose anti-derivatives cannot be expressed in terms of elementary functions.[1] In other situations, we only have numerical data for our original function. In those cases, we can't compute definite integrals using FTC2. We have to compute them by directly estimating the area under the curve or otherwise summing the different contributions.

There are three important numerical methods for estimating definite integrals, all of which are covered in Section 9.3. We have already seen two of them, the **trapezoidal rule** and the **midpoint rule**. While the naive left-endpoint and right-endpoint rules have errors that typically scale[2] as $1/N$, the trapezoidal and midpoint rules have errors that scale as $1/N^2$. By taking a weighted average of the trapezoidal and midpoint rules, we get an estimate whose error scales as $1/N^4$. This is called **Simpson's rule**.

---

[1] Some people even say that "we can't integrate $e^{-x^2}\,dx$". Of course that isn't true, since there *is* a function, namely the accumulation function, whose derivative is e^{-x^2}. We just can't find a *formula* for it!

[2] When we say that an error "scales as" $1/N$ or $1/N^2$ or $1/N^4$, we mean that it is roughly proportional to $1/N$ or $1/N^2$ or $1/N^4$. If we make N ten times larger, then our error will become roughly 10 or 100 or 10,000 times smaller.

9.1. Integration by Substitution

How can we compute

(9.1) $$\int_1^3 2x(x^2 - 1)^3 \, dx \,?$$

We could multiply out $(x^2 - 1)^3$, multiply by $2x$, and then integrate the resulting polynomial term by term, but that's a mess. A better way involves the chain rule.

If u is any function of x, and if $F'(x) = f(x)$, then

(9.2) $$\frac{d}{dx}F(u) = F'(u)\frac{du}{dx} = f(u)\frac{du}{dx}.$$

In other words, $F(u)$ is an anti-derivative of $f(u)\frac{du}{dx}$, so we can write

(9.3) $$\int f(u)\frac{du}{dx} \, dx = F(u) + C = \int f(u) \, du.$$

There's more to this equation than meets the eye! The first indefinite integral means "a function whose derivative with respect to x is $f(u)\frac{du}{dx}$", and the second indefinite integral means "a function whose derivative with respect to u is $f(u)$". Those are very different ideas! However, the equation says that it's legal to replace $\frac{du}{dx}dx$ with du, as if $\frac{du}{dx}$ was actually the ratio of quantities du and dx. So from now on, we will simply write

(9.4) $$du = \frac{du}{dx} \, dx.$$

In our example, we want to find an anti-derivative of $2x(x^2 - 1)^3$. If we let $u = x^2 - 1$, then $\frac{du}{dx} = 2x$, and we have

(9.5) $$\int 2x(x^2 - 1)^3 \, dx = \int u^3 \frac{du}{dx} \, dx = \int u^3 \, du = \frac{u^4}{4} + C = \frac{(x^2 - 1)^4}{4} + C.$$

Let's check that:

(9.6) $$\frac{d}{dx}\frac{(x^2 - 1)^4}{4} = (x^2 - 1)^3 \frac{d(x^2 - 1)}{dx} = 2x(x^2 - 1)^3.$$

Our definite integral is then

(9.7) $$\int_1^3 2x(x^2 - 1)^3 \, dx = \frac{(x^2 - 1)^4}{4}\Big|_1^3 = \frac{8^4 - 0^4}{4} = 1024.$$

Since we substituted the variable u for some expression involving x, this process is called **integration by substitution**. It is also often called u-**substitution**, even if we don't actually call the second variable u! Here are several more examples of how this method works.

- To compute $\int e^{kx} \, dx$, we set $u = kx$, so $du = k\,dx$ and $dx = du/k$. Our integral is then $\int e^u \frac{du}{k} = \frac{e^u}{k} + C = \frac{e^{kx}}{k} + C$. When we take a derivative of e^{kx} we multiply by k. When we integrate $e^{kx}dx$ we divide by k.

- The derivative of $\ln(x)$ is $1/x$, so $\int \frac{dx}{x} = \ln(x) + C$. But that only makes sense when $x > 0$, since we can't take the log of a negative number. What is $\int_{-3}^{-1} \frac{dx}{x}$?

  When $x < 0$, we define $u = -x$, which is positive. Then $du = -dx$ and $\frac{dx}{x} = \frac{du}{u}$, so

$$(9.8) \qquad \int \frac{dx}{x} = \int \frac{du}{u} = \ln(u) + C = \ln(-x) + C.$$

In other words, $\int \frac{dx}{x}$ equals $\ln(x) + C$ when $x > 0$ and equals $\ln(-x) + C$ when $x < 0$. That is, $\int \frac{dx}{x} = \ln(|x|) + C$, no matter whether x is positive or negative. In particular, $\int_{-3}^{-1} \frac{dx}{x} = \ln(|x|)\Big|_{-3}^{-1} = \ln(1) - \ln(3) = -\ln(3)$.

- To compute $\int \tan(x)\, dx = \int \frac{\sin(x)}{\cos(x)}\, dx$, we let $u = \cos(x)$. Then $du = -\sin(x)dx$, and we have

$$(9.9) \qquad \int \tan(x)\, dx = \int \frac{\sin(x)}{\cos(x)}\, dx = \int \frac{-du}{u} = -\ln(|u|) + C$$
$$= -\ln(|\cos(x)|) + C = \ln(|\sec(x)|) + C.$$

In the last step we used the fact that $|\sec(x)| = |\cos(x)|^{-1}$, so $\ln(|\sec(x)|)$ equals $-\ln(|\cos(x)|)$. Another way to get the same result is to let $u = \sec(x)$, so $du = \sec(x)\tan(x)\, dx$, and

$$(9.10) \qquad \int \tan(x)\, dx = \int \frac{\sec(x)\tan(x)}{\sec(x)}\, dx = \int \frac{du}{u}$$
$$= \ln(|u|) + C = \ln(|\sec(x)|) + C.$$

Similarly, we can compute $\int \cot(x)\, dx$ by doing a substitution with $u = \sin(x)$, or by doing a substitution with $u = \csc(x)$. Either way we get

$$(9.11) \qquad \int \cot(x)\, dx = \ln(|\sin(x)|) + C = -\ln(|\csc(x)|) + C.$$

As the last example showed, we can sometimes get the answer in more than one way. *Integration by substitution is a tool, not a recipe.* Likewise, sometimes we can't get the answer by substitution at all. When faced with the integral

$$(9.12) \qquad \int (x^2 - 1)^3\, dx$$

it's tempting to try $u = x^2 - 1$, but that doesn't help at all. The integrand contains u^3, but it isn't $u^3\, du$. It's $u^3\, dx$, which is something else entirely. That's why it's so important to keep track of the dx's and dt's and du's in our integrals. They really matter!

When a method doesn't work, we just have to give up and try something else. The only way to do this integral is to multiply out the expression $(x^2 - 1)^3\, dx = (x^6 - 3x^4 + 3x^2 - 1)\, dx$ and integrate each term separately to get $\frac{x^7}{7} - \frac{3x^5}{5} + x^3 - x + C$.

With the help of integration by substitution, we can expand our table of known integrals; see Table 9.1.

Table 9.1. A revised and expanded table of integrals. The bottom six rows show integration by substitution.

Integrand	Integral	Comment
$x^n \, dx$	$x^{n+1}/(n+1) + C$	for $n \neq -1$
dx/x	$\ln(|x|) + C$	for both $x > 0$ and $x < 0$
$e^x \, dx$	$e^x + C$	world's easiest integral
$e^{kx} \, dx$	$e^{kx}/k + C$	
$\sin(x) \, dx$	$-\cos(x) + C$	note the minus sign
$\cos(x) \, dx$	$\sin(x) + C$	
$\sec^2(x) \, dx$	$\tan(x) + C$	
$\sec(x)\tan(x) \, dx$	$\sec(x)$	
$\tan(x) \, dx$	$\ln(|\sec(x)|) + C$	
$dx/(1 + x^2)$	$\tan^{-1}(x) + C$	
$u^n \, du$	$u^{n+1}/(n+1) + C$	for $n \neq -1$
du/u	$\ln(|u|) + C$	for both $u > 0$ and $u < 0$
$e^u \, du$	$e^u + C$	
$\sin(u) \, du$	$-\cos(u) + C$	note the minus sign
$\cos(x) \, du$	$\sin(u) + C$	
$du/(1 + u^2)$	$\tan^{-1}(u) + C$	

Substitutions and Definite Integrals. We return to our original example of computing $\int_1^3 2x(x^2 - 1)^3 \, dx$. Our anti-derivative was $u^4/4$, which we had to evaluate at our endpoints. The trouble is that 1 and 3 are the starting and ending values of x, not the starting and ending values of u. To avoid mixing apples and oranges, we either have to express everything in terms of x or express everything in terms of u. Our two options are:

- Convert $u^4/4$ into a function of x and plug in $x = 1$ and $x = 3$, or
- Convert $x = 1$ and $x = 3$ into values of u, namely $u = 0$ and $u = 8$.

That is, we can get our answer either as

(9.13)
$$\int_{x=1}^3 2x(x^2 - 1)^3 \, dx = \int_{x=1}^3 u^3 \, du = \frac{u^4}{4}\bigg|_{x=1}^3 = \frac{(x^2-1)^4}{4}\bigg|_{x=1}^3 = \frac{8^4 - 0^4}{4} = 1024,$$

or as

(9.14) $$\int_{x=1}^3 2x(x^2 - 1)^3 \, dx = \int_{x=1}^3 u^3 \, du = \int_{u=0}^8 u^3 \, du = \frac{u^4}{4}\bigg|_{u=0}^8 = \frac{8^4 - 0^4}{4} = 1024,$$

but not as

(9.15) $$\int_1^3 2x(x^2 - 1)^3 \, dx = \int_1^3 u^3 \, du = \frac{u^4}{4}\bigg|_1^3 = \frac{3^4 - 1^4}{4} = 20. \quad \text{(Wrong!)}$$

To avoid that mistake, it's a good idea to explicitly indicate which variable we're talking about in our limits of integration, as on the previous two lines.

We derived the rules for u-substitution by thinking about anti-derivatives and the chain rule. However, we can also understand u-substitution directly in terms of definite integrals and Riemann sums. If $u = x^2 - 1$, then every value of x corresponds to a value of u. If we break the interval $[1, 3]$ into N pieces, with $x_0 = 1 < x_1 < x_2 \cdots < x_N = 3$, then we have also broken the interval $[0, 8]$ into N pieces, with $u_0 = 0 < u_1 < u_2 < \cdots < u_N = 8$, where each $u_i = (x_i^2 - 1)$. We then have

(9.16)
$$\int_0^8 u^3 \, du \approx \sum_{i=1}^N (u_i^*)^3 \Delta_i u$$
$$\approx \sum_{i=1}^N (u_i^*)^3 u'(x_i^*) \Delta_i x$$
$$= \sum_{i=1}^N 2x_i^* ((x_i^*)^2 - 1)^3 \Delta_i x$$
$$\approx \int_1^3 2x(x^2 - 1)^3 \, dx.$$

The first line is the definition of a definite integral with respect to u. On the second line we used the microscope equation for the function $u(x)$, centered at $x = x_i^*$, to say that $\Delta_i u = u_i - u_{i-1} \approx u'(x_i^*) \Delta_i x = 2x_i^* \Delta_i x$. On the third line we just rewrote everything in terms of x, and the fourth line is the definition of a definite integral with respect to x. All the approximations get better and better as $N \to \infty$, and we get equality in the limit.

In summary, the procedure for integration by substitution is summarized in the formal equation

(9.17)
$$du = \frac{du}{dx} \, dx.$$

We derived this by thinking about the chain rule, and the difference between taking a derivative with respect to u and taking a derivative with respect to x. However, it can also be viewed as a statement about the intervals $\Delta_i u$ and $\Delta_i x$ that appear in our Riemann sums. In fact, that is pretty much how Leibniz thought of (9.17) when he invented the whole dx, du notation. To Leibniz, du and dx represented infinitesimally small intervals in u and x whose ratio is the conversion factor du/dx.

9.2. Integration by Parts

How can we integrate $xe^x \, dx$? We know that $\int x \, dx = \frac{x^2}{2} + C$ and that $\int e^x \, dx = e^x + C$, but we don't have a product rule for integrals. There is no obvious combination of x, x^2 and e^x whose derivative is xe^x.

The Method. Instead of going straight to the answer, let's go back to *close is good enough*. We'll guess a function whose derivative is close to xe^x, and then we'll adjust by how much we're off by. Since the derivative of e^x is e^x, we can try xe^x as a function whose derivative is close to xe^x.

The product rule tells us that

(9.18)
$$\frac{d}{dx}xe^x = xe^x + e^x.$$

Our guess has a derivative that is e^x too big. By subtracting off something whose derivative is e^x, we get our answer:

(9.19)
$$\frac{d}{dx}(xe^x - e^x) = xe^x + e^x - e^x = xe^x, \quad \text{so}$$
$$\int xe^x\,dx = xe^x - e^x + C.$$

In general, whenever we are trying to do an integral of the form $\int f(x)g'(x)dx$, we can guess $f(x)g(x)$. By the product rule, the derivative of $f(x)g(x)$ has two parts:

(9.20)
$$\frac{d}{dx}(f(x)g(x)) = f(x)g'(x) + f'(x)g(x).$$

The first part is what we want. The second part is extra, and we have to subtract off something whose derivative is $f'(x)g(x)$. That is, we have the

---

Integration by Parts Formula (long version):

(9.21)
$$\int f(x)g'(x)\,dx = f(x)g(x) - \int g(x)f'(x)\,dx.$$

Integration by parts doesn't solve our integration problem in one step. It just turns one integration problem (computing $\int f(x)g'(x)\,dx$) into another integration problem (computing $\int g(x)f'(x)\,dx$). However, that's often real progress. If the second integral is something that we know how to do, or at least is simpler than the first one, we win.

Integration by parts is often described in terms of expressions u and v instead of $f(x)$ and $g(x)$. If $u = f(x)$ and $v = g(x)$, then $du = f'(x)\,dx$ and $dv = g'(x)\,dx$, and we have the

---

Integration by Parts Formula (short version):

(9.22)
$$\int u\,dv = uv - \int v\,du.$$

Here are some examples of hard problems that can be solved by integration by parts.

- To compute $\int x^2e^x\,dx$, we can take $u = x^2$ and $dv = e^x\,dx$, so $du = 2x\,dx$ and $v = e^x$. (We could also take $v = e^x+3$ or e^x+17, but that's needlessly complicated. We usually pick v to be the simplest integral of dv.) Then

(9.23)
$$\int x^2e^x\,dx = x^2e^x - \int 2xe^x\,dx.$$

We didn't finish the problem, but we reduced it to a simpler problem. To finish the problem we either integrate by parts a second time, or remember equation

(9.19) and get

$$(9.24) \qquad \int x^2 e^x \, dx = x^2 e^x - \int 2x e^x \, dx = x^2 e^x - 2x e^x + 2 e^x + C.$$

- To compute $\int \ln(x) \, dx$ we take $u = \ln(x)$ and $dv = dx$, so $du = dx/x$ and $v = x$. Equation (9.22) becomes

$$(9.25) \qquad \int \ln(x) \, dx = x \ln(x) - \int x \frac{dx}{x} = x \ln(x) - \int dx = x \ln(x) - x + C.$$

- Computing $\int \tan^{-1}(x) \, dx$ is similar. We take $u = \tan^{-1}(x)$ and $dv = dx$, hence $du = \frac{dx}{1+x^2}$ and $v = x$, and get

$$(9.26) \qquad \int \tan^{-1}(x) \, dx = x \tan^{-1}(x) - \int \frac{x \, dx}{1 + x^2}.$$

We can solve the second integral by substitution, setting $w = 1 + x^2$ and $dw = 2x \, dx$. (We already defined u, so we'd better use another letter for our substitution.) Then $\frac{x \, dx}{1+x^2} = \frac{dw}{2w}$. The integral of $\frac{dw}{2w}$ is $\ln(|w|)/2$, so we get

$$(9.27) \qquad \int \tan^{-1}(x) \, dx = x \tan^{-1}(x) - \int \frac{x \, dx}{1 + x^2} = x \tan^{-1}(x) - \frac{1}{2} \ln(1 + x^2) + C.$$

(We could have written $\ln(|1 + x^2|)$, but $1 + x^2$ is always positive, so $|1 + x^2|$ is just $1 + x^2$.)

The Hierarchy. At this point, integration by parts may seem mysterious. Yes, it worked in the examples we just did, but the way that we picked u and dv probably felt like black magic. For instance, when integrating $x e^x \, dx$, we set $u = x$ and $dv = e^x dx$, so that $du = dx$ and $v = e^x$. But we could have tried $u = e^x$ and $dv = x \, dx$, in which case we would have had $du = e^x \, dx$ and $v = x^2/2$, and (9.22) would have read

$$(9.28) \qquad \int x e^x \, dx = \frac{x^2}{2} e^x - \int \frac{x^2}{2} e^x \, dx.$$

That's a true statement, but it doesn't do us any good. We've just replaced a hard integral with an even harder one!

In this subsection we'll learn how to avoid that trap by picking u and dv effectively. Our goal is to *make $\int v \, du$ simpler than $\int u \, dv$*. Ideally, we want to pick u and dv so that du is simpler than u and v is simpler than dv. That isn't always possible, since integration usually makes a function more complicated, but we can often set things up so that what we gain by differentiating u outweighs what we lose by integrating dv.

To develop our strategy, we divide functions into four classes:

(1) Some functions either get much simpler when differentiated (such as $\ln(x)$ and $\tan^{-1}(x)$) or get much more complicated when integrated (like $1/x$ and $1/(1+x^2)$). These functions should almost always be part of u.

(2) Some functions, like positive powers of x, get a little simpler when differentiated and a little more complicated when integrated. These are the next priority for including in u.

(3) Some functions, like $\sin(x)$, $\cos(x)$ and e^x, don't care. They look pretty much the same whether integrated or differentiated, and usually belong in dv.

(4) A few functions, like $1/x^2$ and $2xe^{x^2}$, actually get simpler when integrated. These almost always belong in dv.

When dealing with the product of two functions that are at different levels of this hierarchy, we want to make u the higher function and make dv the lower function (times dx). For instance, positive powers of x are higher than e^x, so when computing $\int x^n e^x \, dx$ we want to take $u = x^n$ and $dv = e^x \, dx$. However, powers of x are lower than $\ln(x)$, so when computing $\int x^3 \ln(x) \, dx$ we want to take $u = \ln(x)$ and $dv = x^3 \, dx$. In this case we have $v = x^4/4$ and $du = dx/x$, so

(9.29)
$$\int x^3 \ln(x) \, dx = \frac{x^4}{4} \ln(x) - \int \frac{x^4}{4} \frac{dx}{x} = \frac{x^4}{4} \ln(x) - \int \frac{x^3}{4} \, dx = \frac{x^4}{4} \ln(x) - \frac{x^4}{16} + C.$$

Converting x^3 to $x^4/4$ (making our job a little harder) was a small price to pay for converting $\ln(x)$ to $1/x$ (making our job a lot easier).

Remember that 1 is a function! We can think of $\ln(x)$ as $\ln(x)$ times 1. When integrating functions at the top of the hierarchy, like $\ln(x)$ and $\tan^{-1}(x)$, we can let u be the entire function being integrated, in which case $dv = dx$.

Integration by parts is often combined with other methods, particularly integration by substitution. For instance, if we wanted to compute

(9.30)
$$\int \frac{4x^2}{(2x + 7)^2} \, dx,$$

we might take $u = 4x^2$ and $dv = (2x + 7)^{-2} \, dx$. Computing v requires a substitution with $w = 2x + 7$, with the result that $v = -\frac{1}{2}(2x + 7)^{-1}$. We then have

(9.31)
$$\int \frac{4x^2}{(2x + 7)^2} \, dx = -\frac{2x^2}{2x + 7} - \int \frac{-2x}{2x + 7} \, dx$$
$$= -\frac{2x^2}{2x + 7} + \int \left(1 - \frac{7}{2x + 7}\right) dx$$
$$= -\frac{2x^2}{2x + 7} + x - \frac{7}{2} \ln(|2x + 7|) + C.$$

In the second line we used the fact that $1 - \frac{7}{2x+7} = \frac{2x}{2x+7}$, and then we did another substitution with $w = 2x + 7$.

Going in Circles. Integration by parts can help us when we have functions at different levels of the hierarchy. In those cases we replace the harder integral $\int u \, dv$ with an easier integral $\int v \, du$, which we can attack with all of our integration methods, including additional rounds of integration by parts. However, sometimes we have two functions at the same level of the hierarchy, so $\int v \, du$ is just as hard as $\int u \, dv$. What do we do then?

In those cases, we can often use integration by parts to turn our integral $I = \int u \, dv$ into a multiple of itself. It may seem like we're going in circles, trading a hard integral for the exact same hard integral, but there's a method to our madness. After one or more rounds of integration by parts, we hope to wind up with something like

(9.32) $$I = \left(\text{Some function of } x\right) + kI,$$

where k is some constant. If $k \neq 1$, then we just put kI on the other side of the equation and solve:

$$I = \left(\text{Some function of } x\right) + kI,$$
$$I - kI = \left(\text{Some function of } x\right) + C,$$

(9.33)

$$I = \frac{\text{Some function of } x}{1 - k} + C.$$

Note the placement of the arbitrary constants. On the first line, both I and kI involve arbitrary constants. They aren't necessarily the same, so on the second line we have to include $+C$ on the right hand side. The constant in the third line is really $\frac{1}{1-k}$ times the constant in the second line, but it's still arbitrary, so we just write $+C$.

For instance, to compute $\int \sin^2(x)\,dx$ we can take $u = \sin(x)$ and $dv = \sin(x)\,dx$, so $du = \cos(x)\,dx$ and $v = -\cos(x)$. We then have

$$\int \sin^2(x)\,dx = -\sin(x)\cos(x) - \int -\cos^2(x)\,dx$$

$$= -\sin(x)\cos(x) + \int \left(1 - \sin^2(x)\right) dx$$

(9.34)

$$= x - \sin(x)\cos(x) - \int \sin^2(x)\,dx,$$

$$2 \int \sin^2(x)\,dx = x - \sin(x)\cos(x) + C,$$

$$\int \sin^2(x)\,dx = \frac{x - \sin(x)\cos(x)}{2} + C.$$

In this example we were able to go in a circle in a single step, but it usually takes two, and sometimes takes more. For instance,

$$\int e^x \sin(x)\,dx = e^x \sin(x) - \int e^x \cos(x)\,dx$$

$$= e^x \sin(x) - e^x \cos(x) - \int e^x \sin(x)\,dx,$$

(9.35)

$$2 \int e^x \sin(x)\,dx = e^x(\sin(x) - \cos(x)) + C,$$

$$\int e^x \sin(x)\,dx = \frac{\sin(x) - \cos(x)}{2} e^x + C.$$

In the first line, we did an integration by parts with $u = \sin(x)$ and $dv = e^x\,dx$. In the second line we did an integration by parts with $u = \cos(x)$ and $dv = e^x\,dx$. That is, *we kept going in the direction we started.* In the first round we differentiated $\sin(x)$ to get $\cos(x)$, so in the second round we differentiated $\cos(x)$. In the first round we integrated $e^x\,dx$ to get e^x, so in the second round we integrated $e^x\,dx$. If in the second round we had integrated $\cos(x)\,dx$ and differentiated e^x, that would have undone what we did in the first round.

In the first round, we didn't have to take $u = \sin(x)$ and $dv = e^x\,dx$. We could just as well have taken $u = e^x$ and $dv = \sin(x)\,dx$. But then we would have differentiated e^x and integrated $\cos(x)$ in the second round, and the calculation would have looked like

$$\int e^x \sin(x)\,dx = \qquad -e^x \cos(x) + \int e^x \cos(x)\,dx$$

$$= \quad e^x \sin(x) - e^x \cos(x) - \int e^x \sin(x)\,dx,$$

(9.36)
$$2 \int e^x \sin(x)\,dx = \qquad e^x(\sin(x) - \cos(x)) + C,$$

$$\int e^x \sin(x)\,dx = \qquad \frac{\sin(x) - \cos(x)}{2} e^x + C.$$

A different calculation, but the same answer.

In general, when we have two functions at the same level of the hierarchy, we can start by differentiating either one and integrating the other. It doesn't much matter which we pick. However, once we pick a direction, we have to stick with it until we get back to a multiple of the original integral.

Definite Integrals. So far we have talked about indefinite integrals. To apply integration by parts to definite integrals, we just evaluate at the endpoints.

Integration by Parts for Definite Integrals:

(9.37)
$$\int_a^b f(x)g'(x)\,dx = f(x)g(x)\Big|_a^b - \int_a^b g(x)f'(x)\,dx$$

$$\int_a^b u\,dv = uv\Big|_a^b - \int_a^b v\,du.$$

The term $uv\Big|_a^b$ is called the **boundary term** (or the boundary *terms*, since we have both $u(b)v(b)$ and $u(a)v(a)$), while $-\int_a^b v\,du$ is called the **bulk term**. This terminology comes from physics. In many settings, we want to integrate a quantity $v\,du$ over some region in space, like the bulk of a crystal. The term $uv\Big|_a^b$ doesn't depend on what is happening inside the crystal, just on what is happening at the boundary.

Integration by parts is especially useful when the boundary term is zero, either because u and/or v vanish at the boundary, or because u and v are periodic and take on the same values at a and b. For instance,

(9.38)
$$\int_0^\pi x \cos(x)\,dx = x \sin(x)\Big|_0^\pi - \int_0^\pi \sin(x)\,dx.$$

As $\sin(0) = \sin(\pi) = 0$, the boundary term vanishes and we are left with just the bulk term

(9.39)
$$-\int_0^\pi \sin(x)\, dx = \cos(x)\Big|_0^\pi = -2.$$

In other cases, the boundary term is only approximately zero, and we can say that the original integral and the bulk term are approximately equal. For instance, consider $\int_0^{100} x^5 e^{-x}\, dx$. We can integrate by parts with $u = x^5$ and $dv = e^{-x}\, dx$, hence $du = 5x^4\, dx$ and $v = -e^{-x}$, to get

(9.40)
$$\int_0^{100} x^5 e^{-x}\, dx = -x^5 e^{-x}\Big|_0^{100} + 5\int_0^{100} x^4 e^{-x}\, dx.$$

As $0^5 e^{-0} = 0$ and $100^5 e^{-100} \approx 3.7{\times}10^{-34}$ is incredibly small, $\int_0^{100} x^5\, dx \approx 5\int_0^{100} x^4\, dx$. Similarly,

(9.41)
$$
\begin{aligned}
\int_0^{100} x^4 e^{-x}\, dx &\approx & 4\int_0^{100} x^3 e^{-x}\, dx \\[6pt]
\int_0^{100} x^3 e^{-x}\, dx &\approx & 3\int_0^{100} x^2 e^{-x}\, dx \\[6pt]
\int_0^{100} x^2 e^{-x}\, dx &\approx & 2\int_0^{100} x^1 e^{-x}\, dx \\[6pt]
\int_0^{100} x^1 e^{-x}\, dx &\approx & 1\int_0^{100} e^{-x}\, dx \\[6pt]
&=& 1 - e^{-100} \approx 1, \text{ so} \\[6pt]
\int_0^{100} x^5 e^{-x}\, dx &\approx & 5 \times 4 \times 3 \times 2 \times 1 = 5! = 120.
\end{aligned}
$$

In fact, if b is any large number and n is a positive integer, then $\int_0^b x^n e^{-x}\, dx \approx n!$ and $\lim_{b\to\infty} \int_0^b x^n e^{-x}\, dx = n!$.

From the perspective of bulk and boundary terms, integration by parts is more than just a trick for computing anti-derivatives. It allows us to relate one bulk quantity to another. As such, it is one of the cornerstones of mathematical physics and mathematical analysis.

9.3. Numerical Integration

So far we have been talking about formulas. If somebody gave us a formula for $f(x)$, we wanted to find a formula for $\int f(x)\, dx$, and we developed two powerful techniques, integration by substitution and integration by parts, for finding one. However, these techniques don't always work. No matter how clever we are, and no matter how hard we try, we will never find a formula (involving only elementary functions) for the anti-derivative of e^{-x^2}. If we want to compute $\int_0^1 e^{-x^2}\,dx$, we can't plug the anti-derivative into the Fundamental Theorem of Calculus. We have to do something completely different.

Likewise, sometimes our functions are defined by data and we don't even have formulas. If we measured the speed of a car every minute for an hour (say, by glancing at the speedometer), how well could we estimate the distance traveled in that hour? Similarly, if we didn't know the formula for a function $f(x)$ but only knew $f(x_0)$, $f(x_1)$, ..., $f(x_N)$, at some equally spaced points $a = x_0 < x_1 < \cdots < x_N = b$, how well could we estimate $\int_a^b f(x)\,dx$?

The Trapezoidal Rule. We have already seen several ways to estimate $\int_a^b f(x)dx$. All of these involve dividing the interval $[a, b]$ into N subintervals $[x_{i-1}, x_i]$, each of width $\Delta x = (b - a)/N$, with right endpoints at $x_i = a + i\Delta x$.

(1) We can do a Riemann sum using left endpoints. We take $x_i^* = x_{i-1}$, and our Riemann sum is

$$(9.42) \qquad L_N = \sum_{i=1}^{n} f(x_{i-1})\Delta x = \Delta x(f(x_0) + f(x_1) + \cdots + f(x_{N-1})).$$

The trouble with this method is that it gives too low an estimate for increasing functions, and too high an estimate for decreasing functions. As you can see in Figure 9.1, for increasing functions this estimate misses the area of N (roughly) triangular regions, each of width Δx and height $\Delta y \approx f'(x)\Delta x$, hence with area about $\Delta x \Delta y/2$. The total error is then roughly $\sum(\Delta x \Delta y)/2 = \Delta x(\sum \Delta y)/2 = \Delta x(f(b) - f(a))/2$. Since $\Delta x = (b - a)/N$, this scales as $1/N$. To get twice as

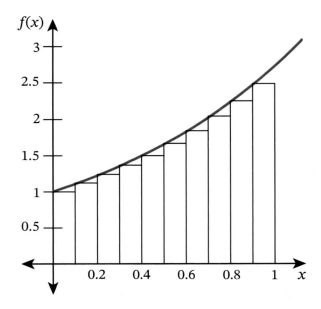

Figure 9.1. The approximation L_{10} with left endpoints

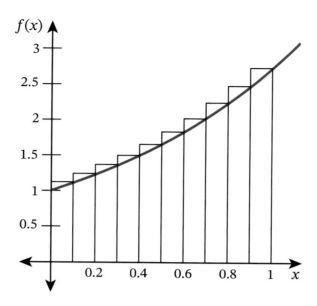

Figure 9.2. The approximation R_{10} with right endpoints

much accuracy, we need twice as many data points. To get 10 times as much accuracy, we need 10 times as many points.

(2) We can do a Riemann sum using right endpoints, as in Figure 9.2. Now we take $x_i^* = x_i$ and our Riemann sum is

(9.43) $$R_N = \sum_{i=1}^{n} f(x_i)\Delta x = \Delta x(f(x_1) + f(x_2) + \cdots + f(x_N)).$$

This has the opposite problem as using left endpoints, as it gives too high an estimate for increasing functions and too low an estimate for decreasing functions. For an increasing function it counts N roughly triangular regions of approximate total area $\Delta x(f(b) - f(a))/2$ that shouldn't have been counted.

Note that the difference between R_N and L_N is

$$
\begin{aligned}
R_N - L_N = \ & \Delta x(\qquad f(x_1) + \cdots + f(x_{N-1}) + f(x_N)) \\
& - \Delta x(f(x_0) + f(x_1) + \cdots + f(x_{N-1}) \qquad) \\
= \ & \Delta x(f(b) - f(a)).
\end{aligned}
$$

(9.44)

(3) If using left endpoints gives an answer that is about $\Delta x(f(b) - f(a))/2$ too low, and using right endpoints gives an answer that is about the same amount too high, then it makes sense to either add $\Delta x(f(b) - f(a))/2$ to the answer with left endpoints, or to subtract $\Delta x(f(b) - f(a))/2$ from the answer with right endpoints, or to average the results of using left and right endpoints. All of these procedures

give exactly the same answer, namely

(9.45)
$$T_N = \sum_{i=1}^{N} \frac{f(x_{i-1}) + f(x_i)}{2} \Delta x$$

$$= \frac{\Delta x}{2} \left(f(x_0) + 2f(x_1) + \cdots + 2f(x_{N-1}) + f(x_N) \right).$$

This is called the **trapezoidal rule** because the area it assigns to the ith subinterval $I_i = [x_{i-1}, x_i]$ is $\Delta x(f(x_{i-1}) + f(x_i))/2$, which is the area of the trapezoid with vertices $(x_{i-1}, 0)$, $(x_{i-1}, f(x_{i-1}))$, $(x_i, 0)$, and $(x_i, f(x_i))$. It does fine with increasing and decreasing functions, but it isn't perfect. When the graph is concave up, as in Figure 9.3, the trapezoids over-estimate the area under the curve, and when the graph is concave down they under-estimate it. The over/under-estimates are small, and are barely visible in the figure, but they're there. We will figure out the size of this error in Exercise 9.44. For now, we merely note that the total error scales as $1/N^2$. Doubling the number of data points gives us 4 times as much accuracy, and multiplying the number of data points by 10 gives us 100 times more accuracy.

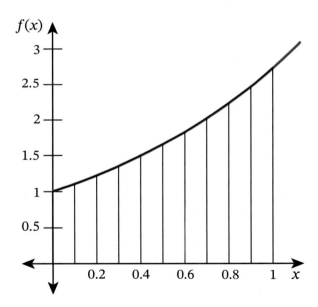

Figure 9.3. The trapezoidal rule T_{10}.

To see how these methods work, let's try them out to compute $\int_0^1 f(x)\,dx$ for a function whose values are listed in Table 9.2. The sum of the first 10 values is 16.337995, the sum of the last 10 values is 18.056277, and $\Delta x = 0.1$, so we have

(9.46) $L_{10} = 1.6337995, \qquad R_{10} = 1.8056277, \qquad T_{10} = 1.7197136.$

Table 9.2. Values of a mystery function

x	$f(x)$
0.0	1
0.1	1.105171
0.2	1.221403
0.3	1.349859
0.4	1.491825
0.5	1.648721
0.6	1.822119
0.7	2.013753
0.8	2.225541
0.9	2.459603
1.0	2.718282

Table 9.3. Comparing different numerical methods of integration

Approximation	Value	Error
$\int_0^1 e^x\,dx$	1.7182818	none
L_{10}	1.6337955	-0.0844868
R_{10}	1.8056277	$+0.0873459$
T_{10}	1.7197136	$+0.0014318$

So which of these is closest? The function in Table 9.2 is actually e^x, and the exact value of the integral is $\int_0^1 e^x\,dx = e^x\Big|_0^1 = e - 1 \approx 1.7182818$. We summarize the different results and their errors in Table 9.3.

It takes slightly more work to compute T_{10} than to compute L_{10} or R_{10}, but the result is a factor of $6N = 60$ better.

The Midpoint Rule and Simpson's Rule. When we introduced Riemann sums we talked about using left endpoints, right endpoints, and midpoints, but there wasn't a single mention of midpoints in the first half of this section. You should be wondering: *What's wrong with midpoints?*

If we have a formula for $f(x)$, there is absolutely nothing wrong with using midpoints! It doesn't have any error associated with slope, and its error from curvature is only half as big as the error from using trapezoids.

For instance, suppose that we were estimating the area under $y = x^2$ from 1 to $1 + h$ using a single interval ($N = 1$), where we imagine that h is a small number, like 0.1. We have $x_0 = 1$, $x_1 = 1 + h$ and our midpoint is $x_1^* = 1 + \frac{h}{2}$. The exact answer (as

Table 9.4. Comparing errors for a small interval

Approximation	Value	Error
$\int_1^{1+h} x^2\,dx$	$h + h^2 + \frac{h^3}{3}$	none
L_1	h	$-h^2 - \frac{h^3}{3}$
R_1	$h + 2h^2 + h^3$	$h^2 + \frac{2h^3}{3}$
T_1	$h + h^2 + \frac{h^3}{2}$	$h^3/6$
M_1	$h + h^2 + \frac{h^3}{4}$	$-h^3/12$

you can work out yourself) is $\int_1^{1+h} x^2\,dx = h + h^2 + \frac{1}{3}h^3$. The other approximations are given in Table 9.4.

While L_1 and R_1 are off by an amount that scales as h^2, T_1 and our midpoint estimate M_1 are only off by amounts proportional to h^3. In fact, T_1 is off by -2 times the amount that M_1 is off. If we can gather data at midpoints as easily as at endpoints, the midpoint rule is a little better than the trapezoidal rule. See Figure 9.4.

The trouble is that when we were integrating our function from a table, we didn't have the values of $f(0.05)$ or $f(0.15)$ or $f(0.25)$! The only way we could implement the midpoint rule was by making our intervals bigger. If we divided $[0, 1]$ into five intervals instead of ten, then the midpoints of the intervals $[0, 0.2]$, $[0.2, 0.4]$, etc. would be 0.1, 0.3, 0.5, 0.7, and 0.9. But while M_5 is a better estimate than T_5, it isn't as good an estimate as T_{10}.

This situation should remind you of what we saw with L_N and R_N. Both L_N and R_N had errors, but by taking their average we eliminated the error that comes from slope. Likewise, M_N and T_N both have errors proportional to $f''(x)/N^2$, with the error in T_N being approximately -2 times the error in M_N. By looking at $(T_N + 2M_N)/3$, we can eliminate that quadratic error entirely. In fact, it turns out that this weighted average eliminates cubic errors, too. All that remains is an error proportional to $1/N^4$. Two extra powers of N is a huge gain in accuracy that more than makes up for only having half as many intervals.

So let's go back to our numerically defined function. The trapezoidal and midterm rules with $N = 5$ and their weighted average $S_{10} = (T_5 + 2M_5)/3$ are

$$T_5 = 0.1\left(f(0) + 2f(0.2) + 2f(0.4) + 2f(0.6) + 2f(0.8) + f(1)\right),$$

$$M_5 = 0.2\left(f(0.1) + f(0.3) + f(0.5) + f(0.7) + f(0.9)\right),$$

(9.47)
$$S_{10} = \frac{0.1}{3}\Big(f(0) + 4f(0.1) + 2f(0.2) + 4f(0.3) + 2f(0.4) + 4f(0.5)$$

$$+ 2f(0.6) + 4f(0.7) + 2f(0.8) + 4f(0.9) + f(1)\Big)$$

$$= 1.7182828.$$

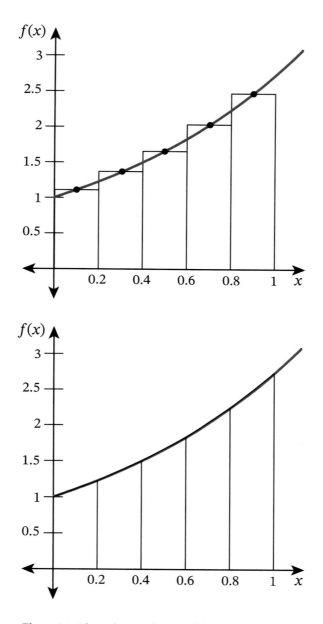

Figure 9.4. The midpoint rule M_5 and the trapezoidal rule T_5

S_{10} is called **Simpson's rule** with $N = 10$. We say $N = 10$ because we're using 10 data points, even though we are only using 5 intervals. In general, if N is even, then

$$S_N = \frac{\Delta x}{3}\Big[f(x_0) + f(x_N) + 2\big(f(x_2) + f(x_4) + \cdots + f(x_{N-2})\big)$$
$$+ 4\big(f(x_1) + f(x_3) \cdots + f(x_{N-1})\big)\Big],$$

(9.48)

where $\Delta x = (b-a)/N$ and each $x_i = a+i\Delta x$. The beginning and end points each count once, all the other even points count twice, the odd points count four times, and the whole thing gets multiplied by $\Delta x/3$. Put another way, the coefficients for Simpson's rule follow the pattern $1, 4, 2, 4, \ldots, 2, 4, 1$.

If $f(x)$ is an arbitrary smooth function, then Simpson's rule gives an answer whose error scales like $1/N^4$. In our example, S_{10} is only 0.000001 too high! This makes S_{10} about 1500 times more accurate than T_{10} and almost 100,000 times more accurate than L_{10} or R_{10}. In summary:

- The trapezoidal rule is easy to remember, easy to implement and has good accuracy, with errors scaling as $1/N^2$.

- Simpson's rule is more complicated, but it gives $1/N^4$ accuracy.

- There do exist more accurate methods, but only hard-core numerical analysts use them. In most situations where you need to compute a numerical integral, Simpson's rule, or even just the trapezoidal rule, is good enough.

9.4. Chapter Summary

The Main Ideas.

- Integration by substitution comes from inverting the chain rule. If $F'(x) = f(x)$, then the rule for integration by substitution can be written in any of the following ways:

(9.49)
$$\int f(g(x))g'(x)\,dx = F(g(x)) + C,$$
$$\int f(u)\frac{du}{dx}\,dx = F(u) + C,$$
$$\int f(u)\frac{du}{dx}\,dx = \int f(u)\,du,$$
$$\frac{du}{dx}dx = du.$$

- Using integration by substitution, we obtain several more "standard" integrals, including:

(9.50)
$$\int \frac{dx}{x} = \ln(|x|) + C \quad \text{even when } x < 0,$$
$$\int \tan(x)\,dx = -\ln(|\cos(x)|) + C = \ln(|\sec(x)|) + C.$$

- When doing definite integrals, the limits of integration have to match the variable being integrated. If $F'(x) = f(x)$, then

(9.51)
$$\int_{x=a}^{b} f(u)\frac{du}{dx}\,dx = F(u(x))\Big|_{x=a}^{b}$$
$$= F(u)\Big|_{u=u(a)}^{u(b)}$$
$$\neq F(u)\Big|_{u=a}^{b}.$$

- Integration by parts comes from inverting the product rule for derivatives. There are two common ways of writing it:

(9.52)
$$\int f(x)g'(x)\,dx = f(x)g(x) - \int g(x)f'(x)\,dx,$$
$$\int u\,dv = uv - \int v\,du.$$

- To decide what to use as u and dv, we pick u from higher in the following hierarchy of functions and dv from lower in the hierarchy, if possible. That makes $\int v\,du$ simpler than $\int u\,dv$.
  - (1) Functions that get much simpler when differentiated or much more complicated when integrated, like logs and inverse trig functions,
  - (2) Functions that get a little simpler when differentiated, like polynomials,
  - (3) Function like $\sin(x)$, $\cos(x)$, and e^x that don't care, and
  - (4) Functions that get simpler when integrated.

- When dealing with products of functions at the same level of the hierarchy, try going in circles.

- Integration by parts for definite integrals

(9.53)
$$\int_a^b u\,dv = uv\Big|_a^b - \int_a^b v\,du$$

is especially useful when the boundary term $uv\Big|_a^b$ is zero or approximately zero.

- When we can't find an anti-derivative, or our function is only defined numerically, we have to evaluate our definite integrals numerically.

- The trapezoidal rule

(9.54)
$$T_N = \frac{\Delta x}{2}\Big(f(x_0) + 2f(x_1) + \cdots + 2f(x_{N-1}) + f(x_N)\Big)$$

is accurate to within an error that scales like $1/N^2$.

- We can get $1/N^4$ accuracy with Simpson's rule with even values of N with:

$$\begin{aligned}
S_N &= \frac{\Delta x}{3}\Big[f(x_0) + f(x_N) + 2\big(f(x_2) + f(x_4) + \cdots + f(x_{N-2})\big) \\
&\quad + 4\big(f(x_1) + f(x_3) \cdots + f(x_{N-1})\big)\Big] \\
&= \frac{\Delta x}{3}\Big[f(x_0) + 4f(x_1) + 2f(x_2) + 4f(x_3) + \cdots \\
&\quad \cdots + 2f(x_{N-2}) + 4f(x_{N-1}) + f(x_N)\Big].
\end{aligned}$$

(9.55)

Expectations. You should be able to:

- Find anti-derivatives of basic functions.
- Use integration by substitution and integration by parts to convert harder integrals into easier ones.
- Apply limits of integration properly to definite integrals.
- Know how to persevere, repeating these methods as many times as necessary to evaluate the harder integrals.
- Also know when to quit. Some integrals, like $\int_0^1 e^{-x^2}\,dx$, can't be evaluated exactly with these or any other methods.
- Estimate integrals numerically with the trapezoidal rule and Simpson's rule.
- Modify the program Riemann (Section 7.2) to instruct a computer to apply the trapezoidal rule and Simpson's rule.

9.5. Exercises

Integration by Substitution

9.1.   Evaluate each integral using substitution.

(a) $\displaystyle\int 3y^2(y^3 - 4)\,dy$

(b) $\displaystyle\int \cos(t)\sqrt{2\sin(t)}\,dt$

(c) $\displaystyle\int \cos(4x)\,dx$

(d) $\displaystyle\int e^t \cos(1 + e^t)\,dt$

(e) $\displaystyle\int \frac{4x^2}{1 + x^2}\,dx$

(f) $\displaystyle\int \tan^2(2t)\sec^2(2t)\,dt$

(g) $\displaystyle\int_0^1 \frac{\sin(\pi\sqrt{x})}{\sqrt{x}}\,dx$

(h) $\displaystyle\int_1^3 \frac{dt}{3t + 1}$

(i) $\displaystyle\int_0^{\ln(e)} \frac{e^x}{e^x + 1}\,dx$

(j) $\displaystyle\int_0^2 \frac{dt}{1 + (t^2/4)}$

(k) $\displaystyle\int_0^{\pi/2} 3\sin(x)\cos^4(x)\,dx$

(l) $\displaystyle\int_0^1 \frac{t}{\sqrt{1 + 2t^2}}\,dt$

9.2. Let $f(x) = x^2 \sin(x^3)$.
 (a) Compute $\int f(x)\,dx$.
 (b) Find an anti-derivative $F(x)$ of $f(x)$ with $F(0) = 8$.
 (c) Find an anti-derivative $\tilde{F}(x)$ of $f(x)$ with $\tilde{F}(\sqrt[3]{\pi}) = 5$.

9.3. An anti-derivative of $f(t)$ is the same thing as the solution to the differential equation $y' = f(t)$. Find three solutions to the differential equation $y' = 4t + 3\ln(t)$.

9.4. **Dialysis.** Dialysis treatment removes urea and other waste products from a patient's blood. When a patient is connected to a dialysis machine, blood flows into the machine, is filtered, and is then returned to the body. The rate at which urea is removed from the blood (in mg/min) can by described by the equation

$$u(t) = \frac{K}{V} u_0 e^{-\frac{Kt}{V}}$$

where K is the rate of flow of blood through the dialysis machine (in mL/min), V is the volume of the patient's blood (in mL), and u_0 is the amount of urea in the blood (in mg) at time $t = 0$. (That is, K, V, and u_0 are parameters.)
 (a) Graph the function $u(t)$ using MATLAB. Use the following parameter values (which are typical for a patient requiring dialysis): $K = 320, V = 5000$, and $u_0 = 8000$.
 (b) What are the units of Kt/V? (This reflects a property that all exponents must have.)
 (c) What does the area between the time axis and $u(t)$ represent?
 (d) Write down the accumulation function that represents the amount of urea removed from the blood.
 (e) Using the parameter values given in part (a), determine approximately how long the dialysis treatment will take to lower the amount of urea in the blood to approximately 3000 mg.

9.5. When buying a car from a mathematically minded car dealer, you are offered the choice of two payment plans. In one, you are required to pay $f(t) = 5000$ dollars per year, continuously, for five years. In the other, you are required to pay $g(t) = 3000e^{0.2t}$ dollars per year for five years. Which of the two plans will cost you more money?

9.6. (a) Sketch the graph of the function $f(x) = x^2 e^{-x^3}$ on the interval $[0, 2]$.
 (b) Find the area under the graph of $f(x)$ between $x = 0$ and $x = 2$.
 (c) Find the area under the graph of $f(x)$ between $x = 0$ and $x = b$. That is, find the accumulation function $A(b) = \int_0^b f(x)\,dx$. Is $A(2)$ the same number you found in part (b)? What are the values of $A(10)$, $A(100)$, and $A(1000)$?
 (d) The area between the graph of $f(x)$ and the entire positive x-axis can be said to be $1/3$. Explain why.

Improper integrals. Even when a function has an unbounded domain, the area under the graph may still be finite. Exercise 9.6 showed us how to handle the integral of such a function. Exercises 9.7–9.10 explore such functions further.

The expression $\int_0^\infty f(x)\,dx$ does *not* mean "take the interval from 0 to ∞, break it into N pieces, build a Riemann sum, and take the limit as $N \to \infty$". That doesn't work, because one of those intervals would have to be infinitely wide. Instead, we set a cutoff at a number b and compute $\int_0^b f(x)\,dx$. That involves breaking the *finite* interval $[0, b]$ into N pieces, setting up a Riemann sum, and taking a limit as $N \to \infty$. Only then do we take a limit as $b \to \infty$. That is,

$$(9.56) \qquad \int_0^\infty f(x)\,dx \quad \text{means} \quad \lim_{b \to \infty} \int_0^b f(x)\,dx.$$

Similarly, we can integrate from $-\infty$ to 0.

$$(9.57) \qquad \int_{-\infty}^0 f(x)\,dx \quad \text{means} \quad \lim_{b \to \infty} \int_{-b}^0 f(x)\,dx.$$

Finally, we can integrate from $-\infty$ to ∞ by combining these operations:

$$(9.58) \qquad \int_{-\infty}^\infty f(x)\,dx = \int_{-\infty}^0 f(x)\,dx + \int_0^\infty f(x)\,dx.$$

Integrals of the form $\int_a^\infty f(x)\,dx$, $\int_{-\infty}^a f(x)\,dx$ and $\int_{-\infty}^\infty f(x)\,dx$ are called **improper integrals** because their definitions are a little different from what we usually mean by integrals. But when $f(x)$ is well-behaved, they still represent the total area under the curve.

In most cases, we can integrate from $-\infty$ to ∞ by setting cutoffs at $-b$ and b:

$$(9.59) \qquad \int_{-\infty}^\infty f(x)\,dx = \lim_{b \to \infty} \int_{-b}^b f(x)\,dx.$$

Be careful, however. The right-hand side sometimes gives an answer even when the improper integral doesn't makes sense. $\int_{-\infty}^\infty x\,dx$ does not exist, since there's an infinite amount of area under the line $y = x$ to the right of the origin, and there's an infinite amount of area over the line to the left of the origin. But $\lim_{b \to \infty} \int_{-b}^b x\,dx = \lim_{b \to \infty} 0 = 0$. The *definition* of $\int_{-\infty}^\infty f(x)\,dx$ is (9.58); equation (9.59) is just a useful-but-not-100%-reliable shortcut.

Improper integrals come up a lot in probability. Many random quantities are described by a **probability density function** $f(x)$, often called a **pdf**. The probability that a particular random quantity is between a and b is then $\int_a^b f(x)\,dx$. Since the quantity has to have *some* value, we must have $\int_{-\infty}^\infty f(x)\,dx = 1$.

9.7. The **exponential distribution** is often used to describe how long a piece of equipment lasts. (We saw a version of this in Exercise 7.28.) Let x be the lifetime of a machine, in hours, and suppose that its pdf is

$$f(x) = \begin{cases} Ce^{-x/1000} & x \geq 0, \\ 0 & x < 0, \end{cases}$$

where C is a constant chosen so that $\int_0^\infty f(x)\,dx = 1$.

(a) Find C.

(b) What is the probability that the machine will last less than 1000 hours? Less than 2000 hours? More than 3000 hours?

(c) Which is more likely, the machine failing in the first 100 hours or failing between the 1000th and 1100th hour?

9.8.   The lifetime of a different piece of equipment is described by a pdf that decays exponentially but at a different rate. Let λ be a positive number, and suppose that

$$f(x) = \begin{cases} Ce^{-x/\lambda} & x \geq 0, \\ 0 & x < 0, \end{cases}$$

where C is a constant chosen so that $\int_0^\infty f(x)\,dx = 1$. As in Exercise 9.7, x is the equipment's lifetime in hours.

(a) Find C in terms of λ.

(b) What is the probability that the equipment will last less than λ hours? Less than 2λ hours? More than 3λ hours?

9.9.   The average lifetime of a piece of equipment is given by the improper integral $\int_0^\infty x f(x)\,dx$. Compute this average for an exponential distribution with parameter λ, as in Exercise 9.8. (*Hint*: Integrate by parts.)

9.10.   If a random quantity is described by a pdf $f(x)$, then the **cumulative probability function**, or **cdf**, is the function $F(X) = \int_{-\infty}^X f(x)\,dx$. This gives the probability that the quantity is X or less. Compute the cdf of an exponential random variable with parameter λ, as in Exercise 9.8. (Note that $f(x) = 0$ if $x < 0$, so we can start our integral at 0 instead of $-\infty$.) If a and b are positive numbers with $a < b$, what is the probability that the random quantity is between $a\lambda$ and $b\lambda$?

Food for Thought: The parameter λ for an exponential distribution is usually called the **mean lifetime**, for reasons that shouldn't surprise you after doing Exercise 9.9. One advantage of the cdf over the pdf is that it is a function of x/λ. If you know the cdf for one quantity that is exponentially distributed, then you essentially know the cdf for every quantity that is exponentially distributed. The **standard exponential distribution** is the distribution with $\lambda = 1$.

Integration by Parts

9.11.   For each integral, pick u and dv so that $\int v\,du$ is simpler than $\int u\,dv$. Indicate what v and du are. You do not have to evaluate $\int v\,du$.

(a) $\displaystyle\int (x^3 + 7)\sin(x)\,dx$

(b) $\displaystyle\int (x^3 + 7)\ln(x)\,dx$

(c) $\displaystyle\int t\tan^{-1}(t)\,dt$

(d) $\displaystyle\int t\sec^2(t)\,dt$

(e) $\displaystyle\int x\sin(x^2)\ln(x)\,dx$

(f) $\displaystyle\int x^3 e^{x^2}\,dx$

9.12.   Evaluate each integral using integration by parts. Some examples may require
        two rounds, or may require a different trick after integrating by parts.

(a) $\int 2xe^{-x}\,dx$

(b) $\int x\cos(x)\,dx$

(c) $\int x^2 e^x\,dx$

(d) $\int x^3 e^{x^2}\,dx$

(e) $\int_{1/2}^{1} x^4 \ln(2x)\,dx$

(f) $\int_{1}^{9} \frac{\ln(x)}{\sqrt{x}}\,dx$

9.13.   Use the going in circles trick discussed in Section 9.2 to evaluate each integral.

(a) $\int e^{-x}\cos(x)\,dx$

(b) $\int_{0}^{\pi/2} \sin(x)\sin(2x)\,dx$

9.14.   Consider the definite integral $\int_{0}^{\pi/2} \sin^n(x)\,dx$, where $n \geq 2$.

(a) Integrate by parts using $u = \sin^{n-1}$ and $dv = \sin(x)\,dx$ to show that
$\int_{0}^{\pi/2} \sin^n(x)\,dx = \frac{n-1}{n} \int_{0}^{\pi/2} \sin^{n-2}(x)\,dx$. This involves both the going
in circles trick and the identity $\cos^2(x) = 1 - \sin^2(x)$.

(b) Using the recursion from part (a), compute $\int_{0}^{\pi/2} \sin^5(x)\,dx$ and
$\int_{0}^{\pi/2} \sin^6(x)\,dx$.

9.15.   The **average value** of the function $f(x)$ over the interval $[a, b]$ is defined to be
$\frac{1}{b-a} \int_{a}^{b} f(x)\,dx$.
(a) Explain in words why that is a reasonable definition.
(b) What is the average value of the function $\ln(x)$ on the interval $[1, e]$?
(c) What is the average value of $\ln(x)$ on $[1, b]$? Express your answer in terms
of b.
(d) Explain why, for large values of b, the average value of $\ln(x)$ on $[1, b]$ is
approximately $\ln(b) - 1$.

9.16.   What is the average value (as defined in Exercise 9.15) of $\sin^2(x)$ on the interval
$[0, 2\pi]$? What is the average value of $\cos^2(x)$ on the same interval?

9.17.   **The Cost of Vigilance.** In the presence of a predator, a prey animal will often modify its behavior to reduce the chance that the predator will find it. For
example, grasshoppers in the presence of wolf spiders will drastically reduce
their foraging in order to lower the chance of a spider seeing them. However,
there is a trade-off for this increased vigilance—less time spent foraging reduces the rate $a(t)$ at which the prey encounters its own food. (This is called
the **prey encounter rate**.) For grasshoppers, this increase in vigilance has
been shown to extend the amount of time spent in a particular instar (stage
of development). Vigilance levels vary considerably throughout the day. For
example, a grasshopper's rate of encounter with wolf spiders is greater in the

morning and evening, because wolf spiders are less active during the heat of the day.

(a) Let $v(t)$ represent the prey's level of vigilance ($0 \le v(t) \le 1$) at time t, with $v = 0$ representing no vigilance and $v = 1$ representing maximal vigilance. Let a_{max} represent the maximum prey encounter rate. Explain why it would be reasonable to use

$$a(t) = a_{max}\left(1 - v(t)\right)$$

as a model for the prey encounter rate.

(b) On warm summer days, grasshoppers are generally active from approximately 6:00 a.m. to 9:00 p.m., while spiders are generally active from 5:00 a.m. to 11:00 a.m. and again from 6:00 p.m. to 9:00 p.m. Explain why it would be reasonable for a grasshopper to have a vigilance function $v(t) = \cos^2\left(\frac{\pi}{14}t\right)$ during its 14 hours of activity, where t is measured in hours and $t = 0$ means 6:00 a.m.

(c) Use MATLAB to graph $a(t)$ when $v(t) = \cos^2\left(\frac{\pi}{14}t\right)$.

(d) Approximately how many encounters with food will a grasshopper have during the 14-hour day if $a_{max} = 3$ and $v(t) = \cos^2\left(\frac{\pi}{14}t\right)$?

9.18. The yeast in a tank at a brewery is producing alcohol at a rate of

$$R(t) = 80te^{-0.2t}$$

liters per day, where t is measured in days. Find the total amount of alcohol produced in the first week.

9.19. Consider the integral $\displaystyle\int_0^{\pi/2} \frac{dx}{1 + \sin^2(x)}$.

(a) Approximate this integral numerically using the trapezoidal rule and $N = 4$.

(b) Approximate this integral numerically using Simpson's rule and $N = 4$.

Degree Days. Temperature plays a major role in the growth and development of plants and insects. For example, if the temperature is unusually cold, a grasshopper is unable to progress through its instar stages quickly and is more susceptible to predation and disease. For food crops, too cool a growing season prevents plants from maturing before the first freeze and drastically reduces the yield of the crop and the nutritional value of the grain.

Each plant or animal must accumulate a certain number of heat units, or **degree days**, in order to reach maturity. As long as the temperature T is above a lower temperature threshold, L, measured in degrees Fahrenheit, the plant or insect will accumulate degree days at a rate $T - L$. However, if the temperature goes above an upper temperature cutoff U, then the organism pretty much shuts down and does not accumulate degree days. That is, the organism accumulates degree days at rate

$$d = \begin{cases} T - L & L < T < U, \\ 0 & \text{otherwise.} \end{cases}$$

For example, suppose that a particular insect has a lower temperature threshold of $L = 70°$ F and an upper temperature cutoff of $U = 85°$ F. Assume that during a 30-minute time interval during a day, the temperature remains at a constant $80°$ F, which is between the two thresholds. Then, during that 30-minute time interval (1/48th of a day), the insect would accumulate

$$\frac{1}{48} \times 5° \text{ or } \frac{5}{48} \text{ degree days.}$$

Plant biologists and entomologists have determined the temperature thresholds, cutoffs, and degree days needed to reach maturity for many plants and insects. Corn, for example, has $L = 50°$ F and $U = 86°$ F. In Exercises 9.20–9.22, we will compute the number of degree days accumulated by corn grown in the Austin, Texas area over a 24 hour period, based on the following table of actual temperature data for Austin on July 18–19, 2021.

Time	Temp	Time	Temp	Time	Temp	Time	Temp
05:53	72	12:53	88	19:53	88	02:53	76
06:53	70	13:53	90	20:53	81	03:53	76
07:53	74	14:53	90	21:53	79	04:53	75
08:53	77	15:53	92	22:53	80	05:53	74
09:53	80	16.53	92	23:53	80		
10:53	84	17:53	92	00:53	78		
11:53	85	18:53	91	01:53	77		

9.20.   (a) Make a table of the degree day accumulation rate $d(t)$ from 5:53 a.m. on July 18 to 5:53 a.m. on July 19.
        (b) Estimate the number of degree days accumulated over that 24-hour period using
            (i) 24 intervals and left endpoints.
            (ii) 24 intervals and right endpoints.
            (iii) 12 intervals and midpoints.
            (iv) 24 intervals and the trapezoidal rule.
            (v) 24 intervals and Simpson's rule.

9.21.   The are many ways to approximate degree days. One of the more accurate methods is to fit a sine curve to the temperature data. If we let $t = 0$ correspond to 5:53 a.m. on July 18th and $t = 1$ correspond to 5:53 a.m. on July 19th, we can approximate the data in the table by

$$T(t) = 82 + 10 \sin\left[2\pi\left(t - \frac{1}{4}\right)\right], \ 0 \le t \le 1$$

where t is now measured in days and T is measured in degrees Fahrenheit.
        (a) Use MATLAB to graph $T(t)$, $L = 50$, and $U = 86$ for $0 \le t \le 1$ all on the same set of axes. Shade the region that corresponds to the degree days accumulated during this one day.

     (b) Write down definite integrals that provide an estimate for the growing degree days accumulated during this single day. The integrals should correspond to the area of your shaded region in part (a).

     (c) Compute the definite integrals in part (b). What units does your integral have, and what does your computed value represent?

9.22. As the climate changes and temperatures rise, the growing degree days accumulated on each date will also change.

     (a) How would you modify the formula for $T(t)$ in Exercise 9.21 to account for a 2-degree Fahrenheit increase in the average temperature for that day? Call your modified function $H(t)$.

     (b) Graph $T(t)$ and $H(t)$ on the same set of axes using MATLAB. Be sure to include $L = 50$, and $U = 86$ and a legend that identifies the two curves.

     (c) Compute the growing degree days using $H(t)$. How does this compare to the answer you found in part (c) of Exercise 9.21?

     (d) Suppose that, instead of the average temperature changing, the high and low temperatures become more extreme. Modify $T(t)$ to account for a 3-degree increase in the high temperature for the day and a 3-degree decrease in the low temperature for the day. Call your modified equation $E(t)$. Repeat parts (b) and (c) using $E(t)$ in place of $H(t)$.

Population Biology. Exercises 9.23–9.25 explore the related ideas of mortality, survivorship, and reproductive output.

9.23. **The Survivorship Function.** Population biologists often track a particular population (cohort) of a certain species, such as all the members that are born in a particular place in a particular year. As time proceeds, members of the cohort die of various causes, such as predation, disease, or weather events. Let H_0 be the number of individuals at time $t = 0$, and let $H(t)$ be the number of individuals still alive at time t. The **per capita mortality rate** at time t, also known as the **hazard function**, is defined to be $m(t) = -H'(t)/H(t)$. For simplicity, suppose that this is equal to a constant m.

     (a) Write down an IVP for $H(t)$ and solve it.

     (b) The **survivorship function** is defined to be

$$S(t) = \frac{H(t)}{H_0}$$

     Explain the significance of $S(t)$.

     (c) Use $S(t)$ to determine the fraction of the cohort that die before age T.

     (d) Now suppose that the per capita mortality rate $m(t)$ is *not* constant. Find an integral expression for the survivorship function $S(t)$. (*Hint*: What is the derivative of $\ln(H)$?)

     (e) What do you think the human survivorship function looks like?[3]

9.24. **Reproductive Output.** Population biologists keep track of the number of offspring produced by an average individual during its lifetime. (Since offspring

---

[3]In developed countries, $m(t)$ is very small for children and younger adults, after which it increases exponentially from $m(40) \approx 1/1000$ to $m(100) \approx 1/3$.

have two parents, mothers and fathers are each given credit for half of their off-spring.) This quantity is referred to as the average lifetime reproductive output, L_R.

(a) Why would we intuitively expect a population to grow if $L_R > 1$, to shrink if $L_R < 1$, and to stay roughly constant if $L_R = 1$?

(b) The rate at which an individual produces offspring normally depends on age. Consider a cohort of animals (say, grasshoppers) that all hatch out at approximately the same time. Let $\beta(a)$ be the average rate at which an *individual* of age a produces offspring. Depending on the species being considered, age could be measured in days (aphids), weeks (house-flies), months (grasshoppers), or years (humans). Based on your personal knowledge, sketch a graph of $\beta(a)$ for humans.

(c) We can determine $\beta(a)$ by computing

$$\beta(a) = \frac{B(a)}{N(a)},$$

where $B(a)$ represents the *total offspring* produced by a *cohort* whose $N(a)$ members are all age a. Explain why this gives $\beta(a)$, the average rate at which an *individual* of age a produces offspring.

(d) Discuss why the accumulation function

$$C(a) = \int_0^a B(x)\,dx = \int_0^a \beta(x)N(x)\,dx$$

gives the *total offspring* between the age of 0 and a produced by the cohort.

(e) Discuss why the average lifetime reproductive output per individual is given by

$$L_R = \lim_{a \to \infty} \frac{C(a)}{N(0)} = \int_0^\infty \beta(x)\left[\frac{N(x)}{N(0)}\right]dx.$$

(*Note*: The quantity $S(x) = N(x)/N(0)$ is the **survivorship function** discussed in Exercise 9.23.)

9.25. In Exercise 9.24 we found that the average lifetime reproductive output, L_R, can be computed using

$$L_R = \int_0^\infty \beta(x)S(x)\,dx,$$

where $\beta(x)$ is the age dependent per capita reproduction rate and $S(x)$ is the survivorship function. We now consider two special cases.

(a) Consider an animal that has a constant per capita mortality rate m, starts to reproduce at age a_r, and thereafter produces offspring at an age-independent per capita rate β_0.

  (i) What are $S(x)$ and $\beta(x)$ for this animal?

  (ii) Show that

$$L_R = \frac{\beta_0}{m}e^{-ma_r}.$$

(b) Now consider an animal that always lives to age a_0, at which time it dies. As in part (a), this animal also starts to reproduce at age a_r and thereafter also produces offspring at an age-independent per-capita rate β_0.
  (i) What are $S(x)$ and $\beta(x)$ for this animal?
  (ii) Show that

$$L_R = \beta_0\,(a_0 - a_r).$$

(c) The approaches of part (a) and part (b) are actually two simple ways to model the same animal, whose actual reproductive and survival characteristics are more complicated. Why is it appropriate to use $m = 1/a_0$?

(d) Compare the values for L_R found in parts (a) and (b), assuming that reproduction starts when the animal reaches 70% of its maximum age. That is, let $a_r = 0.70a_0$. Which L_R would result in more total offspring?[4]

9.26. **SIR Survivorship.** Suppose that an epidemic starts with almost everybody susceptible and just a few people infected. What fraction of the population can expect to stay healthy (and remain Susceptible)? This is exactly a question of survivorship, only of the susceptible population $S(t)$ instead of the total cohort population $H(t)$. We start with the usual SIR equations:

$$\begin{aligned}
S'(t) &= -aS(t)I(t)\\
I'(t) &= aS(t)I(t) - bI(t)\\
R'(t) &= bI(t).
\end{aligned}$$

The survivorship function is then $s(t) = S(t)/S(0)$. (Since $S(0)$ is almost the entire population, this is essentially the same as the parameter $s(t)$ in the scaled SIR equations (2.17).) Let $x = 1 - \lim_{t\to\infty} s(t)$ be the fraction of the population that eventually gets infected.

(a) Find an integral formula for $s(t)$ in terms of $I(t)$.

(b) Take a limit as $t \to \infty$ of your formula from part (a) to get x in terms of the integral of $I(t)$.

(c) Use the fact that each sick person is sick for an average of $1/b$ to directly relate $\int_0^\infty I(t)dt$ to x.

(d) Combine your results from parts (b) and (c) to show that

$$1 - x = e^{-R_0 x},$$

where $R_0 = aS(0)/b$ is the basic replication number.

Food for Thought: When $R_0 > 1$, there are two solutions to this equation. One is $x = 0$, corresponding to nobody ever getting sick, and the other is the fraction of the population that will get sick once an epidemic actually starts. The positive solution x is typically much higher than the herd immunity threshold $1 - \frac{1}{R_0}$. For instance, if $R_0 = 2$, then the herd immunity threshold is 1/2, but the positive solution to $1 - x = e^{-2x}$ is about 0.79. The epidemic stops *growing* when half the population has gotten sick, but it keeps on *going* until 79% of the population has been stricken.

---

[4]If you found Exercises 9.23–9.25 interesting, you can find more details in Gurney & Nisbet (1998), *Ecological Dynamics*, Oxford University Press.

Means, Variances, and Standard Deviations. Probability density functions (pdf's) were introduced in Exercises 9.7–9.9. In Exercises 9.27–9.32, we see how we can find the mean and standard deviation of some common random variables from their pdf's, with the help of improper integrals and integration by parts.

If $f(x)$ is the pdf of a random variable, then the **mean** or the **average** of the probability distribution, often denoted μ or $\langle x \rangle$, is given by the integral

$$\mu = \langle x \rangle = \int_{-\infty}^{\infty} x f(x)\, dx.$$

This is a weighted average of each possible value of x, where the weight is the probability $f(x)\,\Delta x$ of being between x and $x + \Delta x$. Likewise, the average value of x^2 is

$$\langle x^2 \rangle = \int_{-\infty}^{\infty} x^2 f(x)\, dx.$$

The **variance**, often written σ^2, is $\langle x^2 \rangle - \langle x \rangle^2$. Finally, the **standard deviation**, usually written σ, is the square root of the variance. The standard deviation measures, in a technical sense, how wide the probability distribution is.

9.27.   The **standard uniform distribution** has the pdf

$$f(x) = \begin{cases} 1 & 0 < x < 1, \\ 0 & \text{otherwise.} \end{cases}$$

This is what people usually mean by "pick a random number between 0 and 1". Find the mean, variance and standard deviation of this distribution.

9.28.   A uniform distribution on the interval $[a, b]$, where $a < b$, has the pdf

$$f(x) = \begin{cases} \frac{1}{b-a} & a < x < b, \\ 0 & \text{otherwise.} \end{cases}$$

This is what people usually mean by "pick a random number between a and b". Find the mean, variance and standard deviation of this distribution. Notice that the standard deviation only depends on the width $b - a$ of the interval.

9.29.   The **standard exponential distribution** has the pdf

$$f(x) = \begin{cases} e^{-x} & x \geq 0, \\ 0 & x < 0. \end{cases}$$

Compute the mean and standard deviation. This requires integrating by parts and using the fact that $\lim_{b \to \infty} b e^{-b} = \lim_{b \to \infty} b^2 e^{-b} = 0$.

9.30.   An exponential random variable with parameter λ has the pdf

$$f(x) = \begin{cases} \dfrac{e^{-x/\lambda}}{\lambda} & x \geq 0, \\ 0 & x < 0. \end{cases}$$

Compute the mean and standard deviation. For this calculation, you can use the fact that $\lim_{b \to \infty} b e^{-b} = \lim_{b \to \infty} b^2 e^{-b} = 0$.

9.31. The **standard normal distribution** has the pdf

$$f(x) = Ce^{-x^2/2},$$

where C is a constant chosen so that $\int_{-\infty}^{\infty} f(x)\, dx = 1$. (The actual value happens to be $C = 1/\sqrt{2\pi}$, but that isn't relevant here.)
  (a) Compute the mean.
  (b) Compute $\langle x^2 \rangle$. You do not need to know C or the anti-derivative of $e^{-x^2/2}$ (which can't be expressed as a formula) to do this! Instead, integrate by parts to convert the integral for $\langle x^2 \rangle$ into a multiple of $\int_{-\infty}^{\infty} f(x)\, dx$, which equals 1.
  (c) Compute the standard deviation.

9.32. The pdf of a normal distributions with mean μ and standard deviation σ is given by the formula

(9.60) $$f(x) = \frac{C}{\sigma} e^{-(x-\mu)^2/(2\sigma^2)},$$

where C is the same as for the standard normal distribution.
  Let X be a random variable described by a normal distribution with mean μ and standard deviation σ, and let Z be a random variable described by a standard normal distribution. Let a and b be real numbers, with $a < b$. Show that the probability of X being between $\mu + a\sigma$ and $\mu + b\sigma$ is the same as the probability of Z being between a and b. (*Hint*: Do a u-substitution with $z = (x - \mu)/\sigma$.)

Food for Thought: Since the anti-derivative of $e^{-x^2/2}$ cannot be expressed as a formula involving elementary functions, most probability textbooks contain tables of the cdf of the standard normal distribution; these tables are easily accessed online. Computing things involving a general normal distribution almost always involves converting to "z-scores" and using the standard normal distribution.

Numerical Integration

9.33. Consider the integral $\int_0^{\pi/2} \dfrac{dx}{1 + \sin^2(x)}$.
  (a) Approximate this integral numerically using the trapezoidal rule and $N = 4$.
  (b) Approximate this integral numerically using Simpson's rule and $N = 4$.

9.34. A bicycle courier needs to deliver an important package quickly! The courier leaves the pick-up location with the package promptly at 9:00 a.m. and rides down a straight road but is slowed by traffic. The courier needs to deliver the package to a location 1.5 miles away by 9:08 a.m. The following table contains GPS estimates of the velocity $v(t)$ (in mph) of the courier at 1-minute intervals with $t = 0$ corresponding to 9:00 a.m.

t	$v(t)$	t	$v(t)$
0	0	5	8
1	12	6	15
2	10	7	20
3	14	8	8
4	10		

Here t is the number of minutes since 9:00 a.m.

(a) Use the Midpoint rule with $\Delta t = 2$ to estimate the distance the courier traveled from 9:00 a.m. to 9:08 a.m. That is, find M_4. Based on this estimate, was the delivery made on time?

(b) Use the Trapezoidal rule with $\Delta t = 2$ to estimate the distance the courier traveled from 9:00 a.m. to 9:08 a.m. That is, find T_4. Based on this estimate, was the delivery made on time?

(c) Use Simpson's rule with $\Delta t = 2$ to estimate the distance the courier traveled from 9:00 a.m. to 9:08 a.m. That is, find S_4. Based on this estimate, was the delivery made on time?

9.35. The area of the shaded region in the graph below represents the definite integral $\int_0^2 f(x)\,dx$.

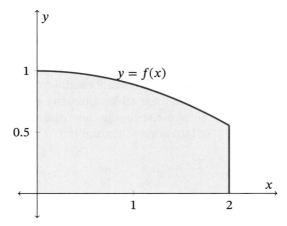

To estimate the definite integral, the left, right, midpoint, and trapezoidal rule approximations were found (using the same N for each estimate). The estimates, from the smallest to largest, were 1.6577, 1.7022, 1.7044, and 1.7466. Which rule produced which estimate?

Numerical Integration by Computer. Exercises 9.36–9.42 involve first creating MATLAB programs to apply the trapezoidal rule and Simpson's rule, and then using those programs to approximate some integrals.

9.36. Create a MATLAB program named Trap that will numerically estimate $\int_a^b f(x)\,dx$ using the trapezoidal rule. You can modify the Riemann program, or you can write your own program from scratch.

9.37. The following is a MATLAB program named Simp that will numerically estimate $\int_a^b f(x)\,dx$ using Simpson's method.

```
function S = Simp(f,a,b,N )
% This function uses Simpson's rule to approximate the integral
% of the function f from a to b where N is an even number
% of equally spaced subintervals.
deltax = (b - a)/N;
S = f(a) + f(b);
for k = 1:2:N-1
    x = a +k*deltax;
    S = S + 4*f(x);
end
for k = 2:2:N-2
    x = a +k*deltax;
    S = S + 2*f(x);
end
S=S*deltax/3;
```

Carefully explain what each line of the program is doing to implement Simpson's rule.

9.38. Consider the integral $\displaystyle\int_0^2 \frac{8}{x^2+4}\,dx$.
   (a) Evaluate this integral exactly, using the substitution $u = x/2$.
   (b) Approximate this integral numerically, using the trapezoidal rule with $N = 1000$. How accurate is the approximation? (Use the "format long" command in MATLAB to display more than four digits of your answer.)
   (c) Approximate this integral numerically, using Simpson's rule with $N = 100$. How accurate is the approximation?

9.39. Consider the integral $\int_0^3 x^2 e^{-x}\,dx$.
   (a) Evaluate this integral exactly, using integration by parts.
   (b) Approximate this integral numerically, using the trapezoidal rule with $N = 100$.
   (c) Approximate this integral numerically, using Simpson's rule with $N = 10$.
   (d) Which numerical approximation is more accurate?

9.40. (a) Use Simp to find numerical estimates for $\int_0^1 f(x)\,dx$ where $f(x) = x$, $f(x) = x^2$, $f(x) = x^3$, and $f(x) = x^4$ when $N = 10$, $N = 100$, and $N = 1000$. Compare these estimates to the exact results found using the Fundamental Theorem of Calculus.
   (b) Why would you expect the answer found using Simp for $f(x) = x^2$ to be the same as the exact answer?

    (c) Repeat part (a), only this time using Trap. Is it true that the error in these estimates is of order $1/N^2$?

    (d) Use Riemann to find L_{10}, L_{100}, and L_{1000} for $f(x) = x^2$. Is it true that the error in these estimates is of order $1/N$?

9.41.   Use Simpson's rule with $N = 6$ to estimate the area under the curve $y = 2^x/x$ from $x = 1$ to $x = 4$.

9.42.   Suppose the probability density function for adult female height (in inches) in a particular country is approximately

$$h(x) = \frac{1}{2.5\sqrt{2\pi}}e^{-(x-63)^2/12.5}.$$

(This is an example of a normal distribution, as in Exercise 9.32.)

    (a) Use Simpson's rule with $N = 12$ to estimate the probability that a woman in that country is between 5 and 6 feet tall.

    (b) Use Simpson's rule to estimate the probability that a woman in that country is over 6.5 feet. You will have to be somewhat creative as Simpson's rule can't be applied to infinite intervals. A way around this is to make the reasonable assumption that there are no women in the country over the height of 7.5 feet.

Improving the Trapezoidal Rule. We noted earlier that the accuracy of the trapezoidal rule goes as $1/N^2$. In Exercises 9.43 and 9.44, we will see why, and we will develop a better approximation, called the **corrected trapezoidal rule**, that provides $1/N^4$ accuracy without the complication of the 1-4-2-4$\cdots$2-4-1 pattern of Simpson's rule.

9.43.   Let $f(x) = c_0 + c_1 + c_2x^2$ be an arbitrary quadratic function, let a and b be arbitrary numbers, and let $\Delta x = b - a$.

    (a) Compute $\int_a^b f(x)dx$ exactly using the Fundamental Theorem of Calculus.

    (b) Compute the area of the trapezoid with vertices $(a, 0)$, $(a, f(a))$, $(b, f(b))$, and $(b, 0)$. (You can assume that $f(a)$ and $f(b)$ are both positive.)

    (c) Show that the area of the trapezoid, minus the integral, is exactly $c_2(\Delta x)^3/6 = f''(a)(\Delta x)^3/12$.

9.44.   Now suppose that $f(x)$ is an arbitrary function, and that we are trying to estimate $\int_a^b f(x)dx$ with the trapezoidal rule using N intervals. Write down a Riemann sum, involving $f''(x)$, that approximates the difference between the trapezoidal rule and the actual integral. (*Hint*: Apply the results of Exercise 9.43 to each subinterval and add up the errors.)

The Corrected Trapezoidal Rule. By now you should be used to approximating integrals with Riemann sums, and approximating Riemann sums with integrals. The answer to Exercise 9.44 is approximately

(9.61) $$\frac{(\Delta x)^2}{12}\int_a^b f''(x)dx = \frac{(\Delta x)^2}{12}f'(x)\Big|_a^b = \frac{(\Delta x)^2}{12}(f'(b) - f'(a)).$$

Since Δx is proportional to $1/N$, this shows that the accuracy of the trapezoidal rule goes as $1/N^2$.

Now that we have a good approximation to the error, we can subtract this (approximate) error to get an even better approximation. The **corrected trapezoidal rule** is

$$(9.62) \qquad C_N = T_N - \frac{(\Delta x)^2}{12}(f'(b) - f'(a)).$$

9.45. Let $f(x) = \cos(x)$. We will approximate $\int_0^{\pi/2} f(x)\,dx$ numerically in several ways: using the left endpoint rule L_N, the right endpoint rule R_N, the trapezoidal rule T_N, Simpson's rule S_N, and the corrected trapezoidal rule C_N. We already know that this integral is exactly $\sin(x)\big|_0^{\pi/2} = 1$, so it's easy to see how accurate each estimate is.

(a) Compute L_2, R_2, T_2, S_2 and C_2.

(b) Now compute L_4, R_4, T_4, S_4 and C_4. How much additional accuracy did we get by doubling the number of points?

(c) For each method, indicate roughly how big N would have to be to give an answer that is accurate to 6 decimal places.

9.46. Let $f(x) = e^{-x}$. We will approximate $\int_0^1 f(x)dx$.

(a) Compute L_{10}, T_{10}, S_{10} and C_{10}. You may want to use the MATLAB programs Riemann, Trap and Simp.

(b) Now compute $\int_0^1 f(x)\,dx$ exactly using the Fundamental Theorem of Calculus. Which approximation was the most accurate, and by how much?

9.47. Consider the integral $\int_{-5}^{5} e^{-x^2}\,dx$. This integral cannot be evaluated exactly, but it is within 10^{-10} of $\sqrt{\pi}$. (See Exercise 10.25 to understand why.)

(a) Evaluate this integral numerically, using the trapezoidal rule with $N = 10$, 100, and 1000.

(b) Evaluate this integral numerically, using Simpson's rule with $N = 10$, 100, and 1000.

(c) The answers using the trapezoidal rule are almost as accurate as the answers using Simpson's rule. Why?

9.48. Suppose that $f(x)$ is a periodic function, with $f(x + 1) = f(x)$, such as $f(x) = \dfrac{1}{1 + \sin^2(2\pi x)}$.

(a) Explain why estimating $\int_0^1 f(x)\,dx$ with left endpoints gives the same answer as using right endpoints, the trapezoidal rule, or the corrected trapezoidal rule.

(b) Explain why, for a fixed value of N, using left endpoints to estimate $\int_0^1 f(x)\,dx$ is just as accurate as using midpoints or Simpson's rule.

9.49. Let $f(x) = c_0 + c_1 x + c_2 x^2 + c_3 x^3$, where c_0, c_1, c_2, and c_3 are arbitrary constants. Show that, when estimating $\int_a^b f(x)dx$, both C_2 and S_2 give *exactly* the right answer. (So do C_N and T_N for larger values of N.)

Food for Thought: The errors for both Simpson's Rule and the corrected trapezoidal rule come from the *fourth* derivative of f and scale roughly as $1/N^4$. It's possible to get even more accurate rules by taking a weighted average of C_N and S_N (specifically, $\frac{1}{5}(C_N + 4S_N)$), or by adding a term proportional to $(f'''(b) - f'''(a))/N^4$ to Simpson's Rule, but that's overkill. In practice, the most common methods for numerical integration are the trapezoidal rule, Simpson's Rule, and the corrected trapezoidal rule. When integrating a smooth periodic function over a complete period, as in Exercise 9.48, all of the methods discussed in this chapter are especially accurate, with errors that shrink faster than any power of N.

Chapter 10

One Variable at a Time

So far we have developed the key tools of calculus, namely derivatives and integrals, for functions of a single variable. The derivative of $f(x)$ is

(10.1)
$$f'(x) = \lim_{h \to 0} \frac{f(x+h) - f(x)}{h}$$

and the definite integral is

(10.2)
$$\int_a^b f(x)\,dx = \lim_{N \to \infty} \sum_{i=1}^{N} f(x_i^*)\Delta_i x.$$

However, many quantities in the real world depend on two or more inputs. The boiling point of salt water depends on your elevation and the salinity of the water. The wind chill on a winter day depends on the temperature and the wind speed. The heat index on a summer day depends on the temperature and the relative humidity. All sorts of weather data, from temperature to atmospheric pressure to average annual rainfall, depend on location, which is described by two coordinates, such as latitude and longitude. The rate of a chemical reaction depends on the temperature and on the concentrations of all of the chemicals involved in the reaction.

The key idea for handling functions of several variables is the 6th and last Pillar of Calculus:

> *One variable at a time!*

When studying how a quantity depends on several variables, we never fiddle with all of the variables at once. Instead, we hold all but one of the variables fixed and see how the quantity depends on the one variable that we are studying. Once we know both how the boiling point of water depends on elevation and how it depends on salinity, we can put the two effects together to approximate the boiling point by using a linear function of elevation and salinity.

The derivative of a function with respect to one variable, while holding all the other variables fixed, is called a **partial derivative**. Partial derivatives are developed in Section 10.1. In Section 10.2 we use partial derivatives to build linear approximations to our functions.

In Section 10.3 we turn to integrals. Not anti-derivatives, but genuine whole-is-the-sum-of-the-parts integrals! If we know the rate of wheat production at every point in a state, how can we compute the total amount of wheat that is grown in the state? As with one-dimensional integrals, we break the state into little pieces, estimate the contribution of each piece, add up the pieces, and take a limit. Since we are working in two dimensions, this is called a **double integral**. In practice, we add the pieces one variable at a time to get something called an **iterated integral**.

10.1. Partial Derivatives

Don't Sweat the Details. The heat index I_H describes how hot it feels as a function of the actual temperature T (usually measured in the USA in degrees Fahrenheit) and the relative humidity R (measured as a percentage between 0 and 100). Since the human body cools itself by the evaporation of sweat, and since sweat evaporates less when the humidity is higher, increased humidity makes it feel hotter, even if the actual temperature doesn't change.[1]

There isn't any formula for the heat index. Instead, it is computed by interpolating from a table produced by the US National Oceanic and Atmospheric Administration (NOAA). A portion of the table is shown in Table 10.1.

Table 10.1. Heat index I_H as a function of relative humidity (R) and temperature (T)

Heat index in °F

R \ T	80	82	84	86	88	90	92	94
40%	80	81	83	85	88	91	94	97
45%	80	82	84	87	89	93	96	100
50%	81	83	85	88	91	95	99	103
55%	81	84	86	89	93	97	101	106
60%	82	84	88	91	95	100	105	110
65%	82	85	89	93	98	103	108	114
70%	83	86	90	95	100	105	112	119
75%	84	88	92	97	103	109	116	124
80%	84	89	94	100	106	113	121	129
85%	85	90	96	102	110	117	126	135

---

[1]Air conditioners both lower the temperature and remove humidity from the air. The hotter it is outside, the more your air conditioner runs to bring the indoor temperature down to the setting on your thermostat, the drier it gets indoors, and the cooler you feel!

For instance, when it is 90° out and the relative humidity is 50%, the heat index is 95°. At that point, raising or lowering the temperature by 2° raises or lowers the heat index by 4°, while raising or lowering the relative humidity by 5% raises or lowers the heat index by 2°.

To study how the heat index I_H depends on humidity, we hold the temperature steady at a single value (say, 90°) and consider the function

(10.3) $g(R) = I_H(R, 90) =$ Heat index at relative humidity R and temperature 90°

We can then approximate $g'(R)$ at a particular value (say, 50% humidity) by a forward difference quotient

(10.4) $$\frac{g(55) - g(50)}{5} = \frac{97 - 95}{5} = 0.4,$$

or by a backward difference quotient

(10.5) $$\frac{g(50) - g(45)}{5} = \frac{95 - 93}{5} = 0.4,$$

or by a centered difference quotient

(10.6) $$\frac{g(55) - g(45)}{10} = \frac{97 - 93}{10} = 0.4.$$

Conveniently, these all give the same answer and we estimate

(10.7) $$g'(50) \approx 0.4° \text{ Fahrenheit/percent humidity.}$$

Similarly, to study the effect of temperature on heat index we could define a function

(10.8) $h(T) = I_H(50, T) =$ Heat index at relative humidity 50% and temperature T

and compute

(10.9) $$h'(90) \approx 2.$$

(The units of h' are degrees/degree, so our answer is dimensionless.)

The derivatives g' and h' give us partial information about how $I_H(R, T)$ depends on R and T, since each one tells us what happens when only one variable is changed. For that reason, g' and h' are called **partial derivatives** of the original function $I_H(R, T)$ and are denoted $\partial I_H/\partial R$ and $\partial I_H/\partial T$.[2] That is,

(10.10)
$$\frac{\partial I_H}{\partial R}(50, 90) = g'(50) = \lim_{h \to 0} \frac{I_H(50 + h, 90) - I_H(50, 90)}{h},$$
$$\frac{\partial I_H}{\partial T}(50, 90) = h'(90) = \lim_{k \to 0} \frac{I_H(50, 90 + k) - I_H(50, 90)}{k}.$$

In the second equation, we used the letter k rather than h to distinguish between a change in R (called h) and a change in T (called k). We could also write $h = \Delta R$, $k = \Delta T$.

---

[2]The symbol ∂ for partial derivatives, usually called "partial", dates to 1770 and is intended to be a stylized "d".

A **partial derivative** of a function of two or more variables is the ordinary derivative of the function that you get by holding all but one variable fixed.

We just computed $\partial I_H/\partial R$ and $\partial I_H/\partial T$ at $(R, T) = (50, 90)$, but the same ideas apply everywhere. At $(45, 82)$ we have $\partial I_H/\partial R \approx 0.2°/\%$ and $\partial I_H/\partial T \approx 1$. At low humidity and (relatively) low temperature, the humidity doesn't have much of an impact and each degree of temperature translates into just one degree of heat index. However, at $(80, 92)$ we have $\partial I_H/\partial R \approx 1°/\%$ and $\partial I_H/\partial T \approx 4$. At high temperatures and very high humidity, small increases in the temperature and humidity can be dangerous, and small decreases can feel wonderful.

Reality Check: Some of the numbers in the heat index table are downright scary. In a city like Houston, Texas, the relative humidity will often hit 80%, and the temperature will usually get into the 90's or 100's in the summer. A temperature of just 94 with 80% humidity would mean a deadly heat index of 129°.

Fortunately, high temperatures and high relative humidities don't usually happen at the same time. The *relative* humidity is the amount of water vapor in the air divided by the maximum amount that the air can hold. Air at 94° can hold about 30% more water vapor than air at 86°. A morning in Houston might well have a temperature of 86° and a relative humidity of 80%, for a heat index of 100°. If the absolute humidity stayed the same in the afternoon when the temperature hit 94°, then the *relative* humidity would drop to about 62%, for a heat index of about 111°. That's plenty hot, but it's nothing like 129°.

On the flip side, falling temperatures make the relative humidity rise. Evening thunderstorms can appear in the summer, seemingly from nowhere, as the relative humidity approaches 100% and the water vapor already in the air coalesces, first into clouds and then into rain.

Partial Derivatives in General. Now that we've endured the heat index, we can return to our air-conditioned classroom and study an arbitrary function $f(x, y)$ of two variables.

Definition of partial derivatives:

(10.11)
$$\frac{\partial f}{\partial x}(a, b) = \lim_{h \to 0} \frac{f(a + h, b) - f(a, b)}{h},$$
$$\frac{\partial f}{\partial y}(a, b) = \lim_{k \to 0} \frac{f(a, b + k) - f(a, b)}{k}.$$

We can also think of $\frac{\partial f}{\partial x}$ and $\frac{\partial f}{\partial y}$ as functions of x and y:

(10.12)
$$\frac{\partial f}{\partial x}(x, y) = \lim_{h \to 0} \frac{f(x + h, y) - f(x, y)}{h},$$
$$\frac{\partial f}{\partial y}(x, y) = \lim_{k \to 0} \frac{f(x, y + k) - f(x, y)}{k}.$$

If we define $g(x) = f(x, b)$ and $h(y) = f(a, y)$, then $\frac{\partial f}{\partial x}(a, b) = g'(a)$ and $\frac{\partial f}{\partial y}(a, b) = h'(b)$. As with ordinary derivatives, the procedure we use for finding the derivative at a point (a *number*) can be applied everywhere at once to get the derivative as a *function* of x and y. To save space, we often denote partial derivatives with subscripts. f_x means $\partial f/\partial x$ and f_y means $\partial f/\partial y$.

Calculating partial derivatives is easy. To compute $\partial f/\partial x$, we *just take the derivative of $f(x, y)$ with respect to x while treating y as a constant.* This is because $\partial f/\partial x = g'(x)$, and $y = b$ is a constant in the function $g(x) = f(x, b)$. Likewise, to compute $\partial f/\partial y$, we treat x as a constant. Since the partial derivatives of $f(x, y)$ are *ordinary* derivatives of $g(x)$ and $h(y)$, all the formulas and rules that we developed for ordinary derivatives can be copied over verbatim for computing partial derivatives.

For instance, suppose that $f(x, y) = e^{xy^2 - 9}$. For any fixed $y = b$, we have

(10.13)
$$\frac{\partial e^{xy^2 - 9}}{\partial x} = \frac{d}{dx} e^{b^2 x - 9} = b^2 e^{b^2 x - 9} = y^2 e^{xy^2 - 9}.$$

Likewise, for any fixed $x = a$, we have

(10.14)
$$\frac{\partial e^{xy^2 - 9}}{\partial y} = \frac{d}{dy} e^{ay^2 - 9} = 2aye^{ay^2 - 9} = 2xye^{xy^2 - 9}.$$

We don't usually write the calculation out in such detail, replacing x and y in turn with constants a and b, taking an ordinary derivative with respect to the other variable, and then replacing a and b with x and y. When taking a partial derivative with respect to x, we just *remember* that y is a constant. When taking a partial derivative with respect to y, we just *remember* that x is a constant. Using the chain rule (specifically, $(e^u)' = e^u u'$), we then write

(10.15)
$$\frac{\partial e^{xy^2 - 9}}{\partial x} = e^{xy^2 - 9} \frac{\partial(xy^2 - 9)}{\partial x} = y^2 e^{xy^2 - 9},$$
$$\frac{\partial e^{xy^2 - 9}}{\partial y} = e^{xy^2 - 9} \frac{\partial(xy^2 - 9)}{\partial y} = 2xye^{xy^2 - 9}.$$

Higher Derivatives: Mixed Partials are Equal! The expressions f_x and f_y are functions in their own right, so we can take their derivatives to get second partial derivatives of $f(x, y)$. There are four of these, each of which can be written in several different ways:

(10.16)
$$f_{xx} = (f_x)_x = \frac{\partial^2 f}{\partial x^2} = \frac{\partial}{\partial x}\left(\frac{\partial f}{\partial x}\right),$$
$$f_{yy} = (f_y)_y = \frac{\partial^2 f}{\partial y^2} = \frac{\partial}{\partial y}\left(\frac{\partial f}{\partial y}\right),$$
$$f_{xy} = (f_x)_y = \frac{\partial^2 f}{\partial y \partial x} = \frac{\partial}{\partial y}\left(\frac{\partial f}{\partial x}\right),$$
$$f_{yx} = (f_y)_x = \frac{\partial^2 f}{\partial x \partial y} = \frac{\partial}{\partial x}\left(\frac{\partial f}{\partial y}\right).$$

Notice that the order of the variables is different in subscript and ∂ notation. f_{xy} means $\frac{\partial^2 f}{\partial y \partial x}$, not $\frac{\partial^2 f}{\partial x \partial y}$. Fortunately, it doesn't really matter, as we'll soon see.

To get a feel for these second derivatives, suppose that $f(x, y)$ is a quadratic function

(10.17) $$f(x, y) = Ax + By + Cx^2 + Dxy + Ey^2,$$

where A, B, C, D, and E, are constants. Then

$$f_x(x, y) = A + 2Cx + Dy,$$
$$f_y(x, y) = B + Dx + 2Ey,$$
(10.18)
$$f_{xx} = 2C,$$
$$f_{yy} = 2E,$$
$$f_{xy} = D,$$
$$f_{yx} = D.$$

Let's take a closer look at what these quantities mean.

- f_x is just the ordinary first derivative of $g(x)$. It tells us how fast $f(x, y)$ is changing when we change x but don't change y. Likewise, f_y is just the ordinary first derivative of $h(y)$. It tells us how fast $f(x, y)$ is changing when we change y but don't change x.

- f_{xx} and f_{yy} are the ordinary second derivatives of $g(x)$ and $h(y)$. If we graphed $g(x)$ and $h(y)$, they would indicate the curvature of the graphs. They also indicate how far $f(x, y)$ is from the linear approximation $Ax + By$ when $x = 0$ or $y = 0$.

- The **mixed partial derivatives** f_{xy} and f_{yx} are more mysterious. To get f_{xy} we first compute f_x by treating y as a constant, but then we take a derivative with respect to y! To get f_{yx} we first compute f_y by treating x as a constant, but then we take a derivative with respect to x. In this example, both are equal to D, and give a sense of the synergy between x and y. Both f_{xy} and f_{yx} capture what happens when we vary both x and y that we don't see by varying x and y separately.

Meanwhile, in our running example $f(x, y) = e^{xy^2 - 9}$,

$$f_{xx} = \left(y^2 e^{xy^2 - 9}\right)_x = y^4 e^{xy^2 - 9},$$
$$f_{yy} = \left(2xy e^{xy^2 - 9}\right)_y = (2x + 4x^2 y^2)e^{xy^2 - 9},$$
(10.19)
$$f_{xy} = \left(y^2 e^{xy^2 - 9}\right)_y = (2y + 2xy^3)e^{xy^2 - 9},$$
$$f_{yx} = \left(2xy e^{xy^2 - 9}\right)_x = (2y + 2xy^3)e^{xy^2 - 9}.$$

The last three calculations all used the product rule, taking the derivative of y^2 or $2xy$ as well as the derivative of $e^{xy^2 - 9}$.

In both of our examples, f_{xy} and f_{yx} happened to be equal. That wasn't a coincidence! The following theorem goes by several names. Some call it Clairaut's theorem, some call it Schwarz's theorem, and some call it Young's theorem, but most people just say "mixed partials are equal".

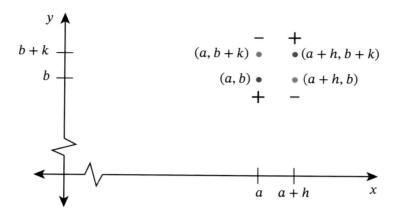

Figure 10.1. $Q(h, k)$ involves taking differences in both the x and y directions.

Theorem 10.1. *If f_{xy} and f_{yx} are continuous in a region around (a, b) then $f_{xy}(a, b) = f_{yx}(a, b)$.*

To see why this is true, we cook up a two-dimensional analogue $Q(h, k)$ of a difference quotient. We'll see that the limit of $Q(h, k)$, as h and k go to zero, is equal to both $f_{xy}(a, b)$ and $f_{yx}(a, b)$, so $f_{xy}(a, b)$ and $f_{yx}(a, b)$ must be equal. See Figure 10.1. We add the values of the function at the blue points marked $+$, subtract the values at the red points marked $-$, and divide by hk to get

$$(10.20) \qquad Q(h, k) = \frac{f(a + h, b + k) - f(a, b + k) - f(a + h, b) + f(a, b)}{hk}.$$

If $f(x, y)$ doesn't depend on y (that is, if $f(x, y)$ is really just a function of x), then $Q(h, k)$ is zero, since then $f(a + h, b + k) = f(a + h, b)$ and $f(a, b + k) = f(a, b)$. $Q(h, k)$ is also zero for functions that don't depend on x, because in that case $f(a + h, b + k) = f(a, b + k)$ and $f(a + h, b) = f(a, b)$. This also means that $Q(h, k) = 0$ when $f(x, y)$ is the sum of a function of x and a function of y.

Before talking about functions in general, let's see how this works for the quadratic function (10.17). The only term that depends on both x and y is Dxy, so to get $Q(h, k)$ for the function (10.17), we can ignore all the other terms and compute

$$
\begin{aligned}
Q(h, k) &= \frac{D(a + h)(b + k) - Da(b + k) - D(a + h)b + Dab}{hk}, \\
&= D\frac{ab + ak + hb + hk - ab - ak - ab - hb + ab}{hk}, \\
&= D\frac{hk}{hk} = D,
\end{aligned}
$$

(10.21)

which is both f_{xy} and f_{yx}.

Returning to a general function $f(x, y)$, we take the limit of $Q(h, k)$ as h and k go to zero. If we first take the limit as $h \to 0$, we get

$$\lim_{h \to 0} Q(h, k) = \frac{1}{k} \lim_{h \to 0} \left(\frac{f(a + h, b + k) - f(a, b + k)}{h} - \frac{f(a + h, b) - f(a, b)}{h} \right)$$

(10.22)

$$= \frac{1}{k} \left(f_x(a, b + k) - f_x(a, b) \right).$$

The limit of *that* as $k \to 0$ is, by definition, $(f_x)_y(a, b)$.

On the other hand, if we first take the limit of $Q(h, k)$ as $k \to 0$, we get

$$\lim_{k \to 0} Q(h, k) = \frac{1}{h} \lim_{k \to 0} \left(\frac{f(a + h, b + k) - f(a + h, b)}{k} - \frac{f(a, b + k) - f(a, b)}{k} \right)$$

(10.23)

$$= \frac{1}{h} \left(f_y(a + h, b) - f_y(a, b) \right).$$

Taking a limit of that as $h \to 0$ gives $(f_y)_x(a, b)$. Since both $f_{xy}(a, b)$ and $f_{yx}(a, b)$ are the limit of $Q(h, k)$ as h and k go to zero, we must have $f_{xy}(a, b) = f_{yx}(a, b)$. Mixed partials are equal!

10.2. Linear Approximations

The Two-Dimensional Microscope Equation. Returning to our running example $f(x, y) = e^{xy^2 - 9}$, suppose we want to use information about the function and its derivatives at $(1, 3)$ to approximate $f(0.99, 3.03)$. We define $g(x) = f(x, 3)$ and $h(y) = f(1, y)$. We have already computed f_x and f_y.

(10.24)
$$\begin{aligned} f_x(x, y) &= y^2 e^{xy^2 - 9}, \\ f_y(x, y) &= 2xy e^{xy^2 - 9}. \end{aligned}$$

Plugging in $(a, b) = (1, 3)$ gives

(10.25)
$$\begin{aligned} f(1, 3) &= 1, \\ f_x(1, 3) &= 9, \\ f_y(1, 3) &= 6. \end{aligned}$$

The microscope equation for $g(x)$ tells us what happens when we decrease x from 1 to 0.99:

(10.26) $f(0.99, 3) = g(0.99) \approx g(1) - 0.01 g'(1) = f(1, 3) - 0.01 f_x(1, 3) = 0.91.$

That is, $f(x, y)$ increases by $f_x(a, b)\Delta x$ (which is negative) when x changes from a to $a + \Delta x$. Likewise, we know what happens when we increase y from 3 to 3.03 without changing x.

(10.27) $f(1, 3.03) = h(3.03) \approx h(3) + 0.03 h'(3) = f(1, 3) + 0.03 f_y(1, 3) = 1.18.$

In this case, $f(x, y)$ increases by $f_y(a, b)\Delta y$.

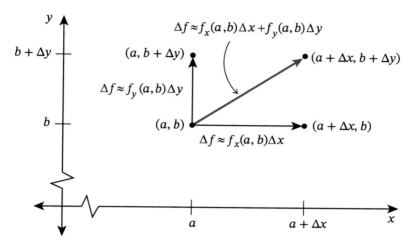

Figure 10.2. The change in $f(x, y)$ as we go along the diagonal arrow is approximately the sum of the changes along the horizontal and vertical arrows.

So what happens when we change both x and y, as in Figure 10.2? We just add up the pieces:

The two-dimensional microscope equation

(10.28)
$$\Delta f \approx f_x \Delta x + f_y \Delta y,$$
$$f(x, y) \approx f(a, b) + f_x(a, b)(x - a) + f_y(a, b)(y - b).$$

In our particular example, with $x \approx 1$ and $y \approx 3$, we have

(10.29)
$$f(x, y) \approx 1 + 9(x - 1) + 6(y - 3).$$

This is the best linear approximation to our function of two variables. Instead of giving the equation of the *line* tangent to the graph $y = f(x)$, it gives the equation of the *plane* tangent to the three-dimensional graph $z = f(x, y)$ at $(1, 3)$, as shown in Figure 10.3.

Of course, equation (10.28) is just an approximation. In our example, $f(0.99, 1.03) \approx 1 + 9(-0.01) + 6(0.03) = 1.09$. In fact, $f(0.99, 1.03) = 1.09318$ to 5 decimal places. There are three sources of error, and these are associated with the three second partial derivatives of $f(x, y)$, namely f_{xx}, f_{yy} and f_{xy}. (We don't consider f_{yx} separately because it equals f_{xy}.)

(1) The microscope equation for $g(x)$ isn't exact. $g''(1) = f_{xx}(1, 3)$ gives a measure of how fast $g(x)$ is curving, and how far off our linear approximation for $g(0.99)$ is.

(2) The microscope equation for $h(y)$ isn't exact, and $f_{yy}(1, 3)$ gives a measure of how far off our linear approximation for $h(1.03)$ is.

(3) $f(a + h, b + k) - f(a, b)$ isn't exactly the sum of $f(a + h, b) - f(a, b)$ and $f(a, b + k) - f(a, b)$. The difference between them is

(10.30)
$$f(a+h, b+k) - f(a+h, b) - f(a, b+k) + f(a, b) = hkQ(h, k)$$
$$\approx hk f_{xy}(a, b).$$

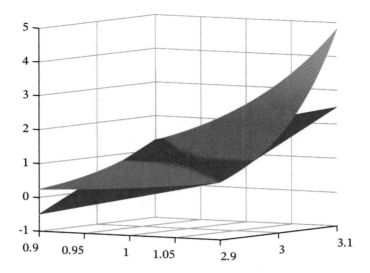

Figure 10.3. The graph of $f(x, y) = e^{xy^2-9}$ in red and the linear approximation (tangent plane) at $(1, 3)$, $z = 1 + 9(x - 1) + 6(y - 3)$ in blue.

If the linear approximation isn't accurate enough for us, we can take these errors into account and obtain a quadratic approximate that is more accurate. If (x, y) is close to (a, b), then the best quadratic approximation is

(10.31)
$$f(x, y) \approx f(a, b) + f_x(a, b)\Delta x + f_y(a, b)\Delta y$$
$$+ \frac{1}{2}f_{xx}(a, b)\Delta x^2 + \frac{1}{2}f_{yy}(a, b)\Delta y^2 + f_{xy}(a, b)\Delta x \Delta y,$$

where $\Delta x = x - a$ and $\Delta y = y - b$. Incidentally, the best quadratic approximation to a function of one variable is

(10.32)
$$f(x) \approx f(a) + f'(a)\Delta x + \frac{f''(a)}{2}\Delta x^2.$$

We will cover quadratic, cubic, and higher order approximations to functions of one variable in Chapter 11.

The Two-Dimensional Chain Rule and the Product Rule Revisited. Suppose that x and y are themselves functions of another variable, such as t. This makes $f(x, y)$ a function of t. What is df/dt? If we know the temperature and humidity, and we know how fast the temperature and humidity are changing, then how fast is the heat index changing?

To figure this out, we can divide the first line of (10.28) by Δt,

(10.33)
$$\frac{\Delta f(x, y)}{\Delta t} \approx f_x(x, y)\frac{\Delta x}{\Delta t} + f_y(x, y)\frac{\Delta y}{\Delta t},$$

and take a limit as $t \to 0$ to get

The two-dimensional chain rule

(10.34)
$$\frac{df(x,y)}{dt} = f_x(x,y)\frac{dx}{dt} + f_y(x,y)\frac{dy}{dt}.$$

The two-dimensional chain rule also tells us things about ordinary derivatives. If we apply the chain rule to the function $f(x,y) = xy$, then we have

(10.35)
$$\frac{d(xy)}{dt} = y\frac{dx}{dt} + x\frac{dy}{dt},$$

as $f_x(x,y) = y$ and $f_y(x,y) = x$. This is our old friend the product rule! The notation is a little different from what we are used to, because usually our functions are $f(t)$ and $g(t)$ (or $f(x)$ and $g(x)$) instead of x and y, but the result is exactly the same. The derivative of a product is the second factor times the derivative of the first, plus the first factor times the derivative of the second.

10.3. Double Integrals and Iterated Integrals

Amber Waves of Grain. Kansas produces a lot of wheat, more than any other state in the nation, earning Kansas the nickname of "the wheat state". Based on its size, can we figure out how much wheat it produces?

The area of Kansas is about 82,300 square miles. The average wheat-growing farm in the United States produces about 50 bushels of wheat[3] per acre per year, or 32,000 bushels per square mile per year. So you might expect Kansas to produce around $32,000 \times 82,300 = 2.63$ billion bushels of wheat per year.

In fact, average annual wheat production in the 2010s was only about 345 million bushels/year. The reason is that not every acre in Kansas produces 50 bushels of wheat per year. The western half of the state is much drier than the eastern half, making it difficult to grow wheat there. (Agriculture in western Kansas is mostly ranching, not farming.) Even in the eastern half, not all farms grow wheat. Some grow corn or soybeans. Some parts of Kansas are forest or park land or urban and don't grow any crops at all. Where wheat is grown, not all farms produce 50 bushels per acre per year. (The average in Kansas is about 40.) To understand wheat production in Kansas, we need to take all that regional variation into account.

Kansas is approximately a rectangle, 400 miles east-west and 210 miles north-south.[4] We can represent it as the rectangle in the x-y plane with vertices at $(0,0)$, $(400,0)$, $(0,210)$ and $(400,210)$, where x and y are measured in miles. Let $f(x,y)$ be the average wheat productivity, in bushels per square mile per year, *as a function of position*.

Given the function $f(x,y)$, we could figure out the average annual wheat production as follows:

---

[3]50 bushels weighs 3,000 pounds and provides about 4.5 million calories.

[4]That makes 84,000 square miles, but about 1700 square miles in the northeast corner of the rectangle are on the other side of the Missouri River and belong to Missouri.

(1) Break Kansas into many little pieces, as in Figure 10.4, which we'll call boxes, each so small that $f(x, y)$ is nearly constant in that box.

(2) Pick a representative point (x_i^*, y_i^*) in each box.

(3) Estimate the wheat production of the ith box by multiplying $f(x_i^*, y_i^*)$ by the area of the ith box.

(4) Add up the contributions of all the boxes.

(5) For greater accuracy, use smaller boxes. For an exact answer, take a limit as the number of boxes goes to ∞ and the size of each box goes to 0.

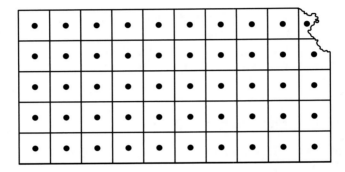

Figure 10.4. Breaking Kansas into smaller pieces

In other words, figuring out a bulk quantity like wheat production in a two-dimensional region follows *exactly* the same slice-and-dice strategy as in one dimension. The only differences are that we multiply our function by area, not by width, and that breaking a two-dimensional region into boxes is more complicated than breaking an interval into pieces.

As in one dimension, the sum is called a **Riemann sum**, the limit of the sum is called an **integral**, and we write

$$(10.36) \qquad \iint_K f(x, y)\, dA = \lim \sum f(x_i^*, y_i^*)\, \Delta A_i.$$

Here K (for Kansas) is the region we are integrating over, ΔA_i is the area of the ith box, and we use the symbol $\iint$ instead of $\int$ to emphasize that we are working in two dimensions instead of one. If we were integrating a function of three variables over a three-dimensional region R we might write $\iiint_R f(x, y, z)\, dV$, where V stands for volume. We also sometimes call $\iint_K f(x, y)\, dA$ a **double integral** and $\iiint_R f(x, y, z)\, dV$ a **triple integral**. In general, we call a function of two (or more) variables integrable over a region K if the limit of the Riemann sum exists and doesn't depend on the choices we make. As in one dimension, all continuous functions are integrable, and many functions that aren't continuous are also integrable.

It's important to remember that double and triple integrals are *integrals*, meaning limits of sums, not *anti-derivatives*. The Fundamental Theorem of Calculus was just a tool for computing one-dimensional integrals, and it doesn't directly apply in two or more dimensions.

Evaluating a Double Integral Numerically. The actual function describing wheat production is complicated, so we'll switch our focus to a simpler (but more contrived) example. Let's estimate the integral

(10.37)
$$\iint_R xe^{-xy}\, dA,$$

where $R = [0, 2] \times [0, 1]$ is the rectangle with vertices at $(0, 0)$, $(0, 1)$, $(2, 0)$, and $(2, 1)$. We divide R into 20 0.4×0.25 boxes, as in Figure 10.5. We could number the boxes from 1 to 20, but it's more practical to label them with two indices i and j. The index i runs from 1 to 5 and says which column a box is in, while j runs from 1 to 4 and says which row it is in. As with our one-dimensional integrals, we pick points $x_0 = 0$, $x_1 = 0.4$, $x_2 = 0.8$, $x_3 = 1.2$, $x_4 = 1.6$, and $x_5 = 2$ that divide the interval $[0, 2]$ into equal sized pieces. We similarly pick points $y_0 = 0$, $y_1 = 0.25$, $y_2 = 0.5$, $y_3 = 0.75$, and $y_4 = 1$. The ijth box is then a rectangle with vertices at (x_{i-1}, y_{j-1}), (x_{i-1}, y_j), (x_i, y_{j-1}), and (x_i, y_y). Each box has width $\Delta x = 0.4$, height $\Delta y = 0.25$, and area $\Delta x \Delta y = 0.1$.

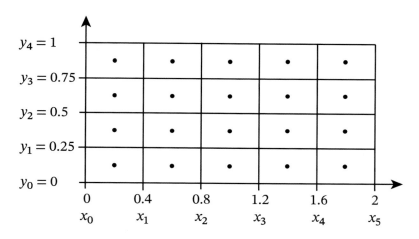

Figure 10.5. Breaking the rectangle R into 20 pieces

For each box, we need to pick a representative point (x^*_{ij}, y^*_{ij}). The simplest way to do this is to make x^*_{ij} depend only on i and y^*_{ij} depend only on j. That is, we pick just one value of x (call it x^*_i) that applies to every box in the ith column, and just one value of y (call it y^*_j) that applies to every box in the jth row. We will use the midpoint rule $x^*_i = (x_{i-1} + x_i)/2$, $y^*_j = (y_{j-1} + y_j)/2$, but we could also use left or right endpoints for x^*_i or use top or bottom endpoints for y^*_j. Our two-dimensional Riemann sum is then

(10.38)
$$\sum_{i=1}^{5}\sum_{j=1}^{4} x^*_i e^{-x^*_i y^*_j}\Delta x \Delta y.$$

To evaluate this sum with the midpoint rule, we use a simple MATLAB program. It is written in a way that can be easily modified for other functions or other rectangles or other numbers of horizontal and vertical divisions.

```
program Riemann2D                              % No input parameters.
sum = 0;                                        % Initialize our sum at 0
a=0; b=2;                                        % x ranges from 0 to 2
c=0; d=1;                                         % y ranges from 0 to 1
N=5; M=4;                                          % 5 columns and 4 rows
deltax = (b - a)/N; deltay=(d-c)/M             % Compute Δx and Δy
for k = 1:N
    xstar = a + (k-1/2)*deltax;                 % x*_k=(x_{k-1}+x_k)/2.
    for j=1:M
        ystar = c + (j-1/2)*deltay;             % y*_j=(y_{j-1}+y_j)/2.
        sum = sum + xstar*exp(-xstar*ystar);    % Add f(x*_{kj},y*_{kj}) to sum
end; end;                                         % Finish the loops
sum = deltax*deltay*sum                          % Multiply sum by ΔxΔy.
```

Note that the program uses k instead of i, since $i = \sqrt{-1}$. The result, to six decimal places, is 1.135906.

Organizing the Boxes: Iterated Integrals. The algorithm encoded in the program has three important features that are good to follow in (almost) all double integral calculations:

(1) The number x^*_{ij} depends only on i, so we can just write x^*_i,

(2) The number y^*_{ij} depends only on j, so we can just write y^*_j,

(3) The sum over all boxes is nested, where we first sum over j for fixed i, and then sum over i.

If we want to divide our region of integration N times horizontally and M times vertically, instead of 5 and 4, we have

$$(10.39) \qquad \text{Riemann sum} = \sum_{i=1}^{N}\sum_{j=1}^{M} f(x^*_i, y^*_j)\Delta y\Delta x.$$

For each value of i, the sum over j gives the contribution of a column of boxes, as shown in Figure 10.6. All the boxes in a column have the same representative x value, and the contribution of the column is proportional to the width of the column, with the proportionality constant being a function of x. Summing over i means adding up all the columns to get the contribution of the entire region of integration.

As with partial derivatives, our slogan is *One variable at a time!* We can rewrite (10.39) to make the iterated sum more explicit.

$$(10.40) \qquad \text{Riemann Sum} = \sum_{i=1}^{N}\left(\sum_{j=1}^{M} f(x^*_i, y^*_j)\Delta y\right)\Delta x$$

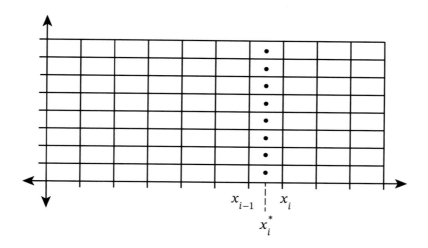

Figure 10.6. Adding up the boxes one column at a time

The sum in parentheses is a sum over y for a fixed value of x, namely x_i^*. As $M \to \infty$ it becomes a one-dimensional *integral* $\int_c^d f(x_i^*, y)\, dy$. The variable of integration is y, and y takes on all values from c to d. We want to do this for many different values of x_i^*, so we define the function

$$(10.41) \qquad g(x) = \int_c^d f(x, y)\, dy,$$

where in the integral we *treat x as a constant*.

For integrating $\iint_R xe^{-xy}\, dA$, our one-dimensional integral becomes

$$(10.42) \qquad g(x) = \int_0^1 xe^{-xy}\, dy.$$

Since the derivative of $-e^{-xy}$ with respect to y (treating x as a constant) is xe^{-xy}, the second Fundamental Theorem of Calculus tells us that

$$(10.43) \qquad g(x) = \int_0^1 xe^{-xy}\, dy = -e^{-xy}\Big|_{y=0}^{1} = 1 - e^{-x}.$$

Holding N fixed (for now), the limit of our two-dimensional Riemann sum as $M \to \infty$ is then

$$(10.44) \qquad \lim_{M \to \infty} \text{Riemann Sum} = \sum_{i=1}^{N}(1 - e^{-x_i^*})\Delta x,$$

where we have simply replaced $\sum_{j=1}^{M} x_i^* e^{-x_i^* y_j^*}\Delta y$ with $g(x_i^*) = \int_0^1 x_i^* e^{-x_i^* y}\, dy = 1 - e^{-x_i^*}$. As $N \to \infty$, the remaining sum also becomes a one-dimensional integral, so

$$(10.45) \qquad \iint_R xe^{-xy}\, dA = \int_0^2 (1 - e^{-x})\, dx = x + e^{-x}\Big|_0^2 = 1 + e^{-2} \approx 1.135335.$$

This answer is exact. The approximate answer of 1.135906 that we got from using 20 boxes and midpoints was accurate to 3 decimal places.

The upshot of this procedure is that we can replace a **double integral** over a two-dimensional region with an **iterated integral**, where we first integrate $f(x, y) \, dy$ for each value of x to get a function $g(x)$, and then integrate $g(x) \, dx$ to get a number. To evaluate those one-dimensional integrals, we can use everything that we previously learned, and in particular we can use the Fundamental Theorem of Calculus. Everything that we did with our example can be summarized as:

$$\iint_R xe^{-xy} \, dA = \int_{x=0}^{2} \left(\int_{y=0}^{1} xe^{-xy} \, dy \right) dx$$

$$= \int_{x=0}^{2} \left(-e^{-xy} \Big|_{y=0}^{1} \right) dx$$

(10.46)

$$= \int_{x=0}^{2} (1 - e^{-x}) \, dx$$

$$= x + e^{-x} \Big|_{x=0}^{2}$$

$$= 1 + e^{-2}.$$

You may (should!) wonder why we summed the boxes in each column and then added up the columns. Why didn't we sum the boxes in each row and then add up the rows? In fact, we could have. In general, either procedure is legal. This is expressed formally as:

Theorem 10.2 (Fubini's Theorem). *If $R = [a, b] \times [c, d]$ is a rectangle, and $f(x, y)$ is an integrable function on R, then the double integral $\iint_R f(x, y) \, dA$ can be expressed as an iterated integral in two ways, and both give the same answer:*

(10.47) $$\iint_R f(x, y) \, dA = \int_{x=a}^{b} \left(\int_{y=c}^{d} f(x, y) \, dy \right) dx = \int_{y=c}^{d} \left(\int_{x=a}^{b} f(x, y) \, dx \right) dy,$$

where in the integrals inside the parentheses the other variable is treated as a constant.

Sometimes it's easier to integrate first over y and then over x. Other times it's easier to integrate first over x and then over y. Still other times both iterated integrals are equally easy or equally hard. In our example, we could have written

(10.48) $$\iint_R f(x, y) \, dA = \int_{y=0}^{1} \left(\int_{x=0}^{2} xe^{-xy} \, dx \right) dy.$$

This is 100% correct, but is almost useless. The inner integral requires integration by parts and works out to be $\frac{1}{y^2} - \left(\frac{1}{y^2} - \frac{2}{y} \right) e^{-2y}$. Integrating *that* from 0 to 1 requires even more advanced methods. Integrating first over y and then over x is much, much easier.

To recap: Fubini's theorem says that we always *can* write a double integral as an iterated integral, integrating first over one variable and then over the other one. However, this doesn't mean that we always *should* do that. Most of the time we should, but not always. Sometimes it makes more sense to organize the boxes differently, or to use boxes that aren't even rectangles.

For instance, suppose that our region of integration is a dartboard, as in Figure 10.7, and that the function being integrated is the value of a throw. In that case, we should pick our boxes to be exactly the regions defined by the rules of the game! A similar division applies to any region of integration that has rotational symmetry. Instead of having columns and rows, we have pie wedges and rings. Instead of getting iterated integrals where you integrate first over y and then over x (or vice-versa), this scheme gives us iterated integrals where we first integrate over the radius r and then over the angle θ, or vice-versa. We will explore integration over such round regions in Exercises 10.24–10.26.

Figure 10.7. A region of integration that shouldn't be divided into rectangles. Photo Xygo_bg]/ iStock/Getty Images Plus, via Getty Images.

The details of how to do a double or triple integral depend on the problem, but the general philosophy is always the same. Organize your boxes so that you can do your Riemann sums, and ultimately your integrals, *one variable at a time*.

10.4. Chapter Summary

The Main Ideas. If $f(x, y)$ is a function of two variables, then

- Almost all problems involving $f(x, y)$ are best handled *one variable at a time*.
- The partial derivative $f_x = \partial f / \partial x$ is the derivative of f with respect to x, treating y as a constant. It is also the derivative of $g(x) = f(x, b)$, where y literally is a constant. Similarly, $f_y = \partial f / \partial y$ is the derivative of $h(y) = f(a, y)$.
- Computing partial derivatives is easy. Just use all the rules you already know, treating the other variable as a constant.

- f_x and f_y give us our best linear approximation to $f(x, y)$:

(10.49) $f(x, y) \approx f(a, b) + f_x(a, b)(x - a) + f_y(a, b)(y - b)$.

- If $z = f(x, y)$, and both x and y are functions of some other variable t, then

(10.50) $\dfrac{dz}{dt} = \dfrac{\partial f}{\partial x}\dfrac{dx}{dt} + \dfrac{\partial f}{\partial y}\dfrac{dy}{dt}.$

  This is the two-dimensional chain rule.

- f_x and f_y are functions in their own right, whose derivatives f_{xx}, f_{xy}, f_{yx} and f_{yy} are called the second partial derivatives of f.

- Mixed partials are equal. If f_{xy} and f_{yx} are continuous, then $f_{xy} = f_{yx}$.

- Bulk quantities in two dimensions come from double integrals. $\iint_R f(x, y)\, dA$ is what we get when we break the region R into boxes, add up $f(x, y)$ times the area of each box, and take a limit as the boxes get smaller and smaller.

- By definition, double integrals can be approximated by Riemann sums. The methods are very similar to those for one-dimensional integrals.

- By organizing the boxes intelligently, we can express our double integral as an iterated integral $\int_a^b \left(\int_c^d f(x, y)\, dy \right) dx$ or $\int_c^d \left(\int_a^b f(x, y)\, dx \right) dy$.

- The Fundamental Theorem of Calculus does not apply to double integrals. However, it can be applied to each stage of an iterated integral.

Expectations. You should be able to:

- Explain what partial derivatives, double integrals, and iterated integrals are.
- Explain the difference between double integrals and iterated integrals.
- Compute partial derivatives both numerically and from formulas.
- Use partial derivatives to approximate functions near known points.
- Express bulk quantities as double integrals.
- Approximate double integrals with Riemann sums.
- Program a computer to do the grungy calculations that this involves.
- Express a double integral as an iterated integral in two different ways.
- Use the Fundamental Theorem of Calculus to evaluate iterated integrals.

10.5. Exercises

10.1.   Find the partial derivatives f_x and f_y of each function.
   (a) $f(x, y) = xy^3$
   (b) $f(x, y) = \sqrt{2x + 5y}$
   (c) $f(x, y) = \cos(xy)$
   (d) $f(x, y) = e^{xy}$
   (e) $f(x, y) = 5\dfrac{y^2}{x^3} - x^3 \sin(y) + \pi$
   (f) $f(x, y) = y \tan(x)$
   (g) $f(x, y) = \dfrac{xy + y^2}{x}$
   (h) $f(x, y) = \ln(x^2 + xe^y)$

10.2. In parts (a)–(d), find the second partial derivatives f_{xx}, f_{yy}, f_{xy}, and f_{yx} of each function.

(a) $f(x, y) = y \tan(x)$

(b) $f(x, y) = 5\dfrac{y^2}{x^3} - x^3 \sin(y) + \pi$

(c) $f(x, y) = e^{xy}$

(d) $f(x, y) = xy^3$

(e) What did you notice about f_{xy} and f_{yx}?

10.3. Consider the function $f(x, y) = \ln\left(\dfrac{1 + xy^2}{2}\right)$.

(a) Compute $f(1, 1)$, $f_x(1, 1)$ and $f_y(1, 1)$.

(b) Use the two-dimensional microscope equation to estimate $f(1.03, 0.99)$.

10.4. Consider the function $f(x, y) = \dfrac{\sin(x)}{1 + y^2}$.

(a) Compute $f(\pi/4, 1)$, $f_x(\pi/4, 1)$ and $f_y(\pi/4, 1)$.

(b) Use the two-dimensional microscope equation to estimate $f(\pi/3, 1.1)$.

10.5. **Ideal Gas Law.** The volume V of a given quantity of gas is a function of the temperature T (in **kelvins**) and the pressure p. (Yes, chemists and physicists always use capital letters for temperature and volume but a lower-case p for pressure.) In a so-called ideal gas the relationship between volume and pressure is given by the ideal gas law:

$$pV = nRT, \quad \text{or equivalently} \quad V = nR\frac{T}{p},$$

where where n is the amount of gas (measured in **moles**) and R is a universal constant. In parts (a)–(c) of this exercise, we are thinking of V as a function of T and p, with n and R constant.

(a) Find formulas for the partial derivatives $V_T(T, p)$ and $V_p(T, p)$.

(b) The constant R equals 8.3×10^3 newton-meters per kelvin. (A **newton** is a unit of force in the MKS unit system.) Check that the units in the ideal gas law are consistent if V is measured in cubic meters, T in kelvins, and p in newtons per square meter, also called **pascals**.

(c) Suppose that one mole of gas at 280 kelvins is under a pressure of 30 newtons per square meter. What is its volume?

(d) The ideal gas law can also be used to express pressure as a function of temperature and volume. If the temperature of the gas in part (c) increases by 8 kelvins and the volume increases by 0.5 cubic meter, will the pressure increase or decrease? By about how much?

10.6. Write the formula for a linear function $f(x, y)$ with the following properties:

$$
\begin{aligned}
f_x(x, y) &= -1.5 && \text{for all } x \text{ and } y, \\
f_y(x, y) &= 1.8 && \text{for all } x \text{ and } y, \\
f(2, 1) &= 6.0.
\end{aligned}
$$

10.7.   Suppose that $z = f(x, y)$ and

$$f(100, 100) = 30, \qquad f_x(100, 100) = -4, \qquad f_y(100, 100) = 5.$$

   (a) Estimate $f(99, 100)$, $f(100, 101)$, $f(99, 101)$ and $f(98, 103)$.
   (b) When $x = 100$ and $y = 100$, how much change in z does a 2% increase in x cause? How much does a 2% increase in y cause? Which has a larger effect: a 2% increase in x or a 2% increase in y? Why?

10.8.   A forester who wants to know the height of a tree walks 200 feet from its base, sights to the top of the tree, and finds the resulting angle to be 50 degrees.
   (a) Draw a diagram illustrating the given information.
   (b) Use trigonometry and the forester's measurements to calculate the height of the tree.
   (c) Suppose the 200-foot measurement has an error of up to 2 feet and the angle measurement has an error of up to 4 degrees. Use a linear approximation to estimate the biggest possible error in the calculated height of the tree. (Don't forget to convert your angle measures to radians!)
   (d) Which would improve the accuracy of the forester's measurements more: a line that gives the exact distance to the tree, or a sight that gives the exact angle? Explain.

10.9.   **Blood Flow.** Blood flow through an artery is modeled using Poiseuille's laws. The resistance R of blood flowing through an artery depends on the artery's length L and radius r, according to the formula

$$R(L, r) = C\frac{L}{r^4},$$

   where C is a parameter indicating the viscosity of the blood.
   (a) Calculate the partial derivative $R_L(L, r)$.
   (b) Calculate the partial derivative $R_r(L, r)$.
   (c) Using insight from your answers to parts (a) and (b), answer the following questions.
      (i) Suppose that a patient undergoes surgery that lengthens a particular artery. However, there is no change to the radius of the artery. Would one expect an increase or decrease in the resistance of the blood flowing through that artery?
      (ii) Suppose a patient has a stent placed in a particular length of artery that increases the interior radius r of the artery without changing the length. Would you expect an increase or decrease in the resistance of the blood flowing through that particular artery?

10.10. **Body Mass Index.** The body mass index (BMI) is an estimate of a person's body fat based on height and weight. The formula for the BMI is

$$\text{BMI} = B(w, h) = \frac{w}{h^2},$$

   where w is the weight of the person in kilograms and h is the height of the person in meters.
   (a) Find the BMI of a preteen girl who weighs 60 kg and is 1.5 m tall.

(b) Find the partial derivatives B_w and B_h at $w = 60$ and $h = 1.5$ and provide concise interpretations for them.

(c) Use the two-dimensional microscope equation to approximate the change in the girl's BMI if she loses 2.5 kg (approximately 5.5 lbs) and grows 4 cm.

(d) At the girl's height and weight, what provides a greater change in the BMI, losing 1 kg or growing 1 centimeter? Explain. Your answer should be supported using information gained from partial derivatives.

Mutualism and Competition. In Section 6.5 we considered several models of how two species can interact, including predator-prey systems, mutualism and competition. In Exercises 10.11–10.13, we revisit the concepts of mutualism and competition from the perspective of multivariable calculus.

The interaction between species is often modeled using differential equations. The models are typically of the form

$$\frac{dN_1}{dt} = f_1(N_1, N_2)$$
$$\frac{dN_2}{dt} = f_2(N_1, N_2)$$

where N_1 and N_2 represent the population density of the two species of interest, and the functions f_1 and f_2 govern the interactions between the two species. The system is said to be **mutualistic** if

$$\frac{\partial f_1}{\partial N_2} > 0 \quad \text{and} \quad \frac{\partial f_2}{\partial N_1} > 0,$$

and **competitive** if

$$\frac{\partial f_1}{\partial N_2} < 0 \quad \text{and} \quad \frac{\partial f_2}{\partial N_1} < 0.$$

10.11. (a) Explain how the conditions $\frac{\partial f_1}{\partial N_2} > 0$ and $\frac{\partial f_2}{\partial N_1} > 0$ encode the idea that each species helps the other.

(b) A very important model of interaction between species was developed independently by Alfred Lotka and Vito Volterra. The Lotka-Volterra model, which we used in Chapter 6 to explore the interactions between foxes and rabbits, can be written as

$$\frac{dN_1}{dt} = \alpha N_1 - \beta N_1 N_2$$
$$\frac{dN_2}{dt} = \gamma N_1 N_2 - \delta N_2,$$

where α, β, γ, and δ are all positive parameters. Show that the Lotka-Volterra model *does not* satisfy the conditions for mutualism.

10.12. Consider two species that grow logistically when they are not near each other. If N_1 represents the population density of species 1, and N_2 represents the population density of species 2, and if they don't interact, then we have the uncoupled initial value problems

$$\frac{dN_1}{dt} = r_1 N_1 \left(1 - \frac{N_1}{K_1}\right), \ N_1(0) = N_1^0$$

$$\frac{dN_2}{dt} = r_2 N_2 \left(1 - \frac{N_2}{K_2}\right), \ N_2(0) = N_2^0$$

(a) Explain why it would be reasonable to model interspecific (between species) competition by *subtracting* a term of the form cN_1N_2 from each differential equation, where c is a parameter that represents the competitiveness of the species. Would you expect the parameter c to be the same for each species?

(b) Modifying the original uncoupled differential equations by introducing this extra term, we arrive at a *coupled* system of differential equations

$$\frac{dN_1}{dt} = r_1 N_1 \left(1 - \frac{N_1}{K_1}\right) - c_1 N_1 N_2$$

$$\frac{dN_2}{dt} = r_2 N_2 \left(1 - \frac{N_2}{K_2}\right) - c_2 N_1 N_2.$$

Does this model satisfy the conditions for mutualism? Does it satisfy the conditions for competition?

10.13. Not surprisingly, the model developed in Exercise 10.12 is called the **Lotka-Volterra competition model.** Shown below is a MATLAB program called Competition for solving the Lotka-Volterra competition model numerically. Most of the parameters are either standard for Euler's method or are directly the parameters of the model. The parameters N10 and N20 are the initial conditions $N_1(0)$ and $N_2(0)$.

```
function [N1,N2,t] = Competition(tmax,T,r1,c1,K1,r2,c2,K2,N10,N20)
dt = tmax/T;
N1 = zeros(1,T+1);
N2 = zeros(1,T+1);
N1(1) = N10;
N2(1) = N20;
for n = 1:T
N1(n+1) = N1(n) +r1*N1(n)*(1-N1(n)/K1)*dt-c1*N1(n)*N2(n)*dt;
N2(n+1) = N2(n) +r2*N1(n)*(1-N1(n)/K2)*dt-c2*N1(n)*N2(n)*dt;
end
t = linspace(0,tmax,T+1);
plot(t,N1,'blue',t,N2,'red')
ylabel('Individuals'), xlabel('Time'), title('Competition Model')
legend('Population $N_1$','Population $N_2$','Interpreter','latex')
```

    (a) Run this program with parameter values $r_1 = 0.2$, $K_1 = 100$, $c_1 = 0.001$, $N_1^0 = 20$, and $r_2 = 0.2$, $K_2 = 100$, $c_2 = 0.004$, and $N_2^0 = 20$. What does the model predict? Explore what happens if you vary the parameter values.

    (b) Write down a Lotka-Volterra-style system of differential equations that models mutualism instead of competition. Verify that your model satisfies the conditions for mutualism.

    (c) Modify the MATLAB program Competition.m to explore the predicted effects of mutualism. Use the same parameter values given in part (a). Investigate the behavior for different parameter values.

Functional Response. The Lotka-Volterra predator-prey model supposes that predation is just a matter of chance. Every time a predator runs into its prey, there's a certain chance of the prey being eaten. As a result, the rate at which prey are eaten is proportional to the product of the density of predators and the density of prey. In reality, it's usually more complicated than that. It takes time and effort for a predator to find prey. Once the prey is found, it usually takes time to "handle" (subdue, kill, eat, and digest) it. A lion that has just taken down a gazelle isn't going to hunt again for a while, no matter how many animals it runs into by accident.

    In 1959, C. S. Holling introduced the term **functional response** to describe the relationship between the predation rate Y and the density of prey N. An additional variable is the time T_t that the predator can devote to predation. (This used to be treated as a constant. However, as the climate warms, ectotherms like snakes and lizards are more alert and have more time to spend on hunting.) This makes the functional response a function $Y(T_t, N)$ of two variables. The time period T_t is itself the sum of two pieces: the time T_s spent searching for prey and the time T_h spent handling prey. Exercises 10.14–10.16 examine the three kinds of functional response defined by Holling.

10.14. In a **type-I** functional response (the simplest case), the handling time for prey is assumed to be insignificant ($T_h = 0$), so $T_s = T_t$ and the predation rate Y is jointly proportional to the time available during the day for predation T_t and the prey density N. In this case, the functional response has the form

$$Y(T_t, N) = aT_t N,$$

where the **discovery rate** a is a proportionality constant.

    (a) Find the partial derivatives Y_{T_t} and Y_N.

    (b) Suppose for a particular system, that if $T_t = 10$ and $N = 100$, then $Y = 10$.

       (i) Determine the parameter a.

      (ii) Use the two-dimensional microscope equation to approximate the change in Y if $T_t = 10.2$ and $N = 98$.

     (iii) Suppose for metabolic purposes the predator requires eating at least 10 prey each day and that and T_t changes from 10 to 10.5. Use the result from part (ii) to determine the smallest N can be that will still allow the predator to meet its metabolic needs.

     (iv) Write down the microscope equation for approximating $Y(T_t, N)$ near $Y(10, 100)$.

    (c) When $T_t = 10$ and $a = 0.01$, $Y(T_t, N)$ becomes $Y = 0.1N$. Use MATLAB to graph Y as a function of the single variable N.

10.15. In a **type-II** functional response, we assume that a predator must take a fixed
time h to handle each prey individual. This type of functional response applies
to most predator-prey situations. This makes the total handling time $T_h = hY$,
so $T_t = T_s + hY$, or equivalently $T_s = T_t - hY$. The functional response $Y(T_t, N)$
is no longer aT_tN, but instead is aT_sN.

(a) Show that

$$Y(T_t, N) = \frac{aT_tN}{1 + ahN}.$$

(*Hint:* Substitute $T_s = T_t - hY$ into the function written in terms of T_s and
solve for Y.)

(b) Find the partial derivatives Y_{T_t} and Y_N.

(c) Suppose that $h = 0.7$ and $a = 1$. Then $Y(20, 10) = 25$. Use the two-
dimensional microscope equation to approximate the change in Y if Y_N
increases to 10.5 and T_t decreases to 19.5.

(d) Find the second partial derivatives of $Y(T_t, N)$ at $(20,10)$ and write down
the quadratic approximation for $Y(T_t, N)$ near $Y(20, 10)$.

(e) If T_t is assumed to be constant (i.e., ignoring the effects of climate change),
then Y is a function of the single variable N. Graph $Y(N)$ when $T_t = 20$,
$h = 0.7$, and $a = 1$.

10.16. A **type-III** functional response involves a predator looking for different kinds
of prey in proportion to their frequency. (If there are lots of gazelles and not
many zebras, a lion will hunt for gazelles. If there are lots of zebras and not
many gazelles, it will hunt for zebras.) For each species of prey, this means
replacing the discovery rate a by aN, so

$$Y(T_t, N) = \frac{aT_tN^2}{1 + ahN^2}.$$

(a) Find the partial derivatives Y_{T_t} and Y_N.

(b) Suppose that $h = 0.25$ and $a = 1$. Then $Y(10, 4) = 32$.

(i) Use the two-dimensional microscope equation to approximate the
change in Y if Y_N decreases to 3.8 and T_t increases to 10.5.

(ii) Suppose the predator needs to eat at least 32 prey each day and that
T_t has changed from 10 to 10.5. Use the two-dimensional microscope equation to determine the smallest N that will still allow the
predator to meet its metabolic needs.

(iii) Find the quadratic approximation for $Y(T_t, N)$ near $Y(10, 4)$.

(c) If T_t is assumed to be constant, then Y is a function of the single variable
N. Graph $Y(N)$ when $T_t = 10$, $h = 0.25$, and $a = 1$.

Iterated Integrals. The remaining exercises all involve evaluating double integrals
by expressing them as iterated integrals. Always remember to do your calculations *one
variable at a time.*

10.17. Evaluate each iterated integral.

(a) $\int_{-2}^{1} \int_{0}^{3} 12x^3y^2 \, dy \, dx$

(b) $\int_{0}^{\ln 2} \int_{0}^{2} x^3 e^y \, dx \, dy$

10.18. Use the MATLAB program Riemann2D to estimate the iterated integral $\int_0^3 \int_1^2 y e^{x^2}\, dx\, dy$, using $\Delta x = \Delta y = 0.1$.

10.19. (a) Write down and compute the Riemann sum that approximates

$$\iint_R x^3 + 2xy^2\, dA$$

over the rectangle $R = [0, 3] \times [1, 3]$ by partitioning R into 6 squares $(R_1, R_2, \ldots, R_6)$, each with an area of 1 square unit, and selecting (x_i^*, y_i^*) to be the lower left corner of R_i.

(b) Convert the double integral from part (a) into an iterated integral.

(c) Evaluate the iterated integral from part (b).

10.20. Let R be the region inside the unit circle in the first quadrant. That is, R is defined by $x \geq 0$, $y \geq 0$, and $x^2 + y^2 \leq 1$.

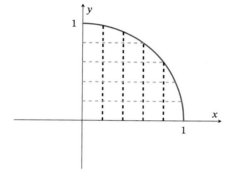

(a) Express $\iint_R xy\, dA$ as an iterated integral, where you integrate first over y and then over x. The subtlety is that our region of integration is not a rectangle. We are still arranging our boxes in columns, summing all the boxes in each column, and then adding up the columns, but the columns don't all have the same height. As a result, the limits of integration for your integral over y will depend on x.

(b) Evaluate the iterated integral from part (a).

Average Values. In Chapter 9 we defined the average value of a function $f(x)$ on an interval $[a, b]$ to be $\frac{1}{b-a} \int_a^b f(x)\, dx$. The same idea works in two dimensions, only replacing the length of the region of integration with its area. If $f(x, y)$ is any function, then the average value of f over a region R, denoted $\bar{f}_R$, is given by

(10.51) $$\bar{f}_R = \frac{1}{\text{Area of } R} \iint_R f(x, y)\, dA.$$

Exercises 10.21–10.23 involve computing average values.

10.21. The state of Wyoming has a rectangular shape, which we model as a rectangle with vertices as $(1,1)$, $(4,1)$, $(1,3)$ and $(4,3)$. Most of Wyoming is desert, but the northwest corner gets a fair amount of rain. We model the average annual rainfall at position (x, y), in inches per year, by the (not very accurate) function

$$W(x, y) = 5(5 - x)^{0.3} y^{0.7}.$$

Find the average rainfall across the state.

10.22. In dry regions, animals tend to live near watering holes. Suppose that a watering hole is located at location $(0,0)$ and that the density of animals (number per square kilometer) is approximately

$$n(x, y) = 2000 - 30(x^2 + y^2),$$

where x and y are measured in kilometers.

(a) What is the average density of animals in an 8 km × 8 km square region centered at the watering hole?

(b) What is the total number of animals in that square?

10.23. The chance of a person with a contagious disease infecting somebody else in a social setting is generally assumed to be a function $f(r)$ of the distance r between them.

(a) Explain why $f(r)$ should be a decreasing function of r.

(b) Suppose you are located at the origin $(0,0)$ of the x, y plane. Explain why $r = \sqrt{x^2 + y^2}$.

(c) For Covid, it was initially believed (without any good empirical data) that the virus could only be transmitted within a distance of $R = 6$ feet. (The parameter R is called the **radius of infectivity**.) Explain why it would be reasonable to use $f(r) = c\left(1 - \frac{r^2}{R^2}\right)$, where c is a constant, for the risk of being a distance r away from an infected person.

(d) For simplicity, take $c = 1$ and $R = 6$ feet. Find the average value of $f(r)$ in the rectangle B_1 defined by $-3 \leq x \leq 3$ and $-1 \leq y \leq 2$.

(e) Find the average value of f in the rectangle B_2 defined by $-2 \leq x \leq 3$ and $1 \leq y \leq 5$.

(f) Which is riskier to your health, talking to an infected person in rectangle B_1 or an infected person in rectangle B_2?

Integration in Polar Coordinates. Exercises 10.24–10.26 involve integrating over circular regions. At the end of this chapter, we saw that when integrating over some regions (like a dartboard), it's a bad idea to integrate first over x and then over y, or vice-versa. Instead, we need different kinds of boxes.

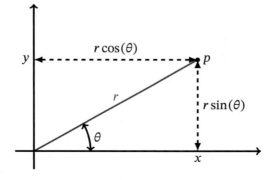

For a circular region centered at the origin, such as the unit disk, it's often useful to describe points by their distance from the origin and their angle above the x axis, instead of by their x and y coordinates. That is, we use **polar coordinates** r and θ, where

(10.52) $$x = r\cos(\theta), \qquad y = r\sin(\theta).$$

We can also express r and θ in terms of x and y:

(10.53) $$r = \sqrt{x^2 + y^2}, \qquad \theta = \tan^{-1}(x/y).$$

The second equation of (10.53) only determines θ up to a multiple of π. For instance, x/y is the same for $x = y = 1$ as for $x = y = -1$, but the first has $r = 1$ and $\theta = \pi/4$, while the second has $r = 1$ and $\theta = 5\pi/4$. The equations (10.52) make sense for any values of r and θ, but we usually restrict our attention to $r \geq 0$ and $0 \leq \theta \leq 2\pi$.

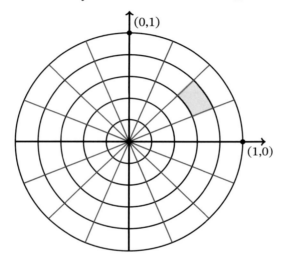

When working with rectangular regions, our boxes are usually smaller rectangles, where each box has a limited range of x-values, of size Δx, and a limited range of y-values, of size Δy. When working with polar coordinates, we divide things into so-called **polar rectangles**. These boxes have a limited range of r-values (running from r_{i-1} to r_i in the ith ring) and a limited range of θ-values (of size $\Delta\theta$). Instead of assembling into columns and rows, they assemble into rings and sectors. When Δr and $\Delta\theta$ are small, each box is roughly rectangular in shape, with the straight edges having length Δr and the (slightly) curved edges having length $r_{i-1}\Delta\theta$ and $r_i\Delta\theta$. This means that the area of the box is between $r_{i-1}\Delta r\Delta\theta$ and $r_i\Delta r\Delta\theta$, and so is approximately $r_i^*\Delta r\Delta\theta$, where r_i^* is our favorite number between r_{i-1} and r_i.

Note that the area of a box is *not* $\Delta r\Delta\theta$! If we are measuring length in meters, then $r_i^*\Delta r\Delta\theta$ has units of square meters, which makes sense for an area. On the other hand, $\Delta r\Delta\theta$ has units of meters, which doesn't make sense. Compared to the formula for integrating $dx\,dy$, there is an extra factor of r in the formula for integration using polar coordinates.

To recap: If we have a function $f(r, \theta)$ that we want to integrate over a polar region, the contribution of each box is approximately

(10.54) $$f(r_i^*, \theta_j^*)r_i^*\,\Delta r\,\Delta\theta.$$

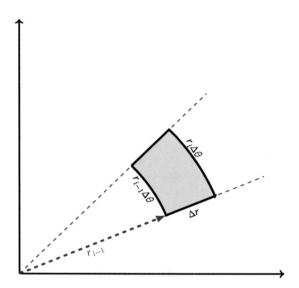

Adding up the boxes in each ring, and then adding up the rings, gives a nested sum

(10.55)
$$\sum_{i=1}^{N} \sum_{j=1}^{M} f(r_i^*, \theta_j^*) r_i^* \, \Delta\theta \, \Delta r.$$

Taking a limit as $M \to \infty$ turns this into

(10.56)
$$\sum_{i=1}^{N} r_i^* \left(\int_0^{2\pi} f(r_i^*, \theta) \, d\theta \right) \Delta r.$$

Taking a limit as $N \to \infty$, this becomes

(10.57)
$$\int_0^1 \int_0^{2\pi} f(r, \theta) r \, d\theta \, dr.$$

If we had nested our sum the other way, adding up all the boxes in each sector and then adding up the sectors, we would have gotten

(10.58)
$$\int_0^{2\theta} \int_0^1 f(r, \theta) r \, dr \, d\theta.$$

That is, if D is the unit disk, then

(10.59)
$$\iint_D f(r, \theta) dA = \int_0^1 \int_0^{2\pi} f(r, \theta) r \, d\theta \, dr$$
$$= \int_0^{2\theta} \int_0^1 f(r, \theta) r \, dr \, d\theta.$$

10.24.   In Exercise 10.23 we used the function $f(r) = 1 - \frac{r^2}{R^2}$, with $R = 6$, to model the risk of infection at distance r from an infected person. We integrated that function over a rectangle to compute an average risk. Now we will compute the risk of infection in two circular regions. Let D_3 be the disk of radius 3 around the origin and let D_6 be the disk of radius 6.

(a) Compute $\iint_{D_3} f(r)\, dA$. Don't forget the factor of r in $r\, dr\, d\theta$! Then find the average value of $f(r)$ in this disk.

(b) Compute $\iint_{D_6} f(r)\, dA$. Then find the average value of $f(r)$ in this disk.

(c) Based on this model, how much riskier is it to talk to somebody within a distance of 3 feet than it is to talk to somebody within a distance of 6 feet?

(d) In general, what fraction of your risk of infection comes from people getting within 3 feet of you?

10.25. Consider the function $f = e^{-(x^2+y^2)} = e^{-r^2}$. We are going to integrate $f\, dA$ over the entire plane $\mathbb{R}^2$ by integrating over a large disk and taking a limit as the radius of that disk goes to infinity.

(a) Compute $\iint_{D_a} e^{-r^2}\, dA$, where D_a is a disk of radius a centered at the origin.

(b) Taking a limit as $a \to \infty$, what is $\iint_{\mathbb{R}^2} e^{-r^2}\, dA$?

(c) Since e^{-r^2} can be written in Cartesian coordinates as $e^{-(x^2+y^2)} = e^{-x^2}e^{-y^2}$, Fubini's theorem tells us that

$$\iint_{\mathbb{R}^2} e^{-r^2}\, dA = \int_{-\infty}^{\infty} \int_{-\infty}^{\infty} e^{-x^2}e^{-y^2}\, dx\, dy.$$

Comparing this to your answer to part (b), what is $\int_{-\infty}^{\infty} e^{-x^2}\, dx$?

10.26. In polar coordinates, the region R of Exercise 10.20 is given by $0 \le r \le 1$, $0 \le \theta \le \pi/2$.

(a) Express the function $f(x, y) = xy$ in polar coordinates.

(b) Using polar coordinates, compute $\iint_R xy\, dA$. Compare your result to your answer to Exercise 10.20.

Chapter 11

Taylor Series

Wouldn't it be nice if all functions were polynomials? Polynomials are easy to evaluate, easy to differentiate, and easy to integrate. The bad news, of course, is that there are plenty of functions that aren't polynomials, including exponentials, logs, trig functions, inverse trig functions, and so on. The good news is that most of those other functions can be *approximated* by polynomials, making calculations much easier.

In this chapter we'll see how to do that. We'll start by looking at ways to represent numbers as infinite sums, or **series**. In Section 11.1 we'll see that adding up an infinite list of numbers requires taking a limit: saying that a number *equals* an infinite sum means that it is *approximated* by a finite sum. Some series have limits and are said to **converge**. Others don't and are said to **diverge**. In particular, we study convergence and divergence of geometric series.

In Section 11.2 we'll look at power series, where each term is a coefficient times a power of x. These are similar to polynomials, except with infinitely many terms, and they give us a practical way to represent many common functions. We'll use what we learned about geometric series to express a number of common functions as power series. In other words, we'll learn how to approximate those functions by polynomials.

In Section 11.3 we'll go beyond those special examples. We'll learn how to approximate arbitrary smooth functions by polynomials, called **Taylor polynomials**, and see how accurate those approximations are. By taking a limit we'll see how to express functions as power series called **Taylor series**. In Section 11.4 we'll turn to the most common and most important Taylor series, for e^x, $\sin(x)$ and $\cos(x)$, and make sense of Euler's strange formula $e^{ix} = \cos(x) + i\sin(x)$.

Finally, we consider questions of convergence. An infinite series that doesn't converge doesn't really make sense. In Section 11.5 we'll develop a lot of tools to determine whether a series converges or not, and in Section 11.6 we'll apply these tools to figure out for which values of x a Taylor series converges.

11.1. What Does $\pi = 3.14159265\cdots$ Mean?

Examples. You have probably seen expressions like

(11.1) $0.333333333333333\cdots$

and

(11.2) $3.14159265358979323846\cdots.$

These are examples of **infinite sums**. A terminating decimal like 0.333 means $3/10 + 3/100 + 3/1000$, so an infinite decimal like $0.333333333\cdots$ means

(11.3) $$\frac{3}{10} + \frac{3}{100} + \frac{3}{1000} + \frac{3}{10,000} + \cdots,$$

where "$+\cdots$" means that we keep adding more and more terms forever. Similarly, $3.14159\cdots$ means

(11.4) $$3 + \frac{1}{10} + \frac{4}{100} + \frac{1}{1000} + \frac{5}{10,000} + \frac{9}{100,000} + \cdots$$

Likewise, we can talk about infinite sums that aren't decimals, like

(11.5) $$1 + \frac{1}{2} + \frac{1}{4} + \frac{1}{8} + \frac{1}{16} + \cdots,$$

(11.6) $1 + 2 + 4 + 8 + 16 + 32 + \cdots,$

and

(11.7) $1 - 2 + 4 - 8 + 16 - 32 + \cdots.$

It takes some work to make sense of any of this, because addition is a *binary* operation. Addition takes two inputs, like 3 and 5, and gives an output of $3 + 5 = 8$. When we add three things we're really adding two things twice. $2 + 3 + 5$ means $(2 + 3) + 5$. (This happens to equal $2 + (3 + 5)$, so we usually don't worry about what order we do the addition.) Likewise, when we add $2 + 3 + 5 + 7 + 11 + 13$, we say something like

> 2 plus 3 makes 5, 5 plus 5 makes 10, 10 plus 7 makes 17, 17 plus 11 makes 28, and 28 plus 13 makes 41, so our answer is 41.

In general, whenever we are faced with a finite sum

(11.8) $a_1 + a_2 + \cdots + a_n$

we generate a sequence of **partial sums**

(11.9)
$$\begin{aligned}
s_1 &= a_1, \\
s_2 &= s_1 + a_2, \\
s_3 &= s_2 + a_3, \\
&\;\;\vdots \\
s_n &= s_{n-1} + a_n,
\end{aligned}$$

and say that the number at the end of the sequence is the sum of all the a_i's:

(11.10) $$s_n = \sum_{i=1}^{n} a_i.$$

But what if the a_i's never end? What do we do then?

Convergence. The answer comes from the 1st Pillar of Calculus: *Close is good enough.* If a sequence of numbers s_1, s_2, ... gets closer and closer to some number S, then we say that the sequence **converges** to S, and we write

(11.11)
$$S = \lim_{n \to \infty} s_n.$$

If those numbers s_n are partial sums as in (11.9), coming from trying to add up an infinite list of numbers, then we say that the infinite sum $a_1 + a_2 + \cdots$ converges to S and we write

(11.12)
$$S = \sum_{i=1}^{\infty} a_i.$$

That is,

(11.13)
$$\sum_{i=1}^{\infty} a_i \quad \text{means} \quad \lim_{n \to \infty} \sum_{i=1}^{n} a_i.$$

An infinite sum is also called an **infinite series**. If it doesn't converge to any number, we say that the series **diverges**.

In our examples, (11.1) converges to 1/3, since by taking enough digits we can get as close to 1/3 as we like. Similarly, (11.2) converges to π and (11.5) converges to 2. The partial sums in (11.5) never actually *reach* 2, but they home in on 2, and that's good enough for us. However, (11.6) just gets bigger and bigger, and doesn't home in on any number at all. The partial sums in the series (11.7) not only get bigger and bigger in size, but they alternate between positive and negative!

It's tempting to say that (11.6) "goes to infinity" and to write something like

(11.14)
$$\sum_{j=0}^{\infty} 2^j = +\infty.$$

That isn't wrong, but please be careful, because ∞ is not a number like 17 or -35! The word "infinity" comes from the Latin words *in*, meaning "not" and *finis*, meaning "finish" or "end", so *infinity* just means "continue without end". When something keeps growing without end, eventually getting bigger than any number you can name, we say it "goes to infinity". When we wrote

(11.15)
$$\int_a^b f(x)\,dx = \lim_{N \to \infty} \sum_{i=1}^{N} f(x_i^*)\Delta x$$

we imagined a process in which N got bigger and bigger and bigger without ever stopping. Likewise, when something gets more and more negative, eventually getting more negative than any number we can name, we say that it "goes to $-\infty$", as in

(11.16)
$$\lim_{x \to 0} \frac{-1}{x^2} = -\infty.$$

We will use the symbols ∞ and $-\infty$, but we must always remember that

> ## *Infinity is a process, not a number!*

In particular, when we write (11.14) we do *not* mean that the series (11.6) converges to infinity. The partial sums don't home in on a number called "infinity". They don't home in on any number at all! The series *diverges*, and "infinity" is just our way of describing *how* it diverges. The series (11.7) also diverges, but in a way that we can't describe as easily.

In summary,

- An **infinite sequence** is an infinite list of numbers $a_1, a_2, \ldots$. We say that the sequence **converges** to a number L if the numbers a_i get closer and closer to L, in which case we write $\lim_{i \to \infty} a_i = L$. If the sequence doesn't converge to a number then it **diverges**. If the sequence diverges by getting more and more positive or more and more negative we might say that it "goes to ∞" or "goes to $-\infty$", but it still diverges.

- An **infinite series** is a sum

$$(11.17) \qquad a_1 + a_2 + a_3 + \cdots$$

  involving infinitely many terms. The only way to make sense of an infinite sum is to think of the sequence of partial sums s_n from (11.9). If the sequence of partial sums converges to a number S we say that the series converges to S and write $\sum_{i=1}^{\infty} a_i = S$. Otherwise the series diverges.

- Representing something as an infinite sum is *exactly* the same as approximating it, better and better, by finite sums.

- Every infinite decimal, and lots of other useful expressions in mathematics, are actually convergent infinite series. The rest of this chapter is about exploring how they work and what they are good for.

Geometric Series. A **geometric series** is a series where the ratio between successive terms is fixed. We have already seen four examples of this, namely (11.1), (11.5), (11.6) and (11.7). We usually call the first term a and the common ratio r, giving us

$$(11.18) \qquad a + ar + ar^2 + ar^3 + \cdots$$

We also usually label the first term as a_0 rather than a_1, so that we can write

$$(11.19) \qquad a_i = ar^i.$$

As long as $r \neq 1$, there is a simple trick to evaluate the partial sums s_n:

$$
\begin{aligned}
s_n &= a + ar + ar^2 + \cdots + ar^n, \\
rs_n &= \qquad\; ar + ar^2 + \cdots + ar^n + ar^{n+1}, \\
(1 - r)s_n &= a \qquad\qquad\qquad\qquad\quad - ar^{n+1}, \\
s_n &= \frac{a - ar^{n+1}}{1 - r} = a\frac{1 - r^{n+1}}{1 - r}.
\end{aligned}
$$

(11.20)

For (11.1) we have $a = 0.3$ and $r = 0.1$, so $1 - r = 0.9$ and $s_n = \frac{1}{3}(1 - 10^{-(n+1)})$. Since this gets closer and closer to 1/3 as $n \to \infty$, the series converges to 1/3. Similarly, for

(11.5) we have $a = 1$ and $r = 1/2$, so $1 - r = 1/2$ and $s_n = 2(1 - 2^{-(n+1)}) = 2 - 2^{-n}$, which approaches 2. The series[1] (11.6) and (11.7) have $r = 2$ and $r = -2$, respectively, so $1 - r^{n+1}$ grows (in magnitude) without bound. Since s_n doesn't approach a definite number, these geometric series diverge.

In general, whenever $|r| < 1$, $1 - r^{n+1}$ approaches 1 as $n \to \infty$, and we have

$$(11.21) \qquad \sum_{i=0}^{\infty} ar^i = \lim_{n \to \infty} a\frac{1 - r^{n+1}}{1 - r} = \frac{a}{1 - r}.$$

When $|r| > 1$, s_n grows without bound in magnitude, and the infinite series $\sum_{i=1}^{\infty} ar^i$ diverges. If $r > 1$, then all terms have the same sign and the sum goes to $\pm\infty$, depending on the sign of a, while if $r < -1$ the partial sums s_n grow in magnitude and alternate in sign, as in (11.7).

When $r = 1$, we have the series

$$(11.22) \qquad a + a + a + a + \cdots$$

which diverges to $\pm\infty$, depending on the sign of a. Finally, if $r = -1$ we have

$$(11.23) \qquad a - a + a - a + a - a + \cdots.$$

The partial sums alternate between $s_n = a$ (when n is even) and $s_n = 0$ (when n is odd). This doesn't approach a *single* fixed number, so the series diverges. We summarize these results as a theorem.

Theorem 11.1 (Geometric series). *Suppose that $a \neq 0$. The geometric series $\sum_{i=0}^{\infty} ar^i$ converges if and only if $|r| < 1$, in which case it converges to $\frac{a}{1-r}$.*

11.2. Power Series

In a geometric series, the parameter r can itself be a variable, or a function of a variable. For instance, as long as $|x| < 1$,

$$(11.24) \qquad 1 + x + x^2 + x^3 + \cdots = \frac{1}{1 - x}$$

and

$$(11.25) \qquad 1 - x^2 + x^4 - x^6 + \cdots = \frac{1}{1 - (-x^2)} = \frac{1}{1 + x^2},$$

since these are just geometric series with $r = x$ or $r = -x^2$. Series like these, where each term is a multiple of a power of x, or of a power of $x - a$ for some fixed constant a, are called **power series**.

Read left-to-right, equations (11.24) and (11.25) say that certain power series add up to certain simple functions. Read right-to-left, they say that certain functions can be represented as power series. (In Sections 11.3 and 11.4, we'll see how to find power series for other functions.)

In particular, equation (11.24) doesn't just tell us that we can add up $1 + x + x^2 + \cdots$ to get $\frac{1}{1-x}$ (as long as $|x| < 1$). It tells us the polynomials $T_n(x)$ that we get by keeping

---

[1]The plural of *series* is *series*.

the first n powers of x in the series are *approximations* to the function $f(x) = \frac{1}{1-x}$. The approximation

(11.26)
$$\frac{1}{1-x} \approx 1 + x$$

is our old friend the microscope equation $f(x) \approx f(a) + f'(a)(x - a)$ with $a = 0$, since $f(0) = 1$ and $f'(0) = 1$. This is the best *linear* approximation to our function near $x = 0$. The best *quadratic* approximation is

(11.27)
$$\frac{1}{1-x} \approx 1 + x + x^2,$$

the best *cubic* approximation is

(11.28)
$$\frac{1}{1-x} \approx 1 + x + x^2 + x^3,$$

and so on. We get a sequence of polynomials

(11.29)
$$T_n(x) = 1 + x + x^2 + \cdots + x^n$$

that approximate $f(x)$ better and better on the interval $(-1, 1)$. These are called Taylor polynomials for the function $f(x) = \frac{1}{1-x}$. The graphs of the first few of these polynomials are shown in Figure 11.1.

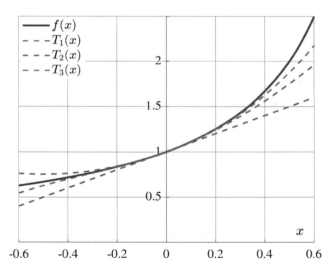

Figure 11.1. The graphs of $f(x) = 1/(1 - x)$ and approximating Taylor polynomials $T_1(x) = 1 + x$, $T_2(x) = 1 + x + x^2$ and $T_3(x) = 1 + x + x^2 + x^3$

Since $T_n(x)$ is a good approximation to $f(x)$, $\int T_n(x)\,dx$ is a good approximation to $\int f(x)\,dx$. Similarly, $T_n'(x)$ is a good approximation to $f'(x)$.[2] This means that $\int f(x)\,dx$ and $f'(x)$ are the limits as $n \to \infty$ of $\int T_n(x)\,dx$ and $T_n'(x)$. But those are

---

[2]This second fact takes some work to prove, but it's still true.

exactly the series that we get by integrating or differentiating every term in the series for $f(x)$. For instance, by differentiating every term in the series (11.24), we get

(11.30)
$$\frac{1}{(1-x)^2} = \frac{d}{dx}\frac{1}{1-x}$$
$$= \frac{d}{dx}\left(1 + x + x^2 + x^3 + x^4 + \cdots\right)$$
$$= 0 + 1 + 2x + 3x^2 + 4x^3 \cdots$$
$$= \sum_{n=0}^{\infty} (n+1)x^n.$$

By integrating each term we get

(11.31)
$$\ln(1-x) = \int_0^x \frac{-ds}{1-s}$$
$$= -\int_0^x \left(1 + s + s^2 + \cdots\right) ds$$
$$= -\left(x + \frac{x^2}{2} + \frac{x^3}{3} + \cdots\right)$$
$$= -\sum_{n=1}^{\infty} \frac{x^n}{n}.$$

Similarly, by integrating each term in the series (11.25), we get

(11.32) $$\tan^{-1}(x) = \int_0^x \frac{ds}{1+s^2} = x - \frac{x^3}{3} + \frac{x^5}{5} - \frac{x^7}{7} + \cdots = \sum_{n=0}^{\infty} \frac{(-1)^n x^{2n+1}}{2n+1}.$$

The formulas for $\ln(1-x)$ and $\tan^{-1}(x)$, which apply whenever $|x| < 1$, give us an effective way to estimate logs and arctangents. For instance, letting $x = \pm 0.1$ we get

(11.33)
$$\ln(0.9) \approx -\left(0.1 + \frac{0.01}{2} + \frac{0.001}{3} + \frac{0.0001}{4}\right) \approx -0.10535833,$$
$$\ln(1.1) \approx 0.1 - \frac{0.01}{2} + \frac{0.001}{3} - \frac{0.0001}{4} \approx .09530833,$$
$$\tan^{-1}(0.1) \approx 0.1 - \frac{0.001}{3} + \frac{0.00001}{5} \approx 0.09966867, \quad \text{and}$$
$$\tan^{-1}(-0.1) \approx -0.1 + \frac{0.001}{3} - \frac{0.00001}{5} \approx -0.09966867.$$

11.3. Taylor Polynomials and Taylor Series

So far we have expressed a number of functions, including $\ln(1-x)$ and $\tan^{-1}(x)$, as power series. However, our methods were completely ad hoc and didn't tell us anything about how to find the series expansions of e^x or $\sin(x)$ or $\cos(x)$ or any other function. In this section we're going to see how to find polynomials that approximate an arbitrary function that has derivatives to all orders. We call such functions **smooth**. We'll also see how to approximate a function as a polynomial in $(x-1)$ or $(x-2)$ or $(x-a)$, which is more useful if we're trying to understand the function near $x = 1$ or $x = 2$ or $x = a$.

Finding the Coefficients. The key idea comes from the Fundamental Theorem of Calculus, specifically FTC2. Suppose that $f(x)$ is a smooth function. Then

$$(11.34) \qquad f(x) - f(a) = \int_a^x f'(s)\, ds, \qquad \text{or} \qquad f(x) = f(a) + \int_a^x f'(s)\, ds.$$

If we can find a good approximation for the function $f'(x)$, we can integrate that approximation to get an approximation for $f(x)$. If our estimate for $f'(s)$ is accurate to within a quantity ϵ for all s between a and x, then our estimate for $f(x)$ will be accurate to within $\epsilon|x - a|$. When x is close to a, this will give us an estimate for $f(x)$ that is more accurate than our estimate for $f'(x)$.

The roughest estimate for $f'(s)$ is simply that $f'(x) \approx f'(a)$ whenever $x \approx a$. That is, we pretend that $f'(x)$ is a constant function. We than have

$$(11.35) \qquad f(x) \approx f(a) + \int_a^x f'(a)\, ds = f(a) + f'(a)(x - a).$$

This is our old friend the microscope equation. We get a better approximation for $f'(x)$ by applying a linear approximation to f':

$$(11.36) \qquad f'(x) \approx f'(a) + f''(a)(x - a), \qquad \text{so} \qquad f'(s) \approx f'(a) + f''(a)(s - a).$$

Integrating this linear approximation for f' gives us a quadratic approximation for f:

$$(11.37)\ \ f(x) \approx f(a) + \int_a^x (f'(a) + f''(a)(s-a))\, ds = f(a) + f'(a)(x-a) + \frac{f''(a)}{2!}(x-a)^2.$$

Likewise, we have a quadratic approximation for $f'(s)$:

$$(11.38) \qquad f'(s) \approx f'(a) + f''(a)(s - a) + \frac{f'''(a)}{2!}(s - a)^2.$$

Plugging this into (11.34) then gives a cubic approximation for $f(x)$:

$$(11.39)$$
$$f(x) \approx f(a) + \int_a^x \left(f'(a) + f''(a)(s - a) + \frac{f'''(a)}{2!}(s - a)^2 \right) ds$$
$$= f(a) + f'(a)(x - a) + \frac{f''(a)}{2!}(x - a)^2 + \frac{f'''(a)}{3!}(x - a)^3.$$

We can continue this bootstrap argument forever. Once we have an nth-order approximation for $f(x)$, we can apply the same procedure to $f'(x)$ to get an nth-order approximation for $f'(x)$, and likewise for $f'(s)$. We then integrate our nth-order approximation for $f'(s)$ to get an $(n+1)$-st order approximation to $f(x)$. In general, the nth-order polynomial that best approximates $f(x)$ is

$$(11.40) \quad T_n(x) = \sum_{k=0}^n \frac{f^{(k)}(a)}{k!}(x - a)^k = f(a) + f'(a)(x - a) + \cdots + f^{(n)}(a)\frac{(x - a)^n}{n!},$$

where the notation $f^{(k)}(a)$ means the kth derivative of $f(x)$, evaluated at $x = a$. In particular, the right hand sides of (11.35), (11.37), and (11.39), which give the best linear, quadratic and cubic approximations to $f(x)$ near $x = a$, are $T_1(x)$, $T_2(x)$ and $T_3(x)$. Taking a limit as $n \to \infty$ we express $f(x)$ as an infinite power series:

$$(11.41) \qquad \sum_{k=0}^{\infty} \frac{f^{(k)}(a)}{k!}(x-a)^k = f(a) + f'(a)(x-a) + \frac{f''(a)}{2!}(x-a)^2 + \cdots$$

The polynomials $T_n(x)$ are called **Taylor polynomials**, after the 18th-century English mathematician Brook Taylor, while the infinite series (11.41) is called a **Taylor series**. When $a = 0$, as with our approximations of $\ln(1-x)$ and $\tan^{-1}(x)$, they are often called MacLaurin polynomials and MacLaurin series, after the 18th-century Scottish mathematician Colin MacLaurin. However, many examples of MacLaurin series were already known to Isaac Newton in the 17th century, and some were even known to Indian mathematicians in the 14th century.

An Alternate Derivation. Another way to understand Taylor polynomials is by matching derivatives. Suppose we want an nth-order polynomial

$$T(x) = c_0 + c_1(x-a) + c_2(x-a)^2 + \cdots + c_n(x-a)^n$$

$$(11.42) \qquad = \sum_{k=0}^{n} c_k(x-a)^k,$$

that does a good job of approximating $f(x)$ near $x = a$. One measure of this is how well the derivatives of $T(x)$ at $x = a$ match the derivatives of $f(x)$ at $x = a$. We start by computing derivatives of $T(x)$:

$$T(x) = c_0 + c_1(x-a) + c_2(x-a)^2 + \cdots$$
$$T'(x) = c_1 + 2c_2(x-a) + 3c_3(x-a)^2 + \cdots$$
$$T''(x) = 2c_2 + 6c_3(x-a) + \cdots$$
$$T'''(x) = 6c_3 + 24c_4(x-a) + \cdots$$

(11.43)
$$\vdots$$

$$T^{(k)}(x) = k!\,c_k + \text{ terms with at least one power of } (x-a)$$

$$\vdots$$

$$T^{(n)}(x) = n!\,c_n.$$

Since $T^{(n)}(x)$ is a constant, all subsequent derivatives are zero. Plugging in $x = a$ gives

$$(11.44) \qquad T^{(k)}(a) = \begin{cases} k!\,c_k & k \leq n \\ 0 & k > n. \end{cases}$$

There's no hope of matching any derivatives higher than nth-order, but we can adjust the coefficients c_k to match the value and first n derivatives of $f(x)$ at $x = a$. Since we want

$$(11.45) \qquad k!\,c_k = f^{(k)}(a),$$

we must have

$$(11.46) \qquad c_k = \frac{f^{(k)}(a)}{k!}.$$

Plugging this back into the second line of equation (11.42) then gives

(11.47)
$$T(x) = \sum_{k=0}^{n} \frac{f^{(k)}(a)}{k!}(x-a)^k.$$

In other words, $T(x)$ is *exactly* what we previously called $T_n(x)$.

A Test Case. As an example of Taylor polynomials, let's approximate $\sqrt{102}$ by hand. (Maybe you left your calculator and phone at home, or maybe you're trying to figure out how an engineer would program a calculator in the first place.) We pick $a = 100$, since that's a place where we understand $f(x) = \sqrt{x}$. We compute the first few derivatives of $f(x)$ at $x = 100$:

k	$f^{(k)}(x)$	$f^{(k)}(100)$	$f^{(k)}(100)/k!$
0	$x^{1/2}$	10	10
1	$\frac{1}{2}x^{-1/2}$	1/20	1/20
2	$-\frac{1}{4}x^{-3/2}$	-1/4000	-1/8000
3	$\frac{3}{8}x^{-5/2}$	3/800,000	1/1,600,000

Plugging these into equation (11.47) gives us our first few Taylor polynomials:

(11.48)
$$T_1(x) = 10 + \frac{x-100}{20},$$
$$T_2(x) = 10 + \frac{x-100}{20} - \frac{(x-100)^2}{8000},$$
$$T_3(x) = 10 + \frac{x-100}{20} - \frac{(x-100)^2}{8000} + \frac{(x-100)^3}{1,600,000}.$$

Setting $x = 102$ then gives

(11.49)
$$T_1(102) = 10 + \frac{2}{20} = \qquad\qquad 10.1,$$
$$T_2(102) = 10 + \frac{2}{20} - \frac{4}{800} = \qquad\qquad 10.0995,$$
$$T_3(102) = 10 + \frac{2}{20} - \frac{4}{800} + \frac{8}{1,600,000} = \quad 10.099505.$$

In fact, $\sqrt{102} = 10.0995049384$ to 10 decimal places. The linear approximation $T_1(102)$ is accurate to within 5×10^{-4}, the quadratic approximation $T_2(102)$ is accurate to within 5×10^{-6}, and the cubic approximation $T_3(102)$ is accurate to within 7×10^{-8}. Each additional term gives us roughly two more decimal places of accuracy.

The bottom line is that the computations needed to use Taylor polynomials to approximate a function near a known data point are very much like using the microscope equation (a.k.a. a linear approximation), except that we compute several derivatives of $f(x)$ at $x = a$ instead of just $f'(a)$, and we get rewarded for our effort with greater accuracy. The linear approximation $T_1(x)$ is good, but $T_2(x)$ is better, $T_3(x)$ is better still, and so on.

Accuracy and Taylor's Theorem. But how much better? Just how accurate *are* Taylor polynomials? The answer is given by:

Theorem 11.2 (Taylor's Theorem). *Suppose that $f(x)$ is a function that is differentiable at least $n + 1$ times on an interval I around a. Then for every $x \in I$ there is a number c, somewhere between a and x, such that*

$$(11.50) \qquad f(x) = T_n(x) + \frac{f^{(n+1)}(c)}{(n+1)!}(x-a)^{n+1}.$$

In other words, when we approximate $f(x)$ by $T_n(x)$ our error looks just like the last term in $T_{n+1}(x)$, except that the derivative is evaluated at a mystery point, somewhere between a and x. Most of the time, $f^{(n+1)}(c)$ is fairly close to $f^{(n+1)}(a)$, and our error is roughly the same size as the next term in our series. In any case, the error is bounded by $\frac{M(x-a)^{n+1}}{(n+1)!}$, where M is the largest value of $|f^{(n+1)}(x)|$ on I.

Note that the error scales like $(x-a)^{n+1}$. When we used $T_3(x)$ to approximate $\sqrt{102}$, we found that our answer was accurate to within 7×10^{-8}. If we had used $T_3(x)$ to approximate $\sqrt{101}$, it would have been $2^4 = 16$ times more accurate, or good to within 5×10^{-9}. If we had used $T_3(x)$ to approximate $\sqrt{106}$, it would have been $3^4 = 81$ times less accurate, or only good to within 6×10^{-6}.

The proof of Taylor's theorem is beyond the scope of this book, involving repeated application of the mean value theorem, which we also aren't covering.

11.4. Sines, Cosines, Exponentials, and Logs

Now that we have a formula (11.41) for the Taylor series of a function $f(x)$, let's find the Taylor series of some common functions.

Compound Interest and the Series for e^x. If $f(x) = e^x$, then $f'(x) = e^x$, $f''(x) = e^x$, and the kth derivative of $f(x)$, evaluated at $x = 0$, is

$$(11.51) \qquad f^{(k)}(0) = e^0 = 1.$$

This makes the nth-order Taylor polynomial

$$(11.52) \qquad T_n(x) = 1 + x + \frac{x^2}{2!} + \cdots + \frac{x^n}{n!},$$

and makes the Taylor series

$$(11.53) \qquad e^x = \sum_{n=0}^{\infty} \frac{x^n}{n!} = 1 + x + \frac{x^2}{2!} + \frac{x^3}{3!} + \cdots.$$

More generally, if A is any expression, then

$$e^A = \sum_{n=0}^{\infty} \frac{A^n}{n!},\tag{11.54}$$

and in particular

$$e^{rt} = 1 + rt + \frac{(rt)^2}{2!} + \frac{(rt)^3}{3!} + \cdots = \sum_{n=0}^{\infty} \frac{(rt)^n}{n!}.\tag{11.55}$$

We can understand each term in this series via compound interest. Suppose that we invest an amount of money P at an interest rate r. How much money will we have at time t? We have already solved this problem with differential equations. If $Y(t)$ represents our bank balance at time t, then $Y(t)$ solves the initial value problem

$$\frac{dY}{dt} = rY, \qquad Y(0) = P.\tag{11.56}$$

The solution to this initial value problem is $Y(t) = Pe^{rt}$, as can be easily checked.

However, there's another way of looking at the problem. Let $Y_0(t)$ be our initial investment, also known as the **principal**. Let $Y_1(t)$ be the interest earned on the principal up to time t. Let $Y_2(t)$ be the interest earned on the interest on the principal, in other words the interest earned by Y_1. Let $Y_3(t)$ be the interest earned by Y_2, and so on. We then have

$$Y(t) = Y_0(t) + Y_1(t) + Y_2(t) + \cdots.\tag{11.57}$$

The function $Y_0(t) = P$ is a constant. Since this generates interest at rate $rY_0(t) = rP$, we have

$$Y_1(t) = \int_0^t rY_0(s)\, ds = \int_0^t rP\, ds = rtP.\tag{11.58}$$

That is, $Y_1(t)$ is the accumulation function for $rY_0(t)$. Similarly,

$$
\begin{aligned}
Y_2(t) &= \int_0^t rY_1(s)\, ds = \int_0^t r^2 sP\, ds = \frac{r^2 t^2}{2} P \\[2mm]
Y_3(t) &= \int_0^t rY_2(s)\, ds = \int_0^t \frac{r^3 s^2}{2} P\, ds = \frac{r^3 t^3}{3!} P \\[2mm]
Y_4(t) &= \int_0^t rY_3(s)\, ds = \int_0^t \frac{r^4 s^3}{3!} P\, ds = \frac{r^4 t^4}{4!} P \\
&\ \ \vdots \\
Y_n(t) &= \int_0^t rY_{n-1}(s)\, ds = \int_0^t \frac{r^n s^{n-1}}{(n-1)!} P\, ds = \frac{r^n t^n}{n!} P. \\[2mm]
Y(t) &= P \sum_{n=0}^{\infty} \frac{(rt)^n}{n!},
\end{aligned}\tag{11.59}
$$

where the $0!$ in the $n = 0$ term means 1. $Y_n(t)$ is exactly the nth term in the Taylor series of Pe^{rt}.

For instance, if we invest $1000 at an interest rate of $r = 6\% = 0.06$, then after one year we will have $Y_0(1) = P = \$1000$. The interest on this $1,000 is $Y_1(1) = 0.06P = \$60$. The interest on the interest is $Y_2(1) = (0.06)^2 P/2! = \$1.80$. The interest on the interest on the interest is $Y_3(1) = (0.06)^3 P/3! = \$0.036$, or less than 4 cents. This, and all subsequent terms, are small enough to be ignored. For all practical purposes, the total interest we earn in one year is

$$(11.60) \qquad \left(0.06 + \frac{(0.06)^2}{2}\right)P = 0.0618P.$$

The multiplier 0.0618 is called the **yield**. That is, an account with interest *rate r* = 0.06 generates a *yield* of approximately $r + \frac{r^2}{2} = 0.0618$. (The exact yield is $e^r - 1$.)

Another way to understand the series (11.53) is through the initial value problem $y' = y$, $y(0) = 1$. If we define $f(x)$ to be the right hand side of equation (11.53), then

$$(11.61) \qquad \begin{aligned} f'(x) &= \frac{d}{dx}\left(1 + x + \frac{x^2}{2!} + \frac{x^3}{3!} + \frac{x^4}{4!} + \cdots\right) \\ &= 0 + 1 + x + \frac{x^2}{2!} + \frac{x^3}{3!} + \cdots \\ &= f(x), \end{aligned}$$

since the derivative of each term of (11.53) is exactly the previous term. Of course $f(0) = 1 + 0 + 0^2/2! + \cdots = 1$, so $f(x)$ satisfies our IVP. Since e^x is the unique function that satisfies this IVP, $f(x)$ must be e^x.

Figure 11.2 shows the graph of e^x and its first few approximating polynomials on two length scales. We have already seen that the linear approximation does well when x is close to 0, and the quadratic approximation does even better. However, they start to fail when we get farther away. To maintain accuracy, we need to keep more and more terms as we get farther and farther away.

Let's see what Taylor's theorem has to say about the approximation

$$(11.62) \qquad e \approx 1 + 1 + \frac{1}{2} + \frac{1}{3!} + \frac{1}{4!} + \cdots + \frac{1}{n!},$$

which we get by evaluating the Taylor polynomial $T_n(x)$ at $x = 1$. Since the $(n + 1)$st derivative of e^x is e^x, and the biggest value of e^x between 0 and 1 is $e^1 = e$, we know that our error is bounded by $e(1 - 0)^{n+1}/(n + 1)! = e/(n + 1)!$. That is, our error as a fraction of e is at most $1/(n + 1)!$. The approximation

$$(11.63) \qquad e \approx 2 + \frac{1}{2} + \frac{1}{6} + \frac{1}{24} = 2.708333$$

is good to one part in $5! = 120$, and the approximation

$$(11.64) \qquad e \approx 2 + \frac{1}{2} + \frac{1}{6} + \frac{1}{24} + \frac{1}{120} + \frac{1}{720} = 2.718056$$

is good to one part in $7! = 5040$.

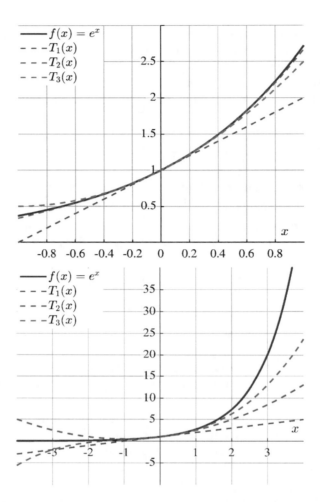

Figure 11.2. The graphs of $f(x) = e^x$ and some approximating polynomials. $T_3(x)$ gives a very good approximation on the interval $[-1, 1]$, but to approximate e^x on $[-4, 4]$ we need to go at least to $T_7(x)$.

An "Impossible" Integral. Can anybody evaluate the indefinite integral $\int e^{-x^2}\, dx$? Integration by substitution doesn't work. Neither does integration by parts, or any other known trick. In fact, the answer cannot be expressed as *any* finite formula involving elementary functions. However, it *can* be expressed as a power series.

Applying (11.54) with $A = -x^2$, we get

$$(11.65) \qquad e^{-x^2} = \sum_{n=0}^{\infty} \frac{(-1)^n x^{2n}}{n!} = 1 - x^2 + \frac{x^4}{2!} - \frac{x^6}{3!} + \frac{x^8}{4!} - \cdots$$

Integrating term by term, an anti-derivative of e^{-x^2} is

$$(11.66) \qquad A(x) = \int_0^x e^{-s^2}\, ds = \sum_{n=0}^{\infty} \frac{(-1)^n x^{2n+1}}{(2n+1)n!} = x - \frac{x^3}{3} + \frac{x^5}{10} - \frac{x^7}{42} + \frac{x^9}{216} - \cdots$$

By evaluating the first few terms, we can accurately approximate quantities such as $\int_{-0.3}^{0.2} e^{-x^2}\,dx$. Approximating integrals on a wider scale, such as $\int_{-3}^{2} e^{-x^2}\,dx$, is similar, except that more terms are needed to be accurate.

Sines and Cosines. We now turn to figuring out the Taylor series for $\sin(x)$ and $\cos(x)$ (around $x = 0$). The first few derivatives of $f(x) = \sin(x)$ are

k	$f^{(k)}(x)$	$f^{(k)}(0)$
0	$\sin(x)$	0
1	$\cos(x)$	1
2	$-\sin(x)$	0
3	$-\cos(x)$	-1
4	$\sin(x)$	0

after which the pattern repeats. Since $f''''(x) = f(x)$, $f^{(4n+k)}(x) = f^{(k)}(x)$, and

(11.67)
$$f^{(4n+k)}(0) = \begin{cases} 0 & k = 0 \text{ or } 2, \\ 1 & k = 1, \\ -1 & k = 3. \end{cases}$$

The Taylor series for $\sin(x)$ with $a = 0$ is then

(11.68)
$$0 + x + 0\frac{x^2}{2!} - \frac{x^3}{3!} + 0\frac{x^4}{4!} + \frac{x^5}{5!} + 0\frac{x^6}{6!} + \cdots.$$

After eliminating the terms that are zero, this simplifies to

(11.69)
$$\sin(x) = \sum_{n=0}^{\infty} \frac{(-1)^n}{(2n+1)!} x^{2n+1} = x - \frac{x^3}{3!} + \frac{x^5}{5!} + \cdots$$

Note that $T_1(x) = T_2(x) = x$. This is the **small angle approximation** that we saw at the end of Chapter 5. The approximation $\sin(x) \approx x$ is accurate for x up to half a radian or so (28.65 degrees). When $x = 1/2$, the error is roughly the size of the next term, namely $(1/2)^3/6 \approx 0.021$, and $\sin(1/2) = 0.479$.

To compute the series for $\cos(x)$ we could compute successive derivatives of $\cos(x)$ and apply equation (11.46). However, since $\cos(x) = (\sin(x))'$, we can get the series for $\cos(x)$ by taking the derivative of each term in the series for $\sin(x)$. The derivative of $(-1)^n x^{2n+1}/(2n+1)!$ is $(-1)^n x^{2n}/(2n)!$, so

(11.70)
$$\cos(x) = \sum_{n=0}^{\infty} \frac{(-1)^n}{(2n)!} x^{2n} = 1 - \frac{x^2}{2!} + \frac{x^4}{4!} + \cdots$$

Here $T_2(x) = T_3(x) = 1 - \frac{x^2}{2}$, which agrees with the small angle approximation $\cos(x) \approx 1 - \frac{x^2}{2}$. Note that the derivative of the series for $\cos(x)$ is minus the series for $\sin(x)$, as of course it must be.

Complex Exponentials and Euler's Formula. The Swiss mathematician Leonhard Euler, after whom Euler's method and the constant e are named, proposed an audacious formula using complex numbers that combines exponentials and trig functions:

$$(11.71) \qquad e^{ix} = \cos(x) + i\sin(x),$$

where $i = \sqrt{-1}$. This is illustrated in Figure 11.3.

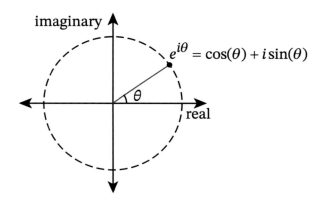

Figure 11.3. In the complex plane, the point $e^{i\theta}$ sits on the unit circle with angle θ.

This looks like nonsense. How can you possibly multiply e by itself an imaginary number of times?! Then again, we had the same issues with $e^{\sqrt{2}}$ and $e^{1/2}$ and e^{-1}, none of which are e multiplied by itself a certain number of times. We managed to define all of those numbers using reciprocals, roots, and limits, but a better method is to simply define e^x to be the unique solution to the initial value problem $\frac{dy}{dx} = y$, $y(0) = 1$. Likewise, for any constant k we can define e^{kx} to be the unique solution to the initial value problem $\frac{dy}{dx} = ky$, $y(0) = 1$.

Let's do that for $k = i$. The derivative of $\cos(x) + i\sin(x)$ is

$$\frac{d}{dx}\Big(\cos(x) + i\sin(x)\Big) = \frac{d(\cos(x))}{dx} + i\frac{d(\sin(x))}{dx}$$
$$(11.72) \qquad\qquad\qquad = -\sin(x) + i\cos(x)$$
$$= i\Big(\cos(x) + i\sin(x)\Big),$$

and

$$(11.73) \qquad \cos(0) + i\sin(0) = 1 + i0 = 1,$$

as required. Since $\cos(x) + i\sin(x)$ solves the initial value problem, we must have $e^{ix} = \cos(x) + i\sin(x)$.

We can also understand Euler's formula (11.71) with Taylor series. Plugging $A = ix$ into the formula (11.54) for e^A, we get

(11.74)
$$e^{ix} = \sum_{n=0}^{\infty} \frac{(ix)^n}{n!}$$

$$= 1 + ix - \frac{x^2}{2} - i\frac{x^3}{3!} + \frac{x^4}{4!} + i\frac{x^5}{5!} + \cdots$$

$$= \left(1 - \frac{x^2}{2} + \frac{x^4}{4!} - \frac{x^6}{6!} + \cdots\right)$$

$$+ i\left(x - \frac{x^3}{3!} + \frac{x^5}{5!} - \frac{x^7}{7!} + \cdots\right)$$

$$= \cos(x) + i\sin(x).$$

Euler's formula allows us to view exponential growth, exponential decay, oscillations like $\sin(x)$ and $\cos(x)$, and oscillations that grow or shrink (such as $e^{\pm x}\cos(x)$) on the same footing. It also gives us a simple way to remember (or derive) the addition of angle formulas for sine and cosine. If A and B are any two angles, then $e^{i(A+B)} = \cos(A + B) + i\sin(A + B)$. However, $e^{i(A+B)}$ also equals e^{iA} times e^{iB}, so

(11.75)
$$\cos(A{+}B) + i\sin(A{+}B) = (\cos(A) + i\sin(A))(\cos(B) + i\sin(B))$$
$$= \cos(A)\cos(B) - \sin(A)\sin(B)$$
$$+ i\big(\cos(A)\sin(B) + \sin(A)\cos(B)\big).$$

This means that

(11.76)
$$\cos(A + B) = \cos(A)\cos(B) - \sin(A)\sin(B) \quad \text{and}$$
$$\sin(A + B) = \cos(A)\sin(B) + \sin(A)\cos(B),$$

since if two complex numbers are equal, then their real and imaginary parts must also be equal.

Logarithms. Finally, we compute the Taylor series for $\ln(x)$ around $x = 1$ in two ways. First, we compute derivatives of $\ln(x)$ at $x = 1$ and apply (11.46) to find the coefficients c_k. If $f(x) = \ln(x)$, then we get the following values.

k	$f^{(k)}(x)$	$f^{(k)}(1)$	c_k
0	$\ln(x)$	0	0
1	x^{-1}	1	1
2	$-x^{-2}$	-1	-1/2
3	$2x^{-3}$	$2!$	1/3
4	$-3!\,x^{-4}$	-3!	-1/4
k	$(-1)^{k-1}(k-1)!\,x^{-k}$	$(-1)^{k-1}(k-1)!$	$(-1)^{k-1}/k$

This gives us

(11.77)
$$\ln(x) = \sum_{k=1}^{\infty} \frac{(-1)^{k-1}}{k}(x-1)^k = (x-1) - \frac{(x-1)^2}{2} + \frac{(x-1)^3}{3} + \cdots$$

The second derivation uses the series we already developed for $\ln(1-x)$ in powers of x, which becomes a series for $\ln(1-A)$ in powers of A for any quantity A. Setting $A = 1 - x$, we then have

$$\ln(x) = \ln(1 - (1-x))$$

$$= -\sum_{n=1}^{\infty}(1-x)^n/n$$

(11.78)

$$= \sum_{n=1}^{\infty}\frac{(-1)^{n-1}}{n}(x-1)^n$$

$$= (x-1) - \frac{(x-1)^2}{2} + \frac{(x-1)^3}{3} + \cdots$$

We summarize the expansions we have derived for common functions in Table 11.1.

Table 11.1. Taylor series for common functions

Function	$\sum$ form	First 3 nonzero terms	Where?
e^x	$\sum_{n=0}^{\infty} x^n/n!$	$1 + x + x^2/2$	All x
$\sin(x)$	$\sum_{n=0}^{\infty}\frac{(-1)^n}{(2n+1)!}x^{2n+1}$	$x - \frac{x^3}{3!} + \frac{x^5}{5!}$	All x
$\cos(x)$	$\sum_{n=0}^{\infty}\frac{(-1)^n}{(2n)!}x^{2n}$	$1 - \frac{x^2}{2!} + \frac{x^4}{4!}$	All x
$\ln(1-x)$	$-\sum_{n=1}^{\infty} x^n/n$	$-x - \frac{x^2}{2} - \frac{x^3}{3}$	$-1 \leq x < 1$
$\tan^{-1}(x)$	$\sum_{n=0}^{\infty}\frac{(-1)^n}{(2n+1)}x^{2n+1}$	$x - \frac{x^3}{3} + \frac{x^5}{5}$	$-1 \leq x \leq 1$
$\ln(x)$	$\sum_{n=1}^{\infty}\frac{(-1)^{n-1}}{n}(x-1)^n$	$(x-1) - \frac{(x-1)^2}{2} + \frac{(x-1)^3}{3}$	$0 < x \leq 2$

11.5. Tests for Convergence

When faced with an infinite series $\sum a_n$, the most basic question is *does it converge?* When faced with a power series $\sum a_n x^n$ or $\sum c_n(x-a)^n$, the question becomes *for what values of x does it converge?* In this section, we start to develop some tests to answer the first question. In the next section we address the second.

The Divergence Test. If a series[3] $\sum_{n=1}^{\infty} a_n$ converges to a value L, then the partial sums s_n all get close to L, and for n large we have

(11.79) $$a_n = s_n - s_{n-1} \approx L - L = 0.$$

That is, in a convergent series the individual terms a_n must approach zero. Conversely, if the individual terms a_n do *not* approach zero, then the series must diverge. For instance, if a geometric series has $|r| \geq 1$, then the series must diverge.

---

[3]Sometimes we label our terms a_i, and sometimes we label them a_n. The name we give to the index doesn't matter.

Note that this does *not* say that a series with $\lim_{n\to\infty} a_n = 0$ automatically converges! It just says that a series whose individual terms don't approach zero automatically *diverges*. For example, the **harmonic series**

(11.80)
$$\sum_{n=1}^{\infty} \frac{1}{n} = 1 + \frac{1}{2} + \frac{1}{3} + \cdots$$

has terms that approach 0, but the sum does not converge. To see this, note that

(11.81)
$$\begin{aligned}
s_{2^n} = {}& 1 + \frac{1}{2} + \left(\frac{1}{3} + \frac{1}{4}\right) \\
& + \left(\frac{1}{5} + \frac{1}{6} + \frac{1}{7} + \frac{1}{8}\right) + \cdots \\
& + \left(\frac{1}{2^{n-1}+1} + \cdots + \frac{1}{2^n}\right).
\end{aligned}$$

The first term in parentheses contains two numbers, each at least 1/4, so it adds up to something at least as big as 1/2. The second term in parentheses contains four numbers, each at least 1/8, so it also adds up to something at least as big as 1/2. The same reasoning applies to each term in parentheses. Since s_{2^n} is $\frac{3}{2}$ plus the sum of $n-1$ such terms, we have $s_{2^n} \geq 1 + \frac{n}{2}$, which is eventually bigger than any number you can name.

Theorem 11.3 (Divergence test). *If* $\lim_{i\to\infty} a_i \neq 0$ *or* $\lim_{i\to\infty} a_i$ *does not exist, then* $\sum_{i=1}^{\infty} a_i$ *diverges.*

The Comparison and Limit Comparison Tests. If the terms a_i of a series are all nonnegative, then the partial sums s_n can only go up. Either they stay bounded and approach a limit, in which case the series converges, or they grow without bound, in which case the series diverges to ∞.

A **positive series** is a series where each term is nonnegative. (Despite the plain meaning of the word "positive", some terms are allowed to be zero.) Not having to worry about oscillations in s_n makes positive series simpler to deal with than arbitrary series.

Suppose that $\sum_{i=1}^{\infty} a_i$ and $\sum_{i=1}^{\infty} b_i$ are positive series, and that for each value of i we have $a_i \leq b_i$. Then the partial sum

(11.82)
$$s_n = \sum_{i=1}^{n} a_i$$

is less than or equal to the partial sum

(11.83)
$$t_n = \sum_{i=1}^{n} b_i.$$

If the s_n's grow without bound, then the t_n's also grow without bound. If the t_n's are bounded, then so are the s_n's. In other words, if $\sum_{i=1}^{\infty} a_i$ diverges, then so does $\sum_{i=1}^{\infty} b_i$, and if $\sum_{i=1}^{\infty} b_i$ converges, then so does $\sum_{i=1}^{\infty} a_i$.

For instance, the series

(11.84) $$\sum_{i=i}^{\infty} \frac{1}{2^i + 1} = \frac{1}{3} + \frac{1}{5} + \frac{1}{9} + \cdots$$

converges, by comparison to the convergent geometric series

(11.85) $$\sum_{i=1}^{\infty} \frac{1}{2^i} = \frac{1}{2} + \frac{1}{4} + \frac{1}{8} + \cdots.$$

Each term in the first series is less than or equal to the corresponding term in the second series, and the second series converges, so the first series converges.

The convergence or divergence of a series doesn't depend on what happens in the first ten terms, or the first thousand terms, or even the first million terms. It depends on what happens in the long run, and we can wait as long as we want for that to happen. We say that a property involving a_i and b_i is **eventually true** if there is a number N such that it is true for all $i > N$. When applying the comparison test, we don't actually need all the a_i's and b_i's to be nonnegative, and we don't actually need to have $a_i \leq b_i$ for every i. We just need the series $\sum a_i$ and $\sum b_i$ to be *eventually* positive, and for $a_i \leq b_i$ to be *eventually* true.

Theorem 11.4 (Comparison test). *Suppose that $\sum_{i=1}^{\infty} a_i$ and $\sum_{i=1}^{\infty} b_i$ are eventually positive series, and that we eventually have $a_i \leq b_i$. Then*

(1) *If $\sum_{i=1}^{\infty} b_i$ converges, then so does $\sum_{i=1}^{\infty} a_i$.*

(2) *If $\sum_{i=1}^{\infty} a_i$ diverges, then so does $\sum_{i=1}^{\infty} b_i$.*

(3) *If $\sum_{i=1}^{\infty} a_i$ converges, or if $\sum_{i=1}^{\infty} b_i$ diverges, that does not tell us anything about the other series.*

If there is a positive constant c such that a_i is eventually less than cb_i, then we can apply the comparison test to $\sum a_i$ and $\sum cb_i$. If $\sum b_i$ converges, then $\sum cb_i$ converges, so $\sum a_i$ converges. If $\sum a_i$ diverges, then $\sum cb_i$ diverges, so $\sum b_i$ diverges. The extra factor of c doesn't make one bit of difference.

For instance, consider the series

(11.86) $$\sum_{i=1}^{\infty} \frac{1}{2^i - 1} = \frac{1}{1} + \frac{1}{3} + \frac{1}{7} + \cdots$$

At first glance, comparing (11.86) to the convergent series (11.85) doesn't seem to work because each term in (11.86) is bigger than the corresponding term in (11.85). In the comparison test, being bigger than a convergent series doesn't tell us anything! However, each term in (11.86) is less than or equal to *twice* the corresponding term in (11.85), and that's enough to show that (11.86) converges.

Next suppose that we have two eventually positive series, $\sum a_i$ and $\sum b_i$, such that the ratio b_i/a_i approaches a positive number L. (For instance, in the last example the ratio of $\frac{1}{2^n-1}$ and $\frac{1}{2^n}$ approached 1.) In that case, b_i/a_i is eventually close to L, and in particular is eventually somewhere between $L/2$ and $2L$. Since b_i is eventually less than $2La_i$, and is eventually greater than $La_i/2$, we can do comparisons between $\sum La_i/2$,

$\sum b_i$, and $\sum 2La_i$. If $\sum a_i$ converges, then $\sum 2La_i$ converges, so $\sum b_i$ converges. On the other hand, if $\sum a_i$ diverges, then $\sum La_i/2$ diverges, so $\sum b_i$ diverges. This observation is summarized in

> **Theorem 11.5** (Limit comparison test). *Suppose that $\sum_{i=1}^{\infty} a_i$ and $\sum_{i=1}^{\infty} b_i$ are eventually positive series, and that $\lim_{i \to \infty} b_i/a_i$ is a positive number L. Then $\sum_{i=1}^{\infty} a_i$ and $\sum_{i=1}^{\infty} b_i$ either both converge or both diverge.*

Absolute Convergence. So far we have been talking about (eventually) positive series. Next we turn to series whose terms are sometimes negative. Of course, no matter what the sign of a_i is, the series $\sum_{i=1}^{\infty} |a_i|$ is a positive series. We say that $\sum_{i=1}^{\infty} a_i$ is **absolutely convergent** if $\sum_{i=1}^{\infty} |a_i|$ converges. In most cases of interest, we can figure out whether $\sum |a_i|$ converges using our other tests, and then apply the following theorem:

> **Theorem 11.6** (Absolute convergence). *If $\sum_{i=1}^{\infty} |a_i|$ converges, then so does $\sum_{i=1}^{\infty} a_i$.*

That is, absolute convergence implies convergence. However, convergence does not imply absolute convergence. When a series $\sum a_i$ converges but $\sum |a_i|$ does not, we say that $\sum a_i$ is **conditionally convergent**.

The proof of Theorem 11.6 uses the comparison test. The quantity $a_i + |a_i|$ is either 0 (if $a_i \leq 0$) or is $2|a_i|$ (if $a_i \geq 0$). This makes $\sum(a_i + |a_i|)$ a positive series that we can compare to $\sum 2|a_i|$. If $\sum |a_i|$ converges, $\sum 2|a_i|$ converges, so $\sum a_i + |a_i|$ converges. Then, since $\sum(a_i + |a_i|)$ and $\sum |a_i|$ both converge, their difference $\sum a_i = \sum(a_i + |a_i| - |a_i|)$ also converges.

Absolute convergence helps us to understand series like

(11.87)
$$\sum_{n=1}^{\infty} \frac{\sin(n\pi/6)}{2^n}.$$

The numerator oscillates, but that doesn't faze us, since the numerator never gets bigger than 1 or less than -1. We always have $|a_n| \leq 2^{-n}$. Since $\sum 2^{-n}$ converges, our series converges absolutely.

The Root and Ratio Tests. The two most powerful and commonly used tests for convergence, called the **root test** and the **ratio test**, come from comparison to a geometric series.

In the ratio test, we look at the limit

(11.88)
$$L = \lim_{n \to \infty} \left| \frac{a_{n+1}}{a_n} \right|.$$

Theorem 11.7 (Ratio test). *Let $\sum_{n=1}^{\infty} a_n$ be an infinite series and let L be defined as in* (11.88).

(1) *If $L < 1$, then $\sum_{n=1}^{\infty} a_n$ converges absolutely.*

(2) *If $L > 1$, or if the ratio $|a_{n+1}/a_n|$ goes to $+\infty$ as $n \to \infty$, then $\sum_{n=1}^{\infty} a_n$ diverges.*

(3) *If the limit fails to exist in some other way, or if $L = 1$, then the test is inconclusive.*

The root test is almost identical, except that we look at the limit

$$(11.89) \qquad \rho = \lim_{n \to \infty} |a_n|^{1/n} = \lim_{n \to \infty} \sqrt[n]{|a_n|}.$$

Theorem 11.8 (Root test). *Let $\sum_{n=1}^{\infty} a_n$ be an infinite series and let ρ be defined as in* (11.89).

(1) *If $\rho < 1$, then $\sum_{n=1}^{\infty} a_n$ converges absolutely.*

(2) *If $\rho > 1$, or if $|a_n|^{1/n}$ diverges to $+\infty$, then $\sum_{n=1}^{\infty} a_n$ diverges.*

(3) *If the limit fails to exist in some other way, or if $\rho = 1$, then the test is inconclusive.*

The ratio test is usually easier to apply, but the root test is more general. If L exists, then ρ also exists and $\rho = L$. However, we will soon see an example where ρ exists but L doesn't. Before proving these theorems, let's apply them to some examples.

- In the series

$$(11.90) \qquad \sum_{n=1}^{\infty} \frac{n^4}{2^n},$$

we have $a_n = n^4/2^n$, so

$$(11.91) \qquad L = \lim_{n \to \infty} \frac{(n+1)^4/2^{n+1}}{n^4/2^n} = \lim_{n \to \infty} \frac{1}{2}\left(\frac{n+1}{n}\right)^4 = \frac{1}{2}.$$

Since $L < 1$, the series converges. After a while, each term is roughly half the previous term, so the series behaves much like the (convergent!) geometric series with $r = 1/2$.

We could also apply the root test to this series, getting

$$(11.92) \qquad \rho = \lim_{n \to \infty} \left(\frac{n^4}{2^n}\right)^{1/n} = \lim_{n \to \infty} \frac{1}{2}(n^{1/n})^4.$$

Evaluating this limit requires knowing that $\lim_{n \to \infty} n^{1/n} = 1$, which is far from obvious (but easy to check numerically). In this example, both the ratio and root tests work, but the ratio test is easier.

- If we were faced with the series $\sum(5n^4 + 37n^3 - 21n^2 + 452n - 1832)2^{-n}$, we could try applying the ratio or root tests directly. However, it is much simpler to apply the limit comparison test, comparing this series to the series $\sum n^4 2^{-n}$ that we just studied. Since $\sum n^4 2^{-n}$ converges, and since the ratio

(11.93)
$$\frac{(5n^4 + 37n^3 - 21n^2 + 452n - 1832)2^{-n}}{n^4 2^{-n}}$$
$$= 5 + 37n^{-1} - 21n^{-2} + 452n^{-3} - 1832n^{-4}$$

approaches 5, the series $\sum(5n^4 + 37n^3 - 21n^2 + 452n - 1832)2^{-n}$ converges.

- Next consider the series

(11.94)
$$\sum_{n=1}^{\infty} \frac{n!}{50^n}.$$

We compute

(11.95)
$$\frac{a_{n+1}}{a_n} = \frac{(n+1)!/50^{n+1}}{n!/50^n} = \frac{n+1}{50}$$

which goes to $+\infty$ as $n \to \infty$. This series therefore diverges. If we tried the root test we would have

(11.96)
$$a_n^{1/n} = (n!)^{1/n}/50.$$

This also goes to $+\infty$, but the fact that $\lim_{n\to\infty}(n!)^{1/n} = \infty$ is not obvious. (See Exercise 11.23 at the end of this chapter.)

- On the other hand, the series

(11.97)
$$\sum_{n=1}^{\infty} \frac{50^n}{n!}$$

has $L = \lim_{n\to\infty} 50/(n+1) = 0$ and $\rho = \lim_{n\to\infty} 50/(n!)^{1/n} = 0$, so the series converges absolutely, despite growing rapidly for the first 50 terms.

- Finally, consider the series

(11.98)
$$\sum_{n=1}^{\infty} \frac{3 + 2(-1)^n}{2^n}.$$

That is, a_n equals 2^{-n} when n is odd and equals 5×2^{-n} when n is even. The ratio a_{n+1}/a_n alternates between 5/2 and 1/10 and has no overall limit, so the ratio test doesn't tell us anything. However, $1^{1/n}$ and $5^{1/n}$ are both close to (or equal to) 1 when n is large, so

(11.99)
$$\rho = \lim_{n\to\infty} \left(\frac{1 \text{ or } 5}{2^n}\right)^{1/n} = \lim_{n\to\infty} \frac{1}{2}(1 \text{ or } 5)^{1/n} = \frac{1}{2}.$$

The ratio test fails, but the root test still works, and it tells us that the series converges.

Proof of the Ratio and Root Tests. We prove Theorems 11.7 and 11.8, the ratio and root tests, starting with Theorem 11.7.

Suppose that $L < 1$. Let M be a number (say, $M = (1 + L)/2$) that is bigger than L but less than 1. Since $|a_{n+1}/a_n|$ gets closer and close to L, it must eventually be less than M. But that means that $|a_n|$ must eventually be less than a constant times M^n. The series $\sum |a_n|$ then converges by comparison to the geometric series $\sum M^n$, so $\sum a_n$ converges absolutely.

If $L > 1$, or if $|a_{n+1}/a_n|$ goes to $+\infty$, then the terms a_n are eventually growing in size, and so can't be approaching zero. By the divergence test, the series $\sum a_n$ diverges.

To see that the test is inconclusive if $L = 1$, we just need examples that converge absolutely, converge conditionally, and diverge. The series $\sum n^2$ has $L = 1$ and diverges. The series $\sum n^{-2}$ has $L = 1$ and converges absolutely. The series $\sum (-1)^n/n$ has $L = 1$ and converges conditionally.

The proof of Theorem 11.8 is very similar. If $\rho < 1$, then we can find a number μ such that $\rho < \mu < 1$. Since $|a_n|^{1/n}$ approaches ρ, it is eventually less than μ, so $|a_n|$ is eventually less than μ^n. Our series then converges absolutely by comparison to $\sum \mu^n$. If $\rho > 1$, or if $|a_n|^{1/n}$ goes to infinity, then $|a_n|^{1/n}$ is eventually bigger than 1, so $|a_n|$ is eventually bigger than 1, and in particular does not approach 0, so the series diverges by the divergence test. The series $\sum n^2$, $\sum n^{-2}$ and $\sum (-1)^n/n$ all have $\rho = 1$, confirming the third statement.

11.6. Intervals of Convergence

When looking at a series $\sum a_n$, the natural question is *does it converge?* When looking at a power series $\sum c_n x^n$ or $\sum c_n(x-a)^n$, the natural question is *where does it converge?* The set of all x's such that the series converges is called the **interval of convergence**. First, we will explore the intervals of convergence of series of the form $\sum c_n x^n$. Once we understand them, the more general case of $\sum c_n(x - a)^n$ will be easy.

Suppose that the series $\sum c_n x^n$ converges when $x = x_0$. This implies that the numbers $c_n x_0^n$ eventually approach zero, and in particular are bounded. That is, there is a number M such that $|c_n x_0^n| < M$ for every natural number n. Then, for every x with $|x| < |x_0|$ we have

$$(11.100) \qquad |c_n x^n| = |c_n x_0^n| \left|\frac{x}{x_0}\right|^n < M \left|\frac{x}{x_0}\right|^n.$$

Since $|x| < |x_0|$, the series $\sum c_n x^n$ converges absolutely by comparison to the geometric series $\sum M|x/x_0|^n$.

In other words, if the interval of convergence contains x_0, then it also contains the entire open interval $(-|x_0|, |x_0|)$. Since the series always converges at $x = 0$ (where it is $c_0 + 0 + 0 + 0 + \cdots)^4$, this leaves only three possibilities for what the interval of convergence is.

---

[4] The notation $\sum_{n=0}^{\infty} c_n x^n$ is a little misleading when $x = 0$, since 0^0 doesn't make sense. A better notation would be $c_0 + \sum_{n=1}^{\infty} c_n x^n$.

Theorem 11.9. *If $\sum c_n x^n$ is a power series, then either*

(1) *The series converges only when $x = 0$.*

(2) *The series converges for every real number x, or*

(3) *There is a positive number R such that the interval of convergence is either $(-R, R)$, $(-R, R]$, $[-R, R)$, or $[-R, R]$.*

The number R in case (3) is called the **radius of convergence** of the series. See Figure 11.4. If $|x| < R$ then the series converges and if $|x| > R$ then it diverges. The situations where $x = \pm R$ are tricky. There are series that converge at both of these endpoints, series that converge at only one, and series that diverge at both endpoints. For instance, $\sum n x^n$, $\sum x^n / n$ and $\sum x^n / n^2$ all have $R = 1$, but the first diverges at $x = \pm 1$, the second converges at $x = -1$ and diverges at $x = 1$, and the third converges at $x = \pm 1$. Depending on what happens at the endpoints of the interval of convergence, R is the largest value of $|x|$ for which the series converges at x, or the smallest value of $|x|$ where it diverges, or both. We also use the language of radius of convergence for the first two cases. In case (1) we say that $R = 0$, and in case (2) we say that R is infinite.

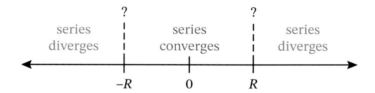

Figure 11.4. A power series $\sum c_n x^n$ with radius of convergence R converges when $|x| < R$ and diverges when $|x| > R$. What happens at $x = \pm R$ is more complicated.

Computing R. So how do we actually compute a radius of convergence? There are many techniques, but the two most common, and the most powerful, are the root and ratio test.

As before, we begin with the ratio test. Theorem 11.7 says that the criterion for $\sum c_n x^n$ converging is for the limit

$$(11.101) \qquad \lim_{n \to \infty} \left| \frac{c_{n+1} x^{n+1}}{c_n x^n} \right|$$

to be less than 1. However, that ratio is just the limit of $\left| \frac{c_{n+1}}{c_n} x \right|$, which is less than 1 precisely when $|x|$ is less than the limit of $\left| \frac{c_n}{c_{n+1}} \right|$. If that last ratio approaches a finite number, then that is our radius of convergence. If it diverges to ∞, then the series converges for all x. If it goes to zero, then the series diverges for all $x \neq 0$.

In other words,

Theorem 11.10 (Ratio test for power series). *If the sequence $|c_n/c_{n+1}|$ has a finite limit, then the radius of convergence of the power series $\sum c_n x^n$ is*

(11.102)
$$R = \lim_{n \to \infty} \left| \frac{c_n}{c_{n+1}} \right|.$$

If the sequence of ratios goes to ∞, then the radius of convergence is also infinite.

Note: Many people look at the limit of $|c_{n+1}/c_n|$ rather than the limit of $|c_n/c_{n+1}|$. When that limit is a positive number, the radius of convergence is equal to the *reciprocal* of the limit of $|c_{n+1}/c_n|$, and is equal to the limit of $|c_n/c_{n+1}|$.

- Let's apply the ratio test to the series for e^x.

(11.103)
$$\lim_{n \to \infty} \left| \frac{c_n}{c_{n+1}} \right| = \lim_{n \to \infty} \frac{1/n!}{1/(n+1)!} = \lim_{n \to \infty} n + 1.$$

Because this goes to ∞, the series for e^x converges for every x. That is, the radius of convergence is infinite.

- The terms in the series for $\sin(x)$ and $\cos(x)$ are either zero or are plus or minus the corresponding terms in the series for e^x. For every x, the series for $\sin(x)$ and $\cos(x)$ converge (absolutely) by comparison to the series for e^x.

- The ratio test also gives us the radius of convergence for the series for $\ln(1 - x)$:

(11.104)
$$R = \lim_{n \to \infty} \left| \frac{c_n}{c_{n+1}} \right| = \lim_{n \to \infty} \frac{-1/n}{-1/(n+1)} = \lim_{n \to \infty} \frac{n+1}{n} = 1.$$

- In the series for $\tan^{-1}(x)$, namely $x - x^3/3 + x^5/5 - x^7/7 + \cdots$, every other coefficient is zero, so the ratio of successive coefficients is either 0 or undefined. However, we can find the radius of convergence by looking at the ratio of successive *nonzero* coefficients, and the answer is 1.

- Finally, we look at the series for $\ln(x)$. The ratio of successive coefficients is 1, as before, so the radius of convergence is 1. However, this does not mean that the series converges when $|x| < 1$! It converges when $|x - 1| < 1$, in other words when $0 < x < 2$. When doing a Taylor series in powers of $x - a$, the interval of convergence runs from $a - R$ to $a + R$, not from $-R$ to R. See Figure 11.5.

The root test is almost identical.

Theorem 11.11 (Root test for power series). *If the sequence $1/\sqrt[n]{|c_n|}$ has a finite limit, then the radius of convergence of the power series $\sum c_n x^n$ is*

(11.105)
$$R = \lim_{n \to \infty} \left| \frac{1}{\sqrt[n]{|c_n|}} \right|.$$

If $1/\sqrt[n]{|c_n|}$ goes to ∞, then the radius of convergence is also infinite.

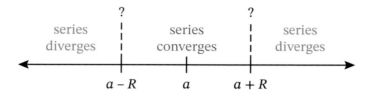

Figure 11.5. A power series $\sum c_n(x-a)^n$ with radius of convergence R converges when $|x-a| < R$ and diverges when $|x-a| > R$. What happens at $x = a \pm R$ is more complicated.

So far we have only looked at the *radius* of convergence. To fully understand the *interval* of convergence, we have to consider what happens at the boundary points $x = a \pm R$. This can be subtle, and it involves tests that are not covered in this text, such as the alternating series test and the integral test, and the fact that $\sum 1/n^p$ (the so-called p-series) converges for $p > 1$ and diverges for $p \le 1$.

In our last example, the series for $\ln(x)$ does not converge at $x = 0$. This should come as no surprise, because $\ln(0)$ does not exist. However, it turns out that the series converges at $x = 2$, and that it converges to

$$(11.106) \qquad \ln(2) = 1 - \frac{1}{2} + \frac{1}{3} - \frac{1}{4} + \frac{1}{5} - \frac{1}{6} + \cdots.$$

The convergence is extremely slow: it takes 1000 terms to get 3 digits of accuracy, 1,000,000 terms to get 6 digits of accuracy, and 10^k terms to get k digits of accuracy. Similarly, the series for $\ln(1-x)$ converges at $x = -1$ and diverges at $x = 1$.

The series for $\tan^{-1}(x)$ converges when $x = \pm 1$. Since $\tan^{-1}(1) = \pi/4$, this gives a curious formula for π:

$$(11.107) \qquad \pi = 4\left(1 - \frac{1}{3} + \frac{1}{5} - \frac{1}{7} + \cdots\right).$$

Like the series for $\ln(2)$, this series converges ridiculously slowly, with over 10^k terms needed to get k digits of accuracy. It's a nice mathematical curiosity, but it's useless for estimating π.

Analytic Functions. Many functions can be represented by convergent Taylor series, just like e^x, $\sin(x)$, $\cos(x)$, $\ln(x)$ and $\tan^{-1}(x)$. Such functions are called **analytic**. In fact, so many of the functions that we encounter in calculus, science, business and engineering are analytic that it sometimes seems that all functions are analytic. However, that isn't true. There are three ways that a function can fail to be analytic.

(1) It may not have a Taylor series. In a Taylor series, $c_n = f^{(n)}(a)/n!$. That only makes sense if the function is n-times differentiable. For instance, the function $f(x) = |x|$ is not differentiable at $x = 0$. Near $x = 0$ there is no linear approximation to the function, much less quadratic, cubic, quartic, etc. approximations.

(2) It may have a Taylor series, but the series may not converge. If the derivatives $f^{(n)}(a)$ grow too quickly with n, then the radius of convergence of the Taylor series may be zero.

(3) It may have a convergent Taylor series, but the series may not converge to the right answer! Sometimes this is because the function changes behavior: you can compute a Taylor series for $f(x) = |x|$ around $a = 1$, getting $1 + (x - 1) + 0(x - 1)^2 + 0(x - 1)^3 + \cdots$, and this converges for every value of x, but it gives the wrong answer for $f(-1)$. However, even smooth functions can have problems. The error in the nth Taylor polynomial is $f^{(n+1)}(c)(x-a)^{n+1}/(n+1)!$. If $f^{(n+1)}(c)$ grows too rapidly with n, this may not go to zero as $n \to \infty$. If the error doesn't go to zero, the series isn't actually converging to $f(x)$.

11.7. Chapter Summary

The Main Ideas.

- An infinite sum $\sum_{k=1}^{\infty} a_k$, also called an infinite series, is the limit of partial sums $s_n = \sum_{k=1}^{n} a_k$. If this limit exists, we say that the series converges. Otherwise it diverges.

- Representing something as an infinite sum is the same as approximating it better and better by finite sums.

- The geometric series $\sum_{n=0}^{\infty} ar^n$ converges to $a/(1 - r)$ if $|r| < 1$, and diverges if $|r| \geq 1$.

- A power series is a series where each term is a coefficient times a power of x or $x - a$. While ordinary series can converge to numbers, power series can converge to functions.

- Many common functions, including e^x, $\sin(x)$, $\cos(x)$, $\ln(x)$ and $\tan^{-1}(x)$, can be expressed as power series. These series are given in Table 11.1 on p. 342.

- The Taylor **series** for a function $f(x)$ around $x = a$ is $\sum_{n=0}^{\infty} c_n(x - a)^n$, where $c_n = f^{(n)}(a)/n!$.

- The Taylor **polynomial** $T_n(x)$ is the first $n+1$ terms of a Taylor series, up through x^n or $(x - a)^n$. It is the nth-order polynomial that best approximates the original function. $T_1(x)$ is the best linear approximation, which is the equation of the tangent line. $T_2(x)$, $T_3(x)$, etc. are improvements on $T_1(x)$. Since we know how to integrate polynomials easily, Taylor polynomials allow us to approximate the integral of a general function.

- The error in approximating $f(x)$ by $T_n(x)$ is

(11.108) $$f(x) - T_n(x) = \frac{f^{(n+1)}(c)}{(n + 1)!}(x - a)^{n+1},$$

where c is a mystery number somewhere between a and x.

- There are many tests for convergence, including the divergence test, the comparison test, the limit comparison test, the absolute convergence test, the ratio test and the root test. The ratio and root tests are based on doing comparisons to geometric series.

- A power series either converges only at $x = a$, or converges everywhere, or converges on an interval of convergence centered at $x = a$. The radius of this interval of convergence is called the radius of convergence.
- We can often compute the radius of convergence from the ratio and root tests. In the ratio test, we look at $\lim_{n \to \infty} |c_n/c_{n+1}|$. If this limit exists, it is the radius of convergence of $\sum c_n(x-a)^n$. The root test is similar, but looks at $\lim_{n \to \infty} 1/\sqrt[n]{|c_n|}$.
- Most common functions are analytic, meaning they can be expressed as convergent power series. However, not every function is analytic.

Expectations. You should be able to:

- Explain what it means for a series to converge, what it means to diverge, and what it means for it to "go to infinity".
- Find the sum of a finite geometric series or a convergent infinite geometric series.
- Recite from memory the series expansions for $1/(1 - x)$, e^x, $\sin(x)$, and $\cos(x)$.
- Manipulate series to get expansions for additional functions, e.g., getting the series for e^{-2x^2} by taking the series for e^x and replacing x with $-2x^2$.
- Differentiate and integrate series term by term.
- Relate the kth coefficient c_k in a Taylor series to the kth derivative of the function being approximated.
- Use Taylor polynomials to approximate functions near points where the first few derivatives are known, and to approximate integrals such as $\int_{-0.1}^{0.2} e^{-x^2}\, dx$.
- Use a variety of tests to determine whether a series converges or diverges.
- Use the ratio and root tests to determine the radius of convergence of a power series.

11.8. Exercises

Geometric Series

11.1. Find the sum of each infinite series, if it exists, or indicate that the series diverges.
   (a) $16 + 12 + 9 + \cdots$
   (b) $1 + \frac{3}{4} + \frac{9}{16} + \cdots$
   (c) $\frac{3}{2} - \frac{3}{4} + \frac{3}{8} - \cdots$
   (d) $7 + 1.6 + 1.06 + 1.006 + \cdots$
   (e) $0.9 + 0.09 + 0.009 + 0.0009 + \cdots$

11.2. The sum of an infinite geometric series is 16 and the common ratio is $1/3$. Find the first three terms of the series.

11.3. The sum of an infinite geometric series is 24 and the common ratio is $-1/4$. Find the first three terms of the series.

11.4. Alex and Charlie have a pet bird, who we'll call Birdie. One day, Alex and Charlie rode bikes toward each other, starting 30 miles apart, with each one riding at 10 mph. At the beginning of the ride, Birdie was next to Alex. Birdie then flew at 20 mph toward Charlie. When Birdie reached Charlie, Birdie turned around and flew at 20 mph back to Alex, then turned around and flew to Charlie, and so on until the three of them met in the middle, where they shared a picnic of sandwiches and birdseed.

   (a) How long did it take Birdie to reach Charlie the first time? Where were Alex, Birdie, and Charlie at that time, relative to Alex's starting point? How far did Birdie fly on that first leg?

   (b) How long did it take Birdie to fly back to Alex? Where was everybody when Birdie reached Alex? How far did Birdie fly on that second leg?

   (c) How far did Birdie fly on the nth leg?

   (d) Write down an infinite series for the total distance that Birdie flew. Then find the sum of the series.

11.5. Here's a completely different way to understand the total distance that Birdie flew in Exercise 11.4. How long did it take for Alex and Charlie to meet? How far did Birdie fly in that time?

11.6. Arquala and Nathan are driving a 10-foot fence post into the ground using a sledgehammer. The first strike of the hammer sinks the post 10 inches, but every subsequent strike only drives the post 60% as far as the previous strike. For example, the second strike drives it only 6 inches deeper. Arquala argues that the depth of the post will eventually converge, with approximately 8 feet of the post remaining above ground, which is plenty to attach a tall fence to. Nathan argues that with each strike the post is being driven deeper. Since the post has a finite length (10 feet), it will eventually be completely driven into the ground. Who (if either) is right, and why?

11.7. **Aquarium Cleaning.** Fish living in an aquarium produce waste. It isn't feasible to clean a large aquarium thoroughly each day. Instead, part of the water is removed at the end of each day and replaced by clean water. Let P be the fraction of the water (and fraction of the existing waste) that is removed. Let W be the daily waste output from the fish in the tank. Starting with a clean tank, the amount W_1 of waste in the tank after one day is

$$W_1 = W(1 - P).$$

Similarly, the waste remaining after two days is

$$W_2 = (W + W_1)(1 - P) = W(1 - P) + W(1 - P)^2.$$

The first term is today's waste, while the second term is what's left of yesterday's waste.

   (a) Write down an expression for the amount of waste in the tank after five days.

   (b) Write down an expression using sigma notation for the amount of waste in the tank after n days.

(c) If a third of the water is removed each day, how much waste is in the tank after 10 days?

(d) For the fish to stay healthy, the waste in the tank after each cleaning must remain under $3W$. What proportion of the water must be replaced each day to achieve this?

Taylor Polynomials

11.8.   Let $f(x) = \tan(x)$ and let $a = \pi/4$.

(a) Compute $f(a)$, $f'(a)$, $f''(a)$ and $f'''(a)$.

(b) Find the first three Taylor polynomials $T_1(x)$, $T_2(x)$ and $T_3(x)$ around $x = a$.

(c) Use the polynomials from part (b) to estimate $\tan(50°)$. (Don't forget to convert 50° to radians!)

(d) Now use the polynomials from part (b) to estimate $\tan(45.5°)$. How does the rate at which these estimates approach the true value compare to the rate for the estimates in part (c)?

11.9.   Let $f(x) = e^{x^3}$.

(a) Find the Taylor polynomial $T_9(x)$ for $f(x)$ around $x = 0$. (*Hint*: this is easier than it looks. You do not have to compute nine derivatives of $f(x)$!)

(b) Estimate $\int_0^{0.5} e^{x^3}\, dx$ to 4 decimal places.

11.10. Let $f(x) = e^{-x^2}$. Find $f^{(50)}(0)$ and $f^{(51)}(0)$. No, you aren't expected to compute 51 derivatives of $f(x)$! Instead, look at the coefficients of x^{50} and x^{51} in the MacLaurin series for $f(x)$.

11.11. Consider the function $f(x) = \sqrt{x+4}$. We will compare the MacLaurin polynomials (Taylor polynomials around $x = 0$) to Taylor polynomials around $x = 45$.

(a) Compute the third-order MacLaurin polynomial $M_3(x) = a_0 + a_1 x + a_2 x^2 + a_3 x^3$ that approximates $f(x)$ near $x = 0$.

(b) Compute the third-order Taylor polynomial $T_3(x) = b_0 + b_1(x - 45) + b_2(x - 45)^2 + b_3(x - 45)^3$ that approximates $f(x)$ near $x = 45$.

(c) Make a graph showing $f(x)$, $M_3(x)$ and $T_3(x)$ for $0 \le x \le 60$. For which values of x is M_3 better than T_3? For which values is T_3 better than M_3?

11.12. Find the sixth-order MacLaurin polynomial $M_6(x)$ for each function.

(a) $f(x) = x^2 \sin x$

(b) $f(x) = e^{\pi x}$

(c) $f(x) = \ln(1 + x)$

(d) $f(x) = \frac{1}{1+x}$

Convergence and Divergence

11.13. Determine whether each series converges or diverges by using either the comparison test or the limit comparison test.

(a) $\displaystyle\sum_{k=1}^{\infty} \frac{5}{k - \frac{1}{2}}$

(b) $\displaystyle\sum_{k=1}^{\infty} \frac{2}{2^k + 2}$

(c) $\displaystyle\sum_{k=1}^{\infty} \frac{\ln k}{k}$

11.14. Determine whether each series converges or diverges by using either the root test or the ratio test.

(a) $\displaystyle\sum_{k=1}^{\infty} \frac{k!}{k^3}$

(b) $\displaystyle\sum_{k=1}^{\infty} \left(\frac{k}{50}\right)^k$

(c) $\displaystyle\sum_{k=1}^{\infty} k\left(\frac{1}{2}\right)^k$

11.15. Consider the series $\displaystyle\sum_{k=1}^{\infty} a_k$, where $a_k = (1 - e^{-k})^k$.

(a) Apply the root test to this series. What does it tell you?
(b) Show that, whenever A is small and positive, $\ln(1 - A)$ is approximately $-A$. In particular, show that $-2A \leq \ln(1 - A) \leq 0$.
(c) Show that, whenever k is big enough, $-2ke^{-k} \leq \ln(a_k) \leq 0$.
(d) We already know that $\lim_{k\to\infty} ke^{-k} = 0$. What is $\lim_{k\to\infty} a_k$?
(e) Does $\sum_{k=1}^{\infty} a_k$ converge or diverge? Why?

11.16. Classify each series as absolutely convergent, conditionally convergent, or divergent.

(a) $\displaystyle\sum_{k=1}^{\infty} 5\sin\left(\frac{k\pi}{2}\right)$

(b) $\displaystyle\sum_{k=1}^{\infty} \frac{\cos(k\pi)}{2k}$

(c) $\displaystyle\sum_{k=1}^{\infty} \frac{(-1)^k}{k!}$

11.17. Find the radius and interval of convergence for each power series.

(a) $\displaystyle\sum_{k=1}^{\infty} 3^k x^k$

(b) $\displaystyle\sum_{k=1}^{\infty} \frac{(-1)^k x^k}{k!}$

(c) $\displaystyle\sum_{k=1}^{\infty} \left(\frac{3}{4}\right)^k (x + 5)^k$

Singularities. In general, the radius of convergence is never greater than the distance to the nearest **singularity**, or place where the function stops existing. For instance, you already know that the Taylor series (11.77) for $\ln(x)$ around $x = 1$ has a radius of convergence of 1. This is because the log function doesn't make sense at $x = 1 - 1 = 0$. We explore this further in Exercises 11.18 and 11.19.

11.18. Consider the function $f(x) = \sqrt{x + 4}$.

    (a) Find a formula for $f^{(n)}(0)$. You may find the following notation useful. If k is an odd integer, then $k!!$ is the product of all the *odd* integers from 1 to k, so $3!! = 1 \times 3 = 3$, $5!! = 1 \times 3 \times 5 = 15$, and so on. (Note that $k!!$ is not the same as $(k!)!$. For instance, $3!! = 3$ but $(3!)! = 6! = 720$.)

    (b) Use the ratio test to find the radius of convergence of the MacLaurin series.

    (c) Does it make sense to you that the radius of convergence is less than 5? Explain. (*Hint*: What is $f(-5)$?)

11.19. Find the radius of convergence of the MacLaurin series for $f(x) = \frac{1}{1+x^2}$. How does that compare to the distance from the origin to the nearest singularity?

Food for Thought: The precise rule, which we won't prove here, is: if the formula for a function can be extended to complex inputs, then the radius of convergence is the distance to the nearest singularity *in the complex plane*; see Figure 11.3. For instance, the function $f(x) = \frac{1}{1+x^2}$ has a singularity at $x = i$, which is at distance 1 from the origin. However, dealing with complex inputs can be confusing and, well, complex. If we stick to real inputs and *analytic* functions, then the rule is

> The radius of convergence is *at most* the distance to the nearest singularity.

11.20. **Geodetic Surveys.** We saw how to find the height of a nearby tree in Exercise 10.8 of Chapter 10. When computing the height of more distant objects, such as a tree on the opposite shore of an ecologically sensitive lake that you're not allowed to cross, we also have to take into account the Earth's curvature, since an object that is barely visible on the horizon is already some distance h above the ground. Let r be the radius of the Earth, and let d be the distance, along the surface of the Earth, from you to the object whose height you are measuring.

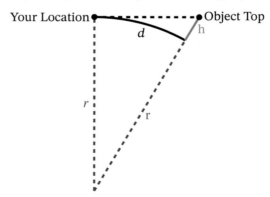

    (a) Show that the height of the object is given by

$$h = r \sec\left(\frac{d}{r}\right) - r.$$

    (b) Find the first two terms in the MacLaurin expansion for $\sec(x)$.

    (c) Explain why $h \approx d^2/2r$.

Food for Thought: The h found in this exercise is called the **curvature correction** in geodetic surveying. If an object at horizontal distance d appears at an angle θ above the horizon, then its true height is approximately $d\tan(\theta) + d^2/2r$. There are also corrections due to the refraction of light, since the index of refraction of air varies with elevation, and due to other factors. Accurate surveying is a complicated business!

11.21. Find the Taylor series around $x = 10$ for the function $f(x) = \ln(x)$.

11.22. Use the first four terms of an appropriate Taylor series for $f(x) = \sqrt[3]{x}$ to approximate $\sqrt[3]{28}$. Compare your answer to the value of $\sqrt[3]{28}$ given by a calculator.

11.23. In the text, we claimed that $\lim_{n\to\infty}(n!)^{1/n} = \infty$. In this exercise we're going to see why.
   (a) When writing $n! = n \times (n-1) \times (n-2) \times \cdots \times 3 \times 2 \times 1$, how many of the factors are $n/2$ or bigger? (*Hint*: The answer depends slightly on whether n is even or odd.)
   (b) Show that, for all positive integers n, $n! \geq (n/2)^{n/2}$.
   (c) Show that, for all positive integers n, $(n!)^{1/n} \geq \sqrt{n/2}$.
   (d) Finally, show that $\lim_{n\to\infty}(n!)^{1/n} = \infty$.

Solving Differential Equations. One use of Taylor (or MacLaurin) polynomials is to find approximate solutions to differential equations. For instance, consider the IVP

(11.109) $y' = \sin(y); \qquad y(0) = 0.3.$

We already know how to get a linear approximation by computing $y'(0) = \sin(0.3) \approx 0.29552$ and setting $y(t) \approx 0.3 + 0.29552t$. To do better than that, we need $y''(0)$, $y'''(0)$, and so on.

We can get $y''(0)$ by taking the derivative of the differential equation itself:

(11.110) $y'' = \cos(y)y' = \cos(y)\sin(y); \qquad y''(0) = \cos(0.3)\sin(0.3) \approx 0.28232.$

Taking another derivative gives us y''':
(11.111)
$$y''' = (\cos(y)\sin(y))' = (\cos^2(y) - \sin^2(y))\,y' = \sin(y)(\cos^2(y) - \sin^2(y)),$$

$$y'''(0) = \sin(0.3)(\cos^2(0.3) - \sin^2(0.3)) \approx 0.24390.$$

Now that we know $y(0)$, $y'(0)$, $y''(0)$ and $y'''(0)$, we can write down a third-order MacLaurin polynomial that approximates $y(t)$:

(11.112) $y(t) \approx 0.3 + 0.29552t + 0.14116t^2 + 0.04065t^3.$

We will apply this method in Exercises 11.24–11.26.

11.24. Find a third-order MacLaurin polynomial that approximates the solution to the initial value problem $y' = e^y$, $y(0) = 0$.

11.25. Recall the SIR model for estimating the progression of a disease.

$$\begin{aligned} S' &= -aSI, \\ I' &= aSI - bI, \\ R' &= bI, \end{aligned}$$

with initial conditions $S(0) = S_0$, $I(0) = I_0$ and $R(0) = R_0$.

(a) Find the third-order MacLaurin polynomial for $S(t)$.

(b) Find the third-order MacLaurin polynomial for $I(t)$.

(c) Find the third-order MacLaurin polynomial for $R(t)$.

11.26. In Exercise 11.25, you found the third-order MacLaurin polynomial for $S(t)$ in the SIR model. Using the parameters $a = -0.0001$, $b = 0.05$, $S_0 = 1000$, and $I_0 = 10$, complete the following steps.

(a) Use MATLAB to plot the third-order MacLaurin approximation for $S(t)$ on the time interval $[0, 100]$.

(b) Use the SIREulers program to graph an approximation for $S(t)$ on the time interval $[0, 100]$.

(c) How do the two approximations in parts (a) and (b) compare? Which one do you think is more accurate? Why?

A **Glucose Tolerance Test** (GTT) is used to screen patients for type-2 diabetes or for gestational diabetes, which may develop during pregnancy. It consists of injecting glucose into the bloodstream, thus inducing an increase in the plasma glucose concentration $G(t)$ and a corresponding increase of the plasma concentration of insulin $I(t)$ secreted by the pancreas. The body's response is modeled by the differential equations

$$\frac{dG(t)}{dt} = -[p_1 + X(t)]G(t) + p_1 G_b,$$

$$\frac{dX(t)}{dt} = -p_2 X(t) + p_3 [I(t) - I_b],$$

$$\frac{dI(t)}{dt} = p_4 [G(t) - G_0]t - p_5 [I(t) - I_b],$$

with initial conditions $G(0) = G_0$, $X(0) = 0$ and $I(0) = I_0$. Here $G(t)$ is the blood glucose concentration after t minutes, $I(t)$ is the blood insulin concentration after t minutes, and $X(t)$ is an artificial function representing the delay in the action of the insulin. The model also includes parameters G_b, I_b, and $p_1, \ldots, p_5$.[5] Exercises 11.27–11.30 refer to this model.

11.27. Find the second-order MacLaurin polynomial for $G(t)$.

11.28. Find the second-order MacLaurin polynomial for $X(t)$.

11.29. Find the second-order MacLaurin polynomial for $I(t)$.

11.30. Use MATLAB to graph the second-order MacLaurin polynomials you found for $G(t)$ and $I(t)$ (in Exercises 11.27 and 11.29) on the same set of axes. Use $p_1 = -0.03$, $p_2 = -0.01$, $p_3 = 0.0001$, $p_4 = 0.0053$, $p_5 = .2$, $G_0 = 250$, $I_0 = 100$, $G_b = 92$, $I_b = 60$, and the time interval $[0, 30]$.

[5]For a full explanation of the model and the various parameters, see: Hyuk Kang, Kyungreem Han & MooYoung Choi (2012) Mathematical model for glucose regulation in the whole-body system, *Islets*, **4**(2), 84–93, DOI: 10.4161/isl.19505.

Index

Selected Published Titles in This Series

For a complete list of titles in this series, visit the
AMS Bookstore at **www.ams.org/bookstore/amstextseries/**.